50 Springer Series in Chemical Physics
Edited by A. L. Schawlow

Springer Series in Chemical Physics

Editors: Vitalii I. Goldanskii   Fritz P. Schäfer   J. Peter Toennies

Managing Editor: H.K.V. Lotsch

Zu-Geng Wang   Hui-Rong Xia

Molecular and Laser Spectroscopy

With 148 Figures

Springer-Verlag

Berlin Heidelberg New York London Paris
Tokyo Hong Kong Barcelona Budapest

Professor Zu-Geng Wang
Professor Hui-Rong Xia
Department of Physics, East China Normal University,
Shanghai 200062, People's Republic of China

Guest Editor:
Professor Arthur L. Schawlow, Ph. D.
Department of Physics, Stanford University, Stanford, CA 94305-4060, USA

Series Editors:

Professor Vitalii I. Goldanskii
Institute of Chemical Physics
Academy of Sciences, Kosygin Street 4
Moscow, 117334, USSR

Professor Dr. Fritz Peter Schäfer
Max-Planck-Institut
für Biophysikalische Chemie
W-3400 Göttingen-Nikolausberg, FRG

Professor Dr. J. Peter Toennies
Max-Planck-Institut für Strömungsforschung
Bunsenstrasse 10, W-3400 Göttingen, FRG

Managing Editor: Dr. Helmut K. V. Lotsch
Springer-Verlag, Tiergartenstrasse 17, W-6900 Heidelberg, Fed. Rep. of Germany

ISBN 3-540-50829-5 Springer-Verlag Berlin Heidelberg New York
ISBN 0-387-50829-5 Springer-Verlag New York Berlin Heidelberg

Library of Congress Cataloging-in-Publication Data. Wang, Zu-Geng, 1939– Molecular and laser spectroscopy / Zu-Geng Wang, Hui-Rong Xia. p. cm.–(Springer series in chemical physics ; v. 50) Includes bibliographical references and index. ISBN 0-387-50829-5 (U.S.) 1. Molecular spectroscopy. 2. Laser spectroscopy. I. Xia, Hui-Rong, 1940– II. Title. III. Series. QC454.M6W36 1991 539'.6–dc20 90-28742

© Springer-Verlag Berlin Heidelberg 1991
Printed in Germany

The use of registered names, trademarks, etc. in this publication does not imply, even in the absence of a specific statement, that such names are exempt from the relevant protective laws and regulations and therefore free for general use.

54/3140-543210 – Printed on acid-free paper

Foreword

The spectra of molecules containing more than one atom are necessarily more complex than those of single atoms. They are correspondingly much richer, not only in the number of spectral lines, but also in qualitatively different phenomena which do not have any counterpart in single atoms.

Historically, molecular spectra have revealed much fundamental physics, such as the connection between nuclear spin statistics. They have provided models of physical systems which have been useful in quite different areas, such as particle physics.

Most especially, molecular spectra are of fundamental importance in understanding chemical bonding. They reveal not only bond lengths but also the strength of the bonding potential between atoms. Moreover, these measurements are obtained for electronic excited states, as well as for the ground state, and for unstable short-lived molecules.

In recent years, tunable lasers have provided powerful tools for the measurement and analysis of molecular spectra. Even before that, molecules were being used in lasers, most notably in the carbon dioxide laser, which finds many industrial applications.

Zu-Geng Wang and Hui-Rong Xia have both worked extensively in molecular and laser spectroscopy, and have made a number of important advances in these fields. It was a great pleasure to collaborate with each of them during the time when they visited my laboratory at Stanford University. They had evidently learned a great deal under the guidance of Professor I-Shan Cheng, and have a very extensive knowledge of these subjects. I am sure that this book will be very useful for anyone interested in any aspect of molecular and laser spectroscopy.

Stanford University   Arthur L. Schawlow
September 1990

Acknowledgements

It is with great pleasure that we acknowledge the many persons who have contributed to our completion of this book.

We would first like to offer our sincere thanks to Professor I.S. Cheng, the director of the Optics Research Group at the East China Normal University (ECNU), for his guidance when we were his students thirty years ago. Having established the group, he led us along the path of spectroscopic research. He made many suggestions for improvements to our previous research publications, which form the background to our scientific experience, and warmly encouraged us to complete this book.

We are specially grateful to Professor A.L. Schawlow of Stanford University, Honorary Professor of ECNU. His great achievements in laser development and laser spectroscopy won our respect and that of all of our colleagues. He visited our university in 1979 and in 1984, and gave us series of lectures on laser spectroscopy. His helpful suggestions and advice propelled forward the research work of our group. We were fortunate to have the opportunity to carry out research at Stanford University under his direction in 1980 and in 1982. We feel greatly honored that he has written the foreword to this book.

We would also like to thank Professor T.W. Hänsch, whose great contributions to the developments of high-resolution spectroscopy show the attractiveness and importance of atomic and molecular spectroscopy. One of the authors (Hui-Rong Xia) attended his lectures on quantum electronics and laser spectroscopy at Stanford. Both of the authors benefited from his valuable suggestions and enthusiastic help.

We are grateful to Dr. J.L. Hall, who is an advisory professor of ECNU, for his impressive lectures on and insights into precision laser spectroscopy at our campus in 1986. Hui-Rong Xia is deeply grateful for the valuable opportunity to work with him on precision measurements at the Joint Institute for Laboratory Astrophysics, University of Colorado at Boulder, in recent years.

Zu-Geng Wang is very grateful to Professor H. Welling and Professor B. Wellegehausen for their valuable suggestions, discussions and warm assistance during his visit to the Institute of Quantum Electronics at the University of Hannover in 1986. We are also grateful to Dr. D.A. Ramsay for his helpful lectures on polyatomic molecular spectroscopy at the campus of ECNU.

Many thanks go to the dozens of domestic and overseas colleagues and research collaborators without whom we would not be able to carry out our research.

We are grateful to the National Natural Science Foundation of China for support.

Finally, we are specially indebted to Dr. H.K.V. Lotsch, the physics editor of Springer-Verlag, for his valuable suggestions and sincere efforts towards the publication of this book.

East China Normal University   Zu-Geng Wang
Boulder, CO   Hui-Rong Xia
October 1990

Contents

1. Introduction

Molecular spectroscopy is an essential tool in establishing the nature of substances. It has had an intimate association with the development of various fundamental branches of science, including Physics, Chemistry, Astrophysics, Metrology and Biology, and has also made substantial contributions to advances of technology ranging from industrial processing to monitoring and controll engineering.

Molecular spectroscopy has had a long history, which was accelerated by the establishment of quantum mechanics six decades ago. Milestones of this history were the publication of the monumental books by *Herzberg* [1.1-4], and by *Townes* and *Schawlow* [1.5], where almost all of the fundamental regularities for molecular electronic, vibrational, rotational and hyperfine spectra as well as their distortions due to inner or external interactions of molecules, and their relations with molecular structures, have been elucidated.

However, molecular energy-level structures are extremely complicated compared with those of atoms. As *Schawlow* put it, a diatomic molecule is a molecule with one atom too many. Although the achievements have been considerable even for large and complicated molecules, it is true that even the knowledge about a molecule as small and simple as a sodium dimer is far from satisfying. Until the middle of the 1970's, there were only 5 excited electronic states in Na_2 for which the constants had been determined experimentally. The situation was worse for other molecules. Furthermore, to go beyond the mere interest on extremely precise spectral data, it was necessary to analyze dynamic characteristics.

Notwithstanding the large number of systems studied and the huge amount of data collected, the fundamental limitation of covnentional spectroscopy, including spontaneous Raman processes, is *linear* in nature. Even the modern Fourier spectrometer, that ultimate marriage of spectroscopic and computer technology, cannot help to breakthrough the ultimate spectral resolution of Doppler broadening due to molecular thermal motions. Thus for a long time the hyperfine constants of molecular energy levels were limited to those obtained for low rotational levels in the lowest vibrational level of the electronic ground state by means of microwave spectroscopy; the traditional analyses of the high spectral density of molecular lines in an electronic vibration-rotational band system were so difficult, requiring months for an expert of molecular spectroscopy with the help of a fast computer, that research was confined to a small circle of famous laboratories.

1

Times changed once the laser was introduced into the field of molecular spectroscopy. The relationship between laser spectroscopy and conventional molecular spectroscopy was not, and will not become, one of replacement, but rather of complementarity. Laser molecular spectroscopy has become an inseparable part of modern molecular spectroscopy. It is well known that by using a low-power tunable laser such as a diode laser instead of a conventional light source an expensive spectrometer, which was once the heart of a molecular spectroscopy research laboratory, is not required. Even if the interaction of molecules with laser radiation is still dominated by linear processes, the obtainable molecular spectra may be incomparable with conventioanl spectroscopy in terms of such aspects as supersensitive and superfast detection, remote sensing, etc., in addtion to high spectral resolution.

Nonlinear interaction of molecules with a laser field is the key aspect of laser molecular spectroscopy. The nonlinear uncoupling and nonlinear coupling spectral effect which are introduced are entirely new concepts in molecular spectroscopy. Here the so-called nonlinear uncoupling interaction indicates the process of coherent absorption induced by the incident laser field with negligible reemission. The most important properties of such kinds of processes are the possibilities of Doppler-free spectral resolution and multiphoton absorption, promising unprecedented opportunities for the precise and direct measurement of the hyperfine constants of molecular excited electronic states in the optical wavelength region, and for the experimental determination of molecular constants or of other fundamental parameters of high-lying, including traditionally forbidden electronic states and high overtone or combination vibrational levels which used to be inaccessible.

The interaction of molecules with intense laser fields is nonlinear and *coupling*, i.e. the molecular reemission field is no longer negligible. In addition to high resolution spectra in the time domain, the distinguishable processes have provided abundant new lasers or coherent radiation in every wavelength regions by optical pumping of various molecules or by different wave-mixing processes via molecular levels.

A resonant absorption process based on the selective excitation of intense laser radiation would also label the relevant levels of a molecular transition, offering the possibility to do selective simplifications of molecular spectra. Such a greatly simplified spectrum is clear and easy even for a beginner to identify, and is especially useful for the unambiguous study of seriously overlapped states.

The new interaction mechanisms have systematically been utilized in the study of small but complete molecules - such as Na_2 - as illustrated in Fig.1.1 [1.6]. The central frame shows the focus on the traditional parity-forbidden or both parity- and spin-forbidden high-lying electronic states. The extension studies framed at the upper left show the comprehensive identification methods developed for the special, simplified molecular spectrum; the studies framed at the lower left are dominated by the nonlinear uncoupling interaction mechanism which reveals the relation between the spectral structure of an isolated line (line shape) and the energy-level

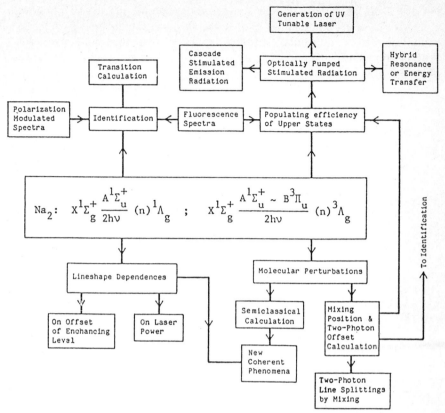

Fig.1.1. The general scheme for the study of high-lying "forbidden" states in Na_2 [1.6]

scheme; to be particularly noted are the frames at the lower right part of the figure which show that new spectral features were observed for the two-photon transitions involving molecular mixing levels; finally the upper right part of the figure represents the sucesssful results in the generation of new lasers and coherent radiation via nonlinear coupling interactions of solidum dimers with incident laser fields, which were also achieved in different kinds of molecules.

There are many excellent books published in the field of laser spectroscopy, including the monographs by *Letokhov* and *Chebotayev* [1.7] and by *Demtröder* [1.8], the books edited by *Bloembergen* [1.9], by *Shimoda* [1.10] and by *Walther* [1.11], etc., in addition to the books on molecular spectroscopy mentioned above. We do not attempt to cover all regularities and applications bridging molecular and laser spectroscopy, but try, taking examples from the spectra of small molecules, to analyse the physical mechanisms and the spectral phenomena for various interactions of laser fields with molecules.

After summarizing the general formulae for energy levels and spectra of molecules, we discuss the coarse and fine spectral structures of molecular two-photon transitions, Doppler-free spectra via nonlinear uncoupling interactions of lasers with molecules, the spectroscopic effects of nonlinear coupling interactions of lasers with molecules and the method of selective simplification of molecular spectra. We hope that this will cover new ground from the books already published in both laser and molecular spectroscopy.

2. Molecular Energy States

In this chapter, starting from the equation of motion for molecules, we briefly summarize the formulas for molecular energy states and levels, including molecular electronic states, vibration-rotational levels and Rydberg states. We treat the energy-level shifts and splittings due to the intramolecular dynamics and angular-momentum couplings as individual perturbations, and list the selection rules for various molecular transitions. Fundamentals and discussions may be found in [2.1, 2].

2.1 The Molecular-Motion Equation, and the Hamiltonian Operator

The Schrödinger equation of a molecule can be written as ·

$$\mathcal{H}\Psi = E\Psi , \tag{2.1}$$

where $\mathcal{H}$ is the total Hamiltonian operator, Ψ is the total wave function, and E is the total energy of the molecule.

The total Hamiltonian operator can be expressed as

$$\mathcal{H} = \mathcal{H}_0 + \mathcal{H}' , \tag{2.2}$$

where $\mathcal{H}_0$ is the part referring to electronic motion as well as to nuclear vibrations and rotations excluding their mutual interactions, whereas $\mathcal{H}'$ represents the effects of all intramolecular interactions. Therefore, with the Born-Oppenheimer approximation we have

$$\mathcal{H}_0 = \mathcal{H}_e + \mathcal{H}_v + \mathcal{H}_r , \tag{2.3}$$

$$\Psi = \Psi_e \, \Psi_v \, \Psi_r , \tag{2.4}$$

$$E = E_e + E_v + E_r , \tag{2.5}$$

where the subscripts e, v and r refer to electronic, vibrational and rotational, respectively. The corresponding term values of the energy levels can be expressed in the form, noting T = E/hc,

$$T = T_e + G(v) + F(J) , \tag{2.6}$$

where v and J are the molecular vibrational and rotational quantum numbers, respectively.

The other part of the Hamiltonian operator $\mathscr{H}$ can be written as

$$\mathscr{H}' = \mathscr{H}'_{reg} + \mathscr{H}'_{irreg} \ . \tag{2.7}$$

The perturbations cause wave function sharing and a shifting of energy levels. The subscript "reg" denotes regular perturbations, which influence all of the energy levels, whereas the subscript "irreg" denotes the irregular perturbations, which affect only some of the levels.

In what follows we will use the double-subscripts to classify the interactions, where s, o, N and I indicate electronic-spin and -orbital motion, nuclear-rotation and -spin, respectively. We will use $\mathscr{H}_{vr}$, for example, to denote the Hamiltonian corresponding to vibrational-rotational interaction, and other symbols analogously. While $\mathscr{H}'_{reg}$ contains

$$\mathscr{H}'_{reg} = \mathscr{H}_{ev} + \mathscr{H}_{er} + \mathscr{H}_{vr} + \mathscr{H}_{so_1} + \mathscr{H}_{NI} + \dots , \tag{2.8}$$

$\mathscr{H}'_{irreg}$ also encompasses many coupling terms

$$\mathscr{H}'_{irreg} = \mathscr{H}_{vv} + \mathscr{H}_{so_2} + \dots . \tag{2.9}$$

Much weaker couplings are omitted.

The regular perturbation shifts the energy levels or splits them systematically. Irregular perturbations, on the other hand, cause unsystematic level shifts and splittings, limited to individual regions where the particular perturbation between certain levels could occur. The latter might also result in changes in the strength of the transitions.

2.2 Molecular Electronic States

Assume a diatomic molecule to be A–B with n electrons, the mass of an electron to be m and the reduced mass of the two nuclei of the molecule to be $\mu = M_a M_b / (M_a + M_b)$ (M_a and M_b are the masses of the two nuclei). Neglecting the various above-mentioned interactions, the Schrödinger equation of the diatomic molecule can be written as

$$\left[-\frac{\hbar^2}{2\mu} \nabla^2 - \frac{\hbar^2}{2m} \sum_{i=1}^{n} \nabla_i^2 + V \right] \Psi = E\Psi \ . \tag{2.10}$$

The molecular potential energy is given by the summation of the electron-electron repulsion energies, the nuclear-nuclear repulsion energy and the electron-nuclear attractive energies, i.e.,

$$V = \sum_{i>j=1}^{n} \frac{e^2}{r_{ij}} + \frac{Z_a Z_b e^2}{R} - \left(\sum_{i=1}^{n} \frac{Z_a e^2}{r_{ia}} + \sum_{i=1}^{n} \frac{Z_b e^2}{r_{ib}} \right), \qquad (2.11)$$

where $Z_a e$ and $Z_b e$ are the charges of the two nuclei, and R is the inter-nuclear distance. If we take the nucleus a to be the origin of the coordinate system, letting r represent the coordinates of all of the electrons and R the coordinates of the nucleus b, then the wave function is in the form of $\Psi = \Psi(r,R)$.

For solving the equation for the energy of the electronic motion, the internuclear distance R can be considered to appear only as a parameter in the wave function and the equation of motion. Therefore, for a given R value, the electronic wave function $\Psi(r,R)$ and the corresponding energy $E(R)$ can be obtained by solving the following equation

$$\left[-\frac{\hbar^2}{2m} \sum_{i=1}^{n} \nabla^2 + V(r,R) \right] \Psi(r,R) = E(R)\Psi(r,R) . \qquad (2.12)$$

Calculating $E(R)$ with progressively varied R values, we can finally obtain a continuous function $E(R)$. As an example, Fig.2.1 shows a part of the electronic states in Na_2. Here the $X^1\Sigma_g^+$ state is the ground state, which is stable with the minimum energy at $R = R_e$. The $a^3\Sigma_u^+$ state shown in the figure is one of the unstable states. A molecule in an unstable state can easily be dissociated. D_e notes the dissociation energy of the molecular ground state. The dissociation of a molecule would occur if its vibrational-rotational energy were larger than D_e. The bond-free atomic states for the atoms dissociated from a sodium dimer have been denoted on the right-hand side of Fig.2.1, such as a $X^1\Sigma_g^+$ molecule can be dissociated into two 3s atoms; a $A^1\Sigma_u^+$ or $B^1\Pi_u$ molecule can be separated into a 3s atom and a 3p atom, etc.

The electrons of a diatomic molecule move in the axially symmetrical electric field. The total electronic orbital angular momentum L precesses around the molecular axis, as shown in Fig.2.2. The axial component of L is equal to $M_L h/2\pi$ with

$$M_L = L, L-1, L-2, ..., -L , \qquad (2.13)$$

where L is the quantum number of the orbital angular momentum. If all of the electrons were to change direction, M_L would become $-M_L$ but the energy remains the same. However, the different values of $|M_L|$ in (2.13) possess unequal energy. Hence the electronic states are specified by $|M_L|$. The states with equal absolute value $|M_L|$ but opposite signs are so-called degenerate states. The quantum number used to represent $|M_L|$ is Λ,

$$\Lambda = |M_L| = 0, 1, 2, 3, ..., L . \qquad (2.14)$$

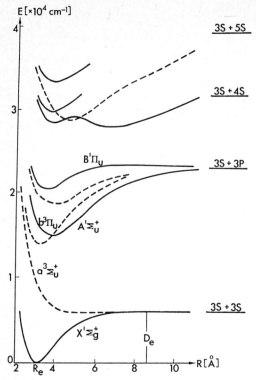

Fig.2.1. Part of the electronic energy states for molecular sodium [2.3]

The molecular electronic states with $\Lambda = 0,1,2,3,...$ are named as Σ, Π, Δ, Φ ... states, respectively. Obviously, all the states except the Σ state in a diatomic molecule have double degeneracies. In addition, the total electronic spin angular momentum S of a diatomic molecule precesses around the direction of the field, as shown in Fig.2.2. The projection Σ of S along the molecular axis is equal to $M_s h/2\pi$, where the quantum number Σ can be taken as

$$\Sigma = M_s = S, S-1, S-2, S-3, ..., -S .\qquad (2.15)$$

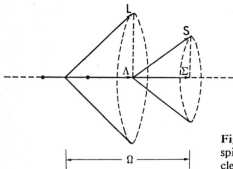

Fig.2.2. Precessions of the orbital and spin angular momenta about the internuclear axis

The total electronic angular momentum along the axis, Ω, is the sum of Λ and Σ. The resultant quantum number Ω is

$$\Omega = |\Lambda + \Sigma| . \qquad (2.16)$$

Due to the magnetic interaction $\mathcal{H}_{so_1}$ in (2.8) between $\mathbf{S}$ and Λ, the electronic term values split into $(2S+1)$ components. Usually, the multiplicity $(2S+1)$ is denoted as an upper-left mark to the electronic-state symbol. For instance, a $\Lambda = S = 0$ state is written as $^1\Sigma$; a $\Lambda=2$, $S=1$ state is denoted by $^3\Delta$. In addition, $(\Lambda+\Sigma)$ is indicated as a lower-right mark (subscript) to the state symbol. For example, with $S = 1$ and $\Lambda = 1$, there are three fine electronic states $^3\Pi_2$, $^3\Pi_1$, $^3\Pi_0$, which correspond to three slightly different potential energy curves.

For a homonuclear diatomic molecule, another mark g or u should be added at the lower-right mark of the electronic-state symbol refering to a positive or negative sign of the total wave function and the nuclear-spin wave function after exchanging the positions of the nuclei. The respective states are then said to have even or odd parity, such as $^1\Sigma_g$ or $^1\Sigma_u$ states.[1]

According to the signs of the electronic wave functions after coordinate inversion through a central point the superscripts + or - are used to mark the electronic state symbol, at the upper-right corner, i.e., $^1\Sigma_g^+$, $^1\Sigma_u^+$, $^1\Sigma_g^-$, $^1\Sigma_u^-$, etc.

The general selection rules of the electronic transitions can be listed as

$$\Delta\Lambda = 0, \pm 1 , \qquad (2.17a)$$
$$\Delta S = 0 , \qquad (2.17b)$$
$$\Delta\Sigma = 0 , \qquad (2.17c)$$
$$\Delta\Omega = 0, \pm 1 . \qquad (2.17d)$$

Of the above rules, (2.17a and b) apply to Hund's coupling cases (a) and (b); rule (2.17c) applies to Hund's case (a); and rule (2.17d) to Hund's case (c) (For details on the Hund cases, see Sect.2.6). The transition between Σ^+ and Σ^- states is forbidden, i.e.,

$$\Sigma^+ \longleftrightarrow \Sigma^+ , \quad \Sigma^- \longleftrightarrow \Sigma^- , \quad \Sigma^+ \longleftrightarrow\!\!\!/ \Sigma^- . \qquad (2.18)$$

For a homonuclear molecule there are some additional selection rules for one-photon transition, namely

$$g \longleftrightarrow u , \quad g \longleftrightarrow\!\!\!/ g , \quad u \longleftrightarrow\!\!\!/ u . \qquad (2.19)$$

---

[1] The conventional notation (g and u) has been derived from the German words gerade (even) and ungerade (odd), respectively.

2.3 Molecular Vibrational Levels

When we consider the nuclear motion (including vibration and rotation) in a certain electronic state, the energy of the electronic motion, E(R), is taken to be the potential energy appearing in the equation of nuclear motion. The total wave function of a molecule can be written as

$$\Psi(\mathbf{r},\mathbf{R}) = \Psi(\mathbf{r},R)\Phi(\mathbf{R}) \ ,$$

where $\Phi(\mathbf{R})$ is the wave function depending only on the nuclear coordinate $\mathbf{R}$. Therefore the Schrödinger equation of the nuclear motion can be separated from (2.10) to be

$$\left[\frac{\hbar^2}{2\mu}\nabla^2 + E(R)\right]\Phi(\mathbf{R}) = E\Phi(\mathbf{R}) \ . \tag{2.20}$$

Introducing the spherical-polar coordinates (R,θ,Φ) to replace the nuclear coordinate $\mathbf{R}$ in the above equation, we have

$$\left\{\frac{-\hbar^2}{2\mu R^2}\left[\frac{\partial}{\partial R}\left(R^2\frac{\partial}{\partial R}\right) + \frac{1}{\sin\theta}\frac{\partial}{\partial\theta}\left(\sin\theta\frac{\partial}{\partial\theta}\right) + \frac{1}{\sin^2\theta}\frac{\partial}{\partial\phi^2}\right] + E(R)\right\}\Phi(R,\theta,\phi)$$

$$= E\Phi(R,\theta,\phi) \ . \tag{2.21}$$

We then divide the wave function of the nuclear motion into a radial part and an angular part, corresponding to molecular vibration and rotation, respectively, as

$$\Phi(R,\theta,\phi) = \Psi_v(R)\Psi_r(\theta,\phi) \ . \tag{2.22}$$

Substituting it into (2.21) we obtain the angular part of the Schrödinger equation of the nuclear motion

$$\frac{-\hbar^2}{2\mu R^2}\left[\frac{1}{\sin\theta}\frac{\partial}{\partial\theta}\left(\sin\theta\frac{\partial}{\partial\theta}\right) + \frac{1}{\sin^2\theta}\frac{\partial^2}{\partial\phi^2}\right]\Psi_r = \frac{J(J+1)\hbar^2}{2\mu R^2}\Psi_r \ , \tag{2.23}$$

and the radial part of the Schrödinger equation

$$\frac{-\hbar^2}{2\mu R^2}\frac{d}{dR}\left[R^2\frac{d\Psi_v}{dR}\right] + \frac{J(J+1)\hbar^2}{2\mu R^2}\Psi_v + E(R)\Psi_v = E\Psi_v \ ,$$

which can be written

$$\frac{-\hbar^2}{2\mu R^2}\frac{d}{dR}\left[R^2\frac{d\Psi_v}{dR}\right] + E(R)\Psi_v = E_v\Psi_v \ . \tag{2.24}$$

For most of the stable molecules the potential energy function E(R) can be expanded in a power series about the equilibrium position R_e, i.e.,

$$E(R) = E(R_e) + \left[\frac{dE}{dR}\right]_{R_e} (R - R_e) + \frac{1}{2}\left[\frac{d^2E}{dR^2}\right]_{R_e} (R - R_e)^2$$

$$+ \frac{1}{3!}\left[\frac{d^3E}{dR^3}\right]_{R_e} (R - R_e)^3 + \dots . \qquad (2.25)$$

Setting to zero the energy at the bottom of the E(R) curve we have $E(R_e) = (dE/dR)_{R_e} = 0$. Since the force constant is given by

$$K = \left[\frac{d^2E}{dR^2}\right]_{R_e} , \qquad (2.26)$$

the potential-energy function E(R) has the form

$$E(R) = \frac{1}{2}K(R - R_e)^2 + \frac{1}{3!}\left[\frac{d^3E}{dR^3}\right]_{R_e} (R - R_e)^3$$

$$+ \frac{1}{4!}\left[\frac{d^4E}{dR^4}\right]_{R_e} (R - R_e)^4 + \dots . \qquad (2.27)$$

Substituting it into (2.24), we obtain the expression for the molecular vibrational energy

$$E_v = hc[\omega_e(v + \tfrac{1}{2}) - \omega_e x_e(v + \tfrac{1}{2})^2 + \omega_e y_e(v + \tfrac{1}{2})^3 + \dots] \quad (v = 0,1,2,3,\dots) , \qquad (2.28)$$

Its term value is

$$G(v) = \omega_e(v + \tfrac{1}{2}) - \omega_e x_e(v + \tfrac{1}{2})^2 + \omega_e y_e(v + \tfrac{1}{2})^3 + \dots , \qquad (2.29)$$

where v is the vibrational quantum number, and the molecular constants obey $\omega_e x_e \ll \omega_e$, $\omega_e y_e \ll \omega_e x_e$, ... Figure 2.3 illustrates the molecular vibrational levels in an electronic state. Let $v = 0$ in (2.29), we see the zero-vibration energy is

$$G(0) = \omega_e/2 - \omega_e x_e/4 + \omega_e y_e/8 + \dots . \qquad (2.30)$$

The molecular dissociation energy counting from the v=0 level is denoted by D_0. The maximum value of G(v) is the uppermost term with the largest v value right before dissociation of the molecule, where v is determined by

$$\frac{d}{dv}G(v) = 0 . \qquad (2.31)$$

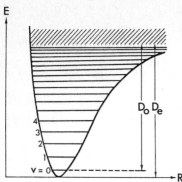

Fig.2.3. Vibrational energy level scheme in the electronic ground state of diatomic molecule

Thus, with the approximation of taking the first two terms in (2.29) the dissociation energy can be evaluated

$$D_e = \omega_e^2/4\omega_e x_e \ . \tag{2.32}$$

The selection rules for the vibrational transitions are

$$\Delta v = \pm 1, \pm 2, \pm 3, \ ... \ . \tag{2.33}$$

Neglecting all of the anharmonic vibrations, i.e., retaining only the first term in (2.28), we get the vibrational energy of a harmonic oscillator

$$E_v = hc\omega_e(v + \tfrac{1}{2}), \quad v = 0,1,2,3,... \ . \tag{2.34}$$

It represents roughly equally spaced vibrational levels.

A polyatomic molecule with N atoms has n = 3N-6 (or n = 3N-5 for a linear molecule) normal vibrational modes. During a certain normal vibration all of the atoms, unless at the molecule's center of mass, move in phase with the same frequency around their own equilibrium positions. Figure 2.4

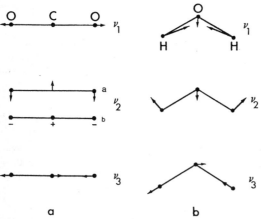

Fig.2.4 The normal modes of CO_2 and H_2O

shows the molecular normal vibrations in CO_2 and H_2O, where the ν_{2a} and ν_{2b} in CO_2 are degenerated.

The vibrational levels of a polyatomic molecule are described approximately by

$$E(v_1, v_2, ... v_n) = h\nu_1(v_1 + \tfrac{1}{2}) + h\nu_2(v_2 + \tfrac{1}{2}) + ... + h\nu_n(v_n + \tfrac{1}{2}) . \quad (2.35)$$

The energy of the lowest vibrational level (the ground state) is

$$E(0,0,...0) = \tfrac{1}{2}h\nu_1 + \tfrac{1}{2}h\nu_2 + ... + \tfrac{1}{2}h\nu_n ; \quad (2.36)$$

the energy of the fundamental levels is

$$E(0,...1,0,...0) = \tfrac{1}{2}h\nu_1 + ... + h\nu_i(1 + \tfrac{1}{2}) + \tfrac{1}{2}h\nu_{i+1} + ... + \tfrac{1}{2}h\nu_n ; \quad (2.37)$$

the energy of the overtone levels is ($v_i = 2, 3, 4, ...$)

$$E(0,...v_i,0,...0) = \tfrac{1}{2}h\nu_1 + ... + h\nu_i(v_i + \tfrac{1}{2}) + \tfrac{1}{2}h\nu_{i+1} + ... + \tfrac{1}{2}h\nu_n , \quad (2.38)$$

and the energy of the combination frequencies is

$$E(0,...v_i,0,...v_j,0,...0) = \tfrac{1}{2}h\nu_1 + ... + h\nu_i(v_i + \tfrac{1}{2}) + \tfrac{1}{2}h\nu_{i+1} + ... + h\nu_j(v_j + \tfrac{1}{2})$$

$$+ \tfrac{1}{2}h\nu_{j+1} + ... + \tfrac{1}{2}h\nu_n , \quad (v_i, v_j \neq 0) . \quad (2.39)$$

Figure 2.5 is the partial vibrational energy level diagram of a triatomic molecule, including the ground state, fundamental level, overtone levels and combination levels under the assumption of $\nu_1 > \nu_2 > \nu_3$.

If a normal frequency ν_i is equal to another normal frequency ν_j ($\nu_i = \nu_j = \nu$), the molecular vibrational energy levels can be written as

$$E_v = h\nu_1(v_1 + \tfrac{1}{2}) + ... + h\nu(v + 1) + h\nu_n(v_n + \tfrac{1}{2}) , \quad (2.40)$$

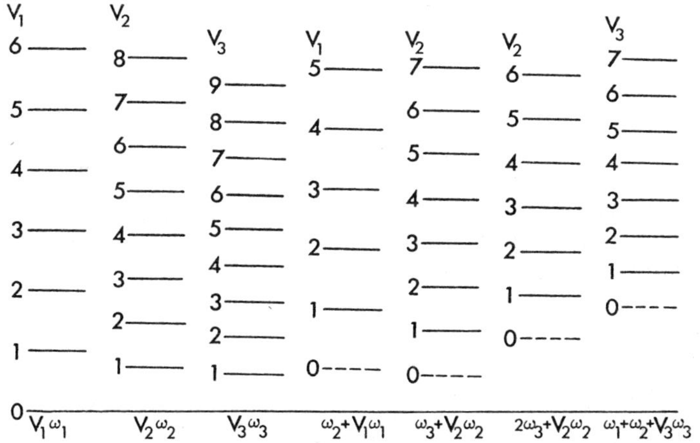

Fig.2.5 Part of the vibrational energy level scheme of a triatomic molecule

where $v = v_i + v_j$. The degree of degeneracy of these vibrational levels is

$$f = v + 1 . \qquad (2.41)$$

It means that the ground state ($v = 0$) relates to only one wave function and is the nondegenerate level; whereas the fundamental level ($v = 1$) with two wave functions is doubly degenerate; there are three wave functions corresponding to the overtone level ($v = 2$), so it is a triply degenerate level.

If three normal vibrations possess the same frequeny, i.e., $\nu_i = \nu_j = \nu_k = \nu$, the vibrational energy levels should be

$$E_v = h\nu_1(v_1 + \tfrac{1}{2}) + ... + h\nu(v + 3/2) + ... + h\nu_n(v_n + \tfrac{1}{2}) , \qquad (2.42)$$

where $v = v_i + v_j + v_k$. The degree of the degeneracy is

$$f = \tfrac{1}{2}(v + 1)(v + 2) . \qquad (2.43)$$

So the ground state ($v = 0$) is still the nondegenerate level, the fundamental level ($v = 1$) is triply degenerate, and the overtone level with $v = 2$ is six-fold degenerate.

Indeed it is necessary to take account of anharmonicities to adequately describe the vibrations of a polyatomic molecule. For instance, the vibrational energy levels of a nonlinear triatomic molecule can be expressed as

$$\begin{aligned}
G(v_1, v_2, v_3) = {} & \omega_1(v_1 + \tfrac{1}{2}) + \omega_2(v_2 + \tfrac{1}{2}) + \omega_3(v_3 + \tfrac{1}{2}) + x_{11}(v_1 + \tfrac{1}{2})^2 \\
& + x_{22}(v_2 + \tfrac{1}{2})^2 + x_{33}(v_3 + \tfrac{1}{2})^2 + x_{12}(v_1 + \tfrac{1}{2})(v_2 + \tfrac{1}{2}) \\
& + x_{13}(v_1 + \tfrac{1}{2})(v_3 + \tfrac{1}{2}) + x_{23}(v_2 + \tfrac{1}{2})(v_3 + \tfrac{1}{2}) + .. , \quad (2.44)
\end{aligned}$$

where ω_1, ω_2 and ω_3 are the three normal vibrational frequencies, or the so-called zero-order frequencies, and x_{ij} are the anharmonic coefficients, which correspond to ω_e and $\omega_e x_e$ in the diatomic molecule case, respectively. Usually the x_{ij}'s are negative. So, in practice, the vibrational levels are lower than they would be if anharmonic corrections were ignored.

2.4 Molecular Rotational Levels

Investigating the rotation of a diatomic molecule with the rigid-rotator model, we can solve the angular part of the Schrödinger equation (2.23) and obtain the level expression for pure rotation

$$F_J = BJ(J + 1) , \quad J = 0,1,2,3,... , \qquad (2.45)$$

where B is the constant of rotation, which is

$$B = h(8\pi^2 cI_e)^{-1} , \qquad \text{where} \qquad (2.46)$$

$$I_e = \mu R_e^2 \; , \tag{2.47}$$

μ is the reduced mass, and R_e is the equilibrium distance. For pure rotational transistions the selection rule of the rotational quantum number is

$$\Delta J = \pm 1 \; . \tag{2.48}$$

However, as a consequence of the centrifugal effect during the molecular rotation, the rotation constant is reduced with increasing J. The rotational energy levels are better expressed in the nonrigid-rotator model

$$F_J = BJ(J + 1) - DJ(J + 1)^2 + ... \; , \tag{2.49}$$

where D is the centrifugal distortion constant determined by

$$D = 4B^3/\nu^2 = \hbar^2(2\mu^2 R_e^6 k)^{-1} \; , \tag{2.50}$$

with a value much smaller than B. The force constant k of a harmonic oscillator is in the form of

$$k = 4\pi^2 \nu^2 c^2 \mu \; . \tag{2.51}$$

Figure 2.6 illustrates the rotational energy level scheme with solid lines for the rigid rotator model and the dashed lines for the nonrigid model of a diatomic molecule.

The simplest model of a rotating polyatomic molecule is a three-dimensional rigid body with three principal axes fixed in the molecule. The labels I_A, I_B and I_C are given to the three principal moments of inertia of the molecule in order of increasing magnitudes ($I_A \le I_B \le I_C$). According to

J

4 ——————

  — — — — — — — — — —

3 ——————

  — — — — — — — — — —

2 ==================

1 ==================   Fig.2.6. Part of the rotational energy level scheme

0 ——————   of a diatomic molecule

the relative values between I_A, I_B, and I_C, and the forms of their respective inertia ellipsoids, the poylatomic molecuels can be classified as:

1) Linear molecule (including the diatomic molecule): $I_B = I_C \neq 0$, $I_A = 0$. The rotational energy term in the rigid-rotator approximation is identical with (2.45) as for diatomic molecule. Regarded as a nonrigid rotator the rotational levels are

$$F_J = BJ(J + 1) - DJ^2 (J + 1)^2 \tag{2.52}$$

with the selection rule for the pure rotational transitions as

$$\Delta J = \pm 1 . \tag{2.53}$$

2) Symmetric top molecules can be further classified as an oblate and a prolate top according to the relative magnitudes of the three principle moments of inertia. For an oblate symmetric top molecule with $I_A = I_B < I_C$ (such as NH_3, BF_3, etc.) the rotational constants are

$$A = B = h(8\pi^2 c I_A)^{-1} = h(8\pi^2 c I_B)^{-1} > C = h(8\pi^2 c I_C)^{-1} . \tag{2.54}$$

In the rigid-rotator approximation the rotational energy levels are

$$F_{JK} = BJ(J + 1) + (C - B)K^2 . \tag{2.55}$$

Here K is the quantum number referring to the component M_c of the total angular momentum along the C axis, that is

$$M_c{}^2 = K^2 \hbar^2 , \quad K = 0,1,2,..,J . \tag{2.56}$$

The selection rules for the infrared spectra are

$$\Delta J = 0, \pm 1 , \quad \Delta K = 0 . \tag{2.57}$$

The solid lines in Fig.2.7 illustrate the rotational energy levels of the rigid-rotator. For each J value $|K|$ can be $(J+1)$ different values. The energy decreases with increasing $|K|$ value.

As a nonrigid-rotator the rotational level is

$$F_{JK} = BJ(J+1) + (C-B)K^2 - D_J J^2(J+1)^2 - D_{JK} J(J+1)K^2 - D_K K^2 . \tag{2.58}$$

Here the correction constants D_J, D_{JK}, D_K, etc. are much smaller than the rotation constants B and C; D_J in (2.58) is always positive. Owing to the effect of the $D_J J^2(J+1)^2$ term all of the levels with J = 1,2,3,... in Fig.2.7 have been shifted down as described in the diatomic molecule case; the larger J the bigger the shifts. On the other hand, due to the existence of the $D_{JK} J(J+1)K^2$ term all of the levels except K = 0 should shift up if D_{JK} is negative; the larger the J and K values the larger the shifts. The dashed lines in Fig.2.7 show the upper shifting case for the three levels with J = K = 1; J = 2, K = 1; and J = K = 2, respectively.

16

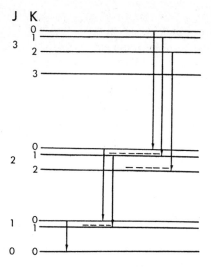

For the prolate symmetric top molecule (such as CH_3F, C_2H_6, etc.) with $I_A < I_B = I_C$ the rotational constants are

$$A = h/8\pi^2 cI_A > B = C = h/8\pi^2 cI_B = h/8\pi^2 cI_C . \qquad (2.59)$$

As a rigid-rotator the resulting rotational energy levels are

$$F_{JK} = BJ(J+1) + (A-B)K^2 . \qquad (2.60)$$

Because $A > B$, the energy is bigger for a level with a larger $|K|$ value than that of a smaller $|K|$ possessing the same J value.

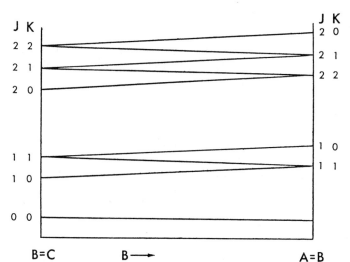

Fig.2.8 Diagrammatic representation of the energy levels of an asymmetric top molecule in relation to the limiting forms of prolate and oblate symmetric top molecules

The spherically symmetric top molecules (such as CH_4, SF_6, etc.) have more than two three-fold symmetric axes and have rotation constants $A = B = C$. Owing to the absence of the permanent dipole moment no pure rotational spectra in the far infrared region could be observed from a spherically symmetric top molecule.

3) Asymmetric top molecules (such as H_2O, C_2H_4, H_2CO, etc.) lack a three-dimension symmetric axis and have the rotation constants as $A>B>C$. One of the limiting cases is the prolate top ($A>B=C$) and the other is the oblate top ($A=B>C$). Figure 2.8 illustrates the energy level relations of the asymmetric top molecules (intermediate values) with the prolate (left scale) and the oblate (right scale) symmetric top molecules.

2.5 Molecular Vibration-Rotational Levels

Ignoring the interactions between molecular vibrations and rotations, the molecular energy in a certain electronic state is simply the sum of the energy of the anharmonic vibration and the nonrigid rotation. But usually this description is not accurate enough to explain experimental phenomena.

In fact, the internuclear distance varies with the molecular vibrations, resulting in a corresponding variation of the rotation constant. This means that the latter is related to the vibrational quantum number and should be denoted by B_v, i.e.,

$$B_v = \frac{h\langle v|1/R^2|v\rangle}{8\pi^2 c\mu}.$$

(2.61)

The calculation gives a power series expansion of B_v on v as

$$B_v = B_e - \alpha_e(v + \tfrac{1}{2}) + \dots ,$$

(2.62)

where B_e is the rotation constant ignoring the vibration-rotation interaction;[2] α_e is much smaller than B_e. By comparison, the following power series is the expansion for the centrifugal distortion constant:

$$D_v = D_e + \beta_e(v + \tfrac{1}{2}) + \dots ,$$

(2.63)

where D_e is the centrifugal distortion constant neglecting the effect of vibration; β_e is much smaller than D_e.

Considering the vibration-rotation interactions, the vibrational-rotational energy of a diatomic molecule is given by

---

[2] In the literature vibration-rotation interaction is frequently abbreviated by *vibronic*. For example, vibronic is used as subtitle in the subject index of Chemical Abstracts

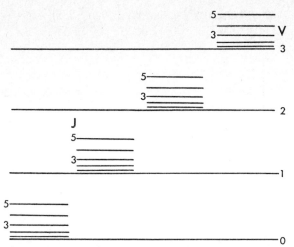

Fig.2.9 The vibration-rotational energy level scheme of a diatomic molecule

$$E_{vr} = hc\omega_e (v + \tfrac{1}{2}) - hc\omega_e x_e (v + \tfrac{1}{2})^2 + ... + hcB_v J(J + 1)$$

$$- hcD_v J^2 (J + 1)^2 + ... , \qquad (2.64)$$

as shown in Fig.2.9.

The vibrational-rotational energy levels of a polyatomic molecule are fairly complicated. Here we list only the simplest expressions, i.e., roughly taking the energy sum of the harmonic oscillator and the rigid rotator for them. For a linear polyatomic molecule it is

$$E_{vr} = \sum_i (v_i + \tfrac{1}{2}d_i) hc\omega_i + hcB_v J(J + 1) , \qquad (2.65)$$

where ω_i is the ith normal vibration frequency; d_i is the degeneracy for ω_i; and B_v is the rotation constant of the vth vibrational level. For a symmetric top molecule, e.g., a prolate top molecule, it is

$$E_{vr} = \sum_i (v_i + \tfrac{1}{2}) hc\omega_i + hcB_v J(J+1) + hc(A_v - B_v)K^2 . \qquad (2.66)$$

2.6 Coupling of Molecular Rotation and Electronic Motion

In the case where the nuclear-spin angular momentum can be neglected the angular momentum of a molecule results from the motion of the electrons in the orbits, the electronic spins and the nuclear rotation.

For most diatomic molecules the total orbital angular momentum and the total spin angular momentum of the electrons are zero in the electronic

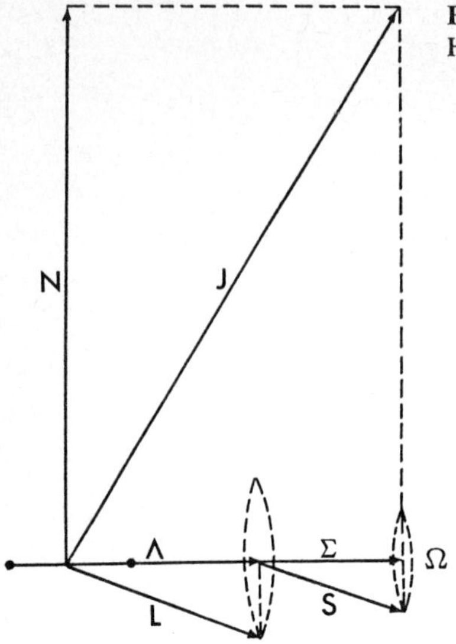

Fig.2.10. Coupling of the angular momenta in Hund's case (a)

ground state. So diatomic molecules usually have a $^1\Sigma$ ground state, where the only source of angular momentum of the molecule is nuclear rotation, the energy of which is determined by the quantum number J. For the electronically excited molecules, however, there is more than one source of angular momentum. It is important then to determine how these various angular momenta could be coupled together to give a resultant angular momentum and to obtain a better expression for the rotational energy of the molecule. Hund's coupling cases (a-e) are the five limiting cases which are enough to approximately explain some main experimental phenomena. However, there are many intermediate cases. The coupling schemes of angular momenta and the respective rotational energies for the Hund case (a-c) are summarized as follows.

Hund's case (a) applies to diatomic molecules with a small internuclear distance. Both the electron orbital and the spin angular momenta ($\mathbf{L}$ and $\mathbf{S}$) process about the internuclear axis, and have the angular momenta $\Lambda h/2\pi$ and $\Sigma h/2\pi$ in the direction of the axis, respectively. Therefore, the total angular momentum along the internuclear axis is $\mathbf{\Omega}$ ($=\Omega h/2\pi$) with the quantum number $\Omega = \Lambda+\Sigma$, as shown in Fig.2.10. The angular momentum $\mathbf{\Omega}$ and the molecular rotational angular momentum $\mathbf{N}$ ($=Nh/2\pi$) could be combined vectorially to form a resultant angular momentum with value of $[J(J+1)]^{1/2}h/2\pi$, here

$$J(J+1) = N(N+1) + \Omega^2 . \tag{2.67}$$

Then the rotational energy of a rigid diatomic molecule is given by

$$E_r = N(N+1)Bhc = [J(J+1) - \Omega^2]Bhc , \tag{2.68}$$

$$J = \Omega, \Omega+1, \Omega+2, ... \tag{2.69}$$

20

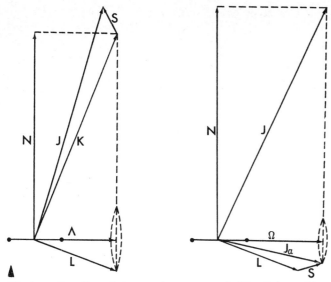

Fig.2.11 Coupling of the angular momenta in Hund's case (b)

Fig.2.12 Coupling of the angular momenta in Hund's case (c)

Obviously, the rotational levels with $J < \Omega$ are absent. For example, the levels with $J = 0$ do not occur for the $^3\Sigma_1$ state. The case where the interaction between S and Λ is negligible compared with that between N and Λ is called Hund's case (b). Here N combines with Λ to form a resultant angular momentum K by the following relation

$$K(K + 1) = N(N + 1) + \Lambda^2 . \tag{2.70}$$

The total angular momentum J is further formed from K and S, as shown in Fig.2.11, and the rotational energy is approximately given by

$$E_r = [K(K+1) - \Lambda^2]Bhc , \tag{2.71}$$

where

$$K = \Lambda, \Lambda+1, \Lambda+2, ... \tag{2.72}$$

For a given K-value, J takes the values

$$J = (K + S), (K+S-1), ... |K-S| . \tag{2.73}$$

It means that each rotational level has $2S+1$ components for a particular K-value.

In Hund's case (c) the internuclear distance is large enough for directly coupling between L and S. A resultant angular momentum J_a and its component Ω ($=\Omega h/2\pi$) along the internuclear axis are shown in Fig. 2.12. The quantum number Ω takes the values

21

$$\Omega = J_a, J_a-1, J_a-2, ... 1/2 \quad \text{or} \quad 0 . \tag{2.74}$$

The Ω and $\mathbf{N}$ vectors then combine to form the resultant angular momentum $\mathbf{J}$. Hund's case (c) has equations similar to (2.67 and 68) in Hund's case (a).

In Hund's cases (a) and (b) the interaction between $\mathbf{N}$ and $\mathbf{L}$ was neglected. For electronic states with $\Lambda \neq 0$, however, each rotational level with a particular J value ($J \neq 0$) splits into two components. The splitting of each rotational level increases with increasing J, and is termed Λ doubling.

2.7 Perturbations

Molecular perturbations arise from various interaction mechanisms. We may consider, as a starting point, the energy levels given by the Born-Oppenheimer approximation. As a result of molecular motion coupling we have such terms as the α_v correction term to account for the vibration-rotation coupling, the Λ doubling to account for the electron-rotation coupling, etc., as mentioned above. With higher spectral resolution much weaker perturbations, such as the interactions of the nuclear quadrupole or the magnetic dipole with other molecular angular momenta, might be observed as the hyperfine splitting or as shifts of a molecular vibration-rotational line, corresponding to the differences of the hyperfine splittings between the two related levels of the transition. This type of perturbation occurs all over the molecular vibration-rotational levels with the amount of shift as well as the number of sublevels depending smoothly on the molecular total angular momentum $\mathbf{J}$.

There is another kind of perturbation, which is characterized by an irregular distribution among molecular energy levels (and is typically recognizable by a J limit). This type of perturbation depends on the energy gap between the mutual perturbation levels, subject to the perturbation selection rules. Therefore one may include the former perturbations in the molecular constants, as discussed in previous sections. The latter perturbations will be briefly discussed in this section.

Introducing an interaction term into the wave equation, a shift of the nominal energy level will arise. The shift depends inversely on the energy gap between a given pair of interacting levels. Their shift directions are repulsive, i.e., the perturbed levels are moved upward and downward from the original higher and lower levels, respectively, by an equal amount. In fact, the terms of the perturbed levels T' are the solution of the following equation:

$$\begin{vmatrix} T_1-T' & H_{12} \\ H_{21} & T_2-T' \end{vmatrix} = 0 , \tag{2.75}$$

where T_1 and T_2 are the unperturbed term levels, and $H_{12} = H_{21}$ are the perturbation matrix elements. The solutions for (2.75) are in the form of

$$T' = \frac{T_1 + T_2}{2} \pm \sqrt{\left[\frac{T_1 - T_2}{2}\right]^2 + H_{12}^2}$$

$$= \frac{(T_1 + T_2)}{2} \pm \left(\frac{T_1 - T_2}{2}\right)\sqrt{1 + \left[\frac{2H_{12}}{T_1 - t_2}\right]^2} . \tag{2.76}$$

The signs of the solutions depend on the related level energies in such a way as to introduce repulsive effects. Assuming, for example, $T_1 > T_2$, we have

$$T_1' = T_1 + \Delta T_p ,$$
$$T_2' = T_2 - \Delta T_p , \tag{2.77}$$

where

$$\Delta T_p = \frac{T_1 - T_2}{2}\left[\sqrt{1 + \left[\frac{2H_{12}}{T_1 - T_2}\right]^2} - 1\right] \tag{2.78}$$

has a positive value. On the other hand, if $T_1 < T_2$, ΔT_p is negative corresponding to the opposite shifts in (2.77) compared with that for the $T_1 > T_2$ case. Equation (2.78) shows that the perturbation shift ΔT_p depends on both the matrix element H_{12} and the energy gap (T_1-T_2). The larger the matrix element H_{12}, and the smaller the energy gap (T_1-T_2) between the mutual perturbation levels, i.e., the smaller the denominator in (2.78), then the larger are the perturbation shifts. Consider two rotational term series $F_1(J)$ and $F_2(J)$ with a common level with respect to the rotational quantum number J_c, as shown by the broken lines in Fig.2.13. Their slopes correspond to the related rotational constants B_i by virtue of the relationship

$$T_i = T_{ei} + G(v)_i + B_i J(J+1) \quad (i = 1,2) \tag{2.79}$$

with $B_2 > B_1$. For the individual J values the energy gaps between these two term series are gradually changed, according to (2.79), as

$$T_1 - T_2 = \Delta T_0 + \Delta B J(J+1) , \tag{2.80}$$

where

$$\Delta T_0 = [T_{e1} + G_1(v)] - [T_{e2} + G_2(v)] ,$$

$$\Delta B = B_1 - B_2 .$$

Substituting (2.80) into (2.78), we arrive at the gradual perturbation shifts described. The deviations of the solid curves $F_i'(J)$ from the dashed lines $F_i(J)$ in Fig.2.13 illustrate the mutual perturbation shifts between the two-

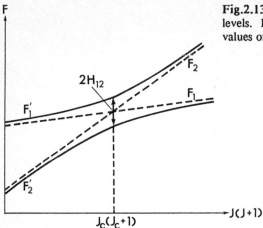

term series from the unperturbed terms as a function of $J(J+1)$, where at the intersection $J = J_c$ the perturbation shifts reach a maximum value, according to (2.78), of

$$\Delta T_{p\,max} = H_{12} \ . \tag{2.81}$$

The curves described by the perturbed rotational term are no longer linear. Instead of (2.79) they are given by

$$T_i' = T_i(J) + \Delta T_p(J) \ . \tag{2.82}$$

As mentioned above, $\Delta T_p(J)$ will change sign from $T_1 > T_2$ to $T_1 < T_2$, thus avoiding a crossing of the perturbed-term curves. The intersection position $J = J_c$ is the so-called perturbation center. The width of the effective perturbation region for J depends on the energy difference in (2.80), and is limited in practice by the maximum energy gap allowed by the detection sensitivity.

The perturbation element of matrix H_{12} is defined by

$$H_{12} = \int \Psi_1^* \mathscr{H}' \Psi_2 dr \ , \tag{2.83}$$

where $\mathscr{H}'$ is the perturbation Hamiltonian operator in (2.2). It has a specific content and form for each of the perturbation sources in (2.8 and 9). Let us consider a few examples.

Spin-orbital interaction is caused by the coupling between molecular electronic spin and orbital motions. The relative Hamiltonian operator can be written as

$$\mathscr{H}_{so} = \sum_k Z_k(r)(\hat{\ell}_k \hat{S}_k) \ , \tag{2.84}$$

$$Z_k(r) = \frac{\hbar e}{2m^2 c^2} \left[\frac{1}{r} \frac{\partial U}{\partial r} \right]_k \, , \tag{2.85}$$

where $\hat{\ell}_k$ and $\hat{S}_k$ are, respectively, the orbital and spin angular momentum operators of the kth electron in a molecule, $Z_k(r)$ is the coupling coefficient, and U is molecular potential function depending on internuclear distance r.

For a definite molecular electronic configuration with certain nonzero total electronic spin and orbital quantum numbers S and Λ, respectively, the spin-orbital coupling operator, corresponding to $\mathcal{H}_{so_1}$ in (2.8), can be expressed in the simple form

$$\mathcal{H}_{so} = A L \cdot S \, , \tag{2.86}$$

which provides nonzero values for all magnitudes of the angular momentum J. A is the coupling coefficient between the total spin and the total orbital angular momenta, which is about 7.85 cm^{-1} for the $b^3\Pi_{\Omega u}$ state in Na$_2$ [2.4]. Obviously, there are three components for S, corresponding approximately to 0 and $\pm A$ separations of the three fine electronic states. Their effects on the molecular energy levels are a shift of the rotational curves up or down almost constantly via the rotational quantum number J, as shown in Fig.2.14, where the fine electronic states are denoted by $\Omega = 0,1,2, \ldots$ ($\Lambda+S$) as subscripts. The constant A may then be considered a molecular energy-level constant.

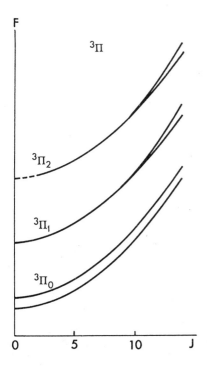

Fig.2.14. Fine splittings of a molecular electronic state due to spin-orbit coupling

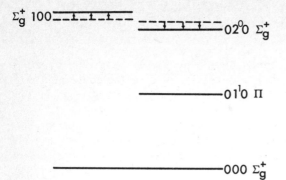

Σ_g^+ 100 ——————— ——————— $02^0 0$ Σ_g^+

——————— $01^1 0$ Π

——————— 000 Σ_g^+

Fig.2.15 Illustrating the Fermi resonance between vibrational levels of the CO_2 molecule

There is another kind of spin-orbital interaction occuring between different electronic configurations. As shown in Fig.2.1, for example, there is a large superimposed region between $A^1\Sigma_u^+$ and $b^3\Pi_{\Omega u}$ states in Na_2. When one substitutes (2.84) into (2.83), the integral would show a nonzero value only if the mutual perturbing states have the same total angular momentum J. This means that the molecular spin-orbital intraction between different electronic configurations belongs to the term $\mathscr{H}_{so_2}$ on the right-hand side in (2.9), corresponding to accidental perturbations via J. The perturbation shift dependence on J is described by the last term in (2.82).

Coriolis interaction arises from the coupling of a molecular frame rotation with the orthogonal molecular vibration. When two levels in different species have a Coriolis interaction, the extent of the Coriolis shifts is governed by the symmetric species. Typically the magnitude of Coriolis coupling as large as a few percent of a wave number, has been measured in CH_3I [2.5]. The present-day high-resolution spectral techniques with infrared diode lasers have enabled fundamental research into molecular physics via these phenomena. Coriolis interaction was included in $\mathscr{H}_{vr}$ in (2.9).

If, in a polyatomic molecule, two vibrational levels belonging to different vibrational modes but with the same species have nearly the same energy (i.e., accidental degeneracy) the perturbation known as *Fermi resonance* will take place. As an example, the Fermi resonance between the levels of (100) and ($02^0 0$) in CO_2 is illustrated in Fig.2.15, where the two broken lines represent the unperturbed levels (100) and ($02^0 0$). The effective rotational constants for the perturbed vibrational levels can be written as functions of the quantum number K. The perturbed terms are also described by (2.76). Fermi resonance was included in $\mathscr{H}_{vv}$ in (2.9).

Moreover, $\mathscr{H}_{vv}$ also includes the perturbation effect when two vibrational levels belonging to different electronic states lie in the neighborhood of the intersection of two pontential curves. The amount of perturbation depends on the degree of overlap between the related vibration wavefunctions. In Fig.2.16, for example, the vibrational levels 3 and 4 or 7 and 8 can influence each other, but no great perturbation will take place between levels 1 and 2 or 5 and 6. A strong perturbation with considerable repulsive

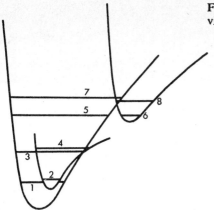

shifts may result in avoided crossing of the two potential curves. There is no strict selection rules for the vibrational quantum numbers.

The perturbation terms $\mathscr{H}_{ev}$ and $\mathscr{H}_{er}$ in (2.8) have been included in energy level expressions, e.g., (2.28 and 68), respectively. The perturbation operator $\mathscr{H}_{NI}$ in (2.8) indicates the spectral effects caused by molecular nuclear spin, producing so-called hyperfine splitting of molecular vibration-rotational levels, as discussed in the following sections. There are still more angular momentum couplings which can perturb molecular levels. The reader is refered to the literature for details [2.6].

2.8 Quadrupole Hyperfine Structure of Molecules

Molecular spectra contain hyperfine structures which depend on the characteristics of the nuclei in the molecule, including the isotope effect, the isomeric nuclear effect due to one of the nuclei in a molecule being excited to a substable state, the electric quadrupole moment effect due to the deviation of the charge distribution in a nucleus from spherical shape, and the magnetic dipole effect due to the motion of the intranuclear charge coupling with electronic orbital motion, and so on.

The quadrupole effect is usually the dominant source of molecular hyperfine structure. The Hamiltonian operator of the nuclear quadrupole interaction is the inner product of the quadrupole-moment operator of a nucleus and the gradient operator of the electric field due to the charge distribution surrounding it, i.e.,

$$\mathscr{H}_Q = -\frac{1}{6}\mathscr{Q} : \nabla\mathscr{E} . \tag{2.87}$$

For a rotating molecule it can be expressed as [2.2]

$$\mathscr{H}_Q = eQ\left[\frac{\partial^2 V}{\partial z_J^2}\right]_{av} \frac{3(\mathbf{I}\cdot\mathbf{J})^2 + 3(\mathbf{I}\cdot\mathbf{J})/2 + |\mathbf{I}|^2|\mathbf{J}|^2}{2I(2I-1)J(2J-1)} , \tag{2.88}$$

27

where e is the charge of a proton, Q is the nuclear quadrupole moment, and

$$g_J \equiv \left(\frac{\partial^2 V_0}{\partial z_J^2} \right)_{av}$$

is the average gradient of the electric field along the direction of the rotational angular momentum. A nucleus with $I = 0$ or $\frac{1}{2}$ has zero quadrupole moment, $Q = 0$; whereas $Q \neq 0$ only for $I > \frac{1}{2}$. If there are n nuclei with $I > \frac{1}{2}$, the total quadrupole moment Hamiltonian will be

$$\mathcal{H}_Q = \sum_{i=1}^{n} \mathcal{H}_{Qi} \tag{2.89}$$

where the individual $\mathcal{H}_{Qi}$ have the form of (2.88).

The quantum number F of the total angular momentum $F = I+J$ takes the values $J+I$, $J+I-1$, $J+I-2$, ..., $|J-I|$. The statistical-average values of $|I|^2$, $|J|^2$ and $I \cdot J$ are $I(I+1)$, $J(J+1)$ and $[F(F+1)-I(I+1)-J(J+1)]/2$, respectively. Then the hyperfine splittings of the molecular energy levels produced by the quadrupole effect of a single nucleus can be expressed as

$$W_Q = \frac{eq_J Q}{2I(2I-1)J(2J-1)} \left[\frac{3}{4} C(C+1) - I(I+1)J(J+1) \right], \tag{2.90}$$

where $eq_J Q$ is the nuclear quadrupole coupling constant, $c = \frac{1}{2}[F(F+1) -I(I+1)-J(J+1)]$.

In a linear molecule the charge distribution is symmetric around the molecular axis. Let z_m indicate the direction of the molecular axis, (2.90) becomes

$$W_Q = \frac{-eq_m Q}{2I(2I-1)(2J-1)(2J+3)} \left[\frac{3}{4} C(C+1) - I(I+1)J(J+1) \right]$$
$$= - e\, q_m Q\, F(I,J,F) , \tag{2.91}$$

where $F(I,J,F)$ is the so-called Casimir's function. When F has its maximum or minimum values, $F(I,J,F)$ is positive; towards intermediate values of F the function is turning negative. Figure 2.17 shows the magnetic dipole (a) and electric quadrupole (b) splittings of the vibration-rotational energy level with $J = 5$ and the nuclear spin of $I = 5/2$. In Fig.2.17b the components corresponding to intermediate values of F are shifted downwards, whereas the components for maximum or near-minimum values of F are shifted upwards.

For a symmetric-top molecule we have

$$W_Q = \frac{eq_m Q \left(\frac{3K^2}{J(J+1)} - 1 \right)}{2I(2I-1)(2J-1)(2J+3)} \left[\frac{3}{4} C(C+1) - I(I+1)J(J+1) \right]. \tag{2.92}$$

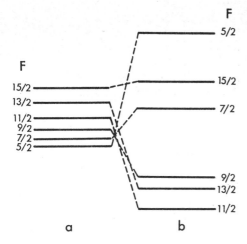

F

F

5/2

15/2

13/2
11/2
9/2
7/2
5/2

15/2

7/2

9/2
13/2
11/2

a   b

Fig.2.17. Molecular rotational level splittings in the forms of (a) magnetic dipole hyperfine structure and (b) electric quadrupole hyperfine structure

Since K may take $(2J+1)$ values, i.e., $K = J$, $J-1$, $J-2$, ..., $-J$, the number of hyperfine components will be more than that in the case of a linear molecule, which corresponds to $K = 0$.

If the quadrupole interaction energy is appreciable compared with the rotational energy, the second-order quadrupole effect should be considered. Of course, the latter effect on the hyperfine splittings would be a few orders smaller than the first-order effect.

2.9 Magnetic Dipole Hyperfine Structure in Molecules

When a molecule has electronic angular momentum (either orbit or spin angular momentum), the magnetic fields associated with this momentum will interact with the nuclear-spin magnetic dipole moment to produce a magnetic dipole hyperfine structure.

In fact, there are several contributions to the intramolecular magnetic field at a nucleus under consideration, in addition to the contributions of the orbital motion and the spin of the unpaired valence electrons. These are the contributions from the spins of other nuclei in the molecule, the weak magnetic field due to molecular frame rotation and the so-called contact field due to the penetrating-like motion of the valence electrons, etc. Owing to the fact that the spin angular momentum of the electrons is three orders of magnitude larger than that of the nucleus the major contribution from the above-mentioned sources is that from unpaired valence electrons. The other sources mentioned might be significant only if the total angular momentum of the valence electrons in a molecule is zero ($^1\Sigma$ state). The frame rotation field is very weak but exists for every molecular energy state, mostly contributing magnetic interaction energy to the molecular $^1\Sigma$ states.

The magnitude of the interaction energy of a given nucleus with molecular magnetic fields depends on the coupling of its spin with other molecular vectors. Just as the three vectors of **L**, **S**, and **J** form different coupling patterns for the Hund cases (a, b) and so on to give different expressions for the rotational energy terms (Sect.2.6), the various coupling schemes between the four vectors **I**, **L**, **S** and **J**, will result in different expressions for the hyperfine splittings as an additional energy term for a given molecular rotational term. Depending on **I** coupled with the molecular axis or with **J**, for example, the coupling pattern for the original Hund's case (a) is denoted by Hund's case (a_α and a_β), respectively. For the original Hund's case (b) there are the cases of ($b_{\beta K}$, $b_{\beta S}$, and $b_{\beta J}$) according to **I** coupled with **K**, **S** and **J**, respectively. Among them, Hund's case ($b_{\beta J}$) is the most common, and **S** is almost always coupled before **I** due to the fact that the electronic magnetic moment is about 1836 times larger than the nuclear one. This implies that Hund's case ($b_{\beta K}$) will, in fact, never be present. For the most common coupling cases the hyperfine energy terms due to the magnetic interactions can be derived from vector models [2.2] as

α) for Hund's case (a_α)

$$W_{mag} = [a\Lambda + (b + c)\Sigma]\Omega_I - \hbar B[\Omega_I + 2(\Lambda + \Sigma)]\Omega_I , \qquad (2.93)$$

β) for Hund's case (a_β)

$$W_{mag} = [a\Lambda + (b + c)\Sigma](\Lambda + \Sigma)\mathbf{I}\cdot\mathbf{J}/J(J + 1) , \qquad (2.94)$$

γ) for Hund's case ($b_{\beta J}$) with $S = \tfrac{1}{2}$

$$W_{mag}(J = K+\tfrac{1}{2}) =$$

$$\left[\frac{2a\Lambda^2}{(K + 1)(2K + 1)} + \frac{b}{(2K + 1)} + \frac{c}{(2K + 1)(2K + 3)}\left(1 + \frac{2\Lambda^2}{K + 1}\right) \right] \mathbf{I}\cdot\mathbf{J} ,$$

$$W_{mag}(J = K-\tfrac{1}{2}) =$$

$$\left[\frac{2a\Lambda^2}{K(2K + 1)} - \frac{b}{2K + 1} - \frac{c}{(2K - 1)(2K + 1)}\left(1 - \frac{2\Lambda^2}{K}\right) \right] \mathbf{I}\cdot\mathbf{J} , \qquad (2.95)$$

and with $S = 1$

$$W_{mag}(J = K+1) =$$

$$\left[\frac{a\Lambda^2}{(K + 1)^2} + \frac{b}{(K + 1)} + \frac{c}{(K + 1)(2K + 3)}\left(1 + \frac{2\Lambda^2}{K + 1}\right) \right] \mathbf{I}\cdot\mathbf{J} ,$$

$$W_{mag}(J = K) =$$

$$\left[\frac{a\Lambda^2(K^2 + K - 1)}{K^2(K + 1)^2} + \frac{b}{K(K + 1)} + \frac{c}{K(K + 1)}\left(1 - \frac{2\Lambda^2}{K(K + 1)}\right) \right] I \cdot J \ , \quad (2.96)$$

$$W_{mag}(J = K-1) = \left[\frac{a\Lambda^2}{k^2} - \frac{b}{K} + \frac{c}{K(2K - 1)}\left(1 - \frac{2\Lambda^2}{K}\right) \right] I \cdot J \ ,$$

where Λ, Σ and Ω_I are the quantum numbers of the projections of L, S, and I, respectively, on the molecular axis, K is the quantum number of N (frame rotation angular momentum); a, b, and c are constants determined by the distances of the electrons from the considered nucleus and by the molecular rotational inertia, and the scalar product $I \cdot J$ is

$$I \cdot J = [F(F + 1) - I(I + 1) - J(J + 1)]/2 \ . \quad (2.97)$$

So even for nonlinear molecules the magnetic hyperfine splitting can simply be expressed as

$$W_{mag} = 2C(I \cdot J) = C[F(F + 1) - I(I + 1) - J(J + 1)] \ , \quad (2.98)$$

where C is the so-called magnetic hyperfine constant, determined by molecular parameters. The lower the value of F, the smaller are the hyperfine shifts from an original rotational level, as shown in Fig.2.17. For large J, (2.98) is approximately given by

$$W_{mag} = CM_I J \ , \quad (2.99)$$

where $M_I = I, I-1, I-2, ..., -I$.

As an example, Table 2.1 lists the electric quadrupole and magnetic dipole hyperfine constants for methyl-halides and the homonuclear molecules with nuclear spin larger than or equal to 1/2. It shows that in the electronic ground state the magnetic hyperfine constant is often an order of magnitude smaller than the quadrupole constant. On the other hand, with one or more excited electrons the magnetic constant may exceed the quadrupole constant, as observed in Table 2.1 for Na_2.

2.10 Isotopic Energy-Level Shifts

Isotopic molecules differ only by the replacement of one or more atoms by its isotope(s), without frame structure changing of the molecule. Having a different reduced mass μ the molecular constants of such molecules, ω_e, $\omega_e x_e$, B_e, D_e and α_e etc., are varied and the energy levels are shifted. Let the subscript o represent the ordinary form of a molecule and i an isotopic

Table 2.1. Examples of molecular hyperfine constants

Molecule	State	eqQ [MHz]	C [KHz]		Ref.
CH_3F	ground	0	30		2.99
CH_3Br	ground	-74.41	~1.5		
CH_3Cl	ground	577.15	-0.72		
CH_3I	ground	-1934	~5		
I_2 [a]	ground	$(-3400\pm400)Q$	$(\leq34\pm9)P$		2.100
	low-lying state B	$(-1000\pm400)Q$	$(\dfrac{1.2\cdot10^5}{4400-E}-12.5)P$		
	ground	0	0		
Na_2	$b^3\Pi_{\Omega u}$ [b]	0	5000-9500	$(\Omega=0)$	
			<160	$(\Omega=1)$	2.101
			-6500	$(\Omega=2)$	2.102
	$(3)^3\Pi_{\Omega g}$ [c]	0	-1600	$(\Omega=0)$	
			~300	$(\Omega=1)$	2.103
			~800	$(\Omega=2)$	

[a] For $^{127}I_2$, Q = 0.79, P = 1.1232; for $^{129}I_2$, Q = 0.55, P = 0.7477; E is the vibrational energy of the upper level in cm^{-1}
[b] For the fine electronic state with $\Omega = 0$ the constant C was found experimentally to be approximately inversely proportional to J
[c] The data were determined for large J of about 60

variant. We then have the ratios between the molecular constants for two isotopic molecules as follows:

$$(\omega_e)_i/(\omega_e)_o \quad = (\mu_o/\mu_i)^{1/2} = \rho\,,$$

$$(\omega_e x_e)_i/(\omega_e x_e)_o = \mu_o/\mu_i = \rho^2\,,$$

$$(B_e)_i/(B_e)_o \quad = \mu_o/\mu_i = \rho^2\,,$$ (2.100)

$$(D_e)_i/(D_e)_o \quad = (\mu_o/\mu_i)^2 = \rho^4\,,$$

$$(\alpha_e)_i/(\alpha_e)_o \quad = (\mu_o/\mu_i)^{3/2} = \rho^3\,.$$

Assume the subscript i to refer to the heavier isotope. The constant ρ is then smaller than 1. This means that the levels of the heavier isotopic molecule lie below the corresponding levels of the lighter one. The smallest value of ρ will occur for isotopic molecules composed only of hydrogen atoms. Among their energy-level constants, on the other hand, the absolute shifts will depend on the magnitudes of the relevant constants according to (2.100). While the isotopic change in the centrifugal constant D_e of D_2 from that of H_2 is only about 0.035 cm^{-1}, the normal vibrational frequency ω_e of D_2 from H_2 shows an isotopic change of more than 1000 cm^{-1}. For the molecules with heavy isotopic elements, however, ρ approaches 1 and produces much smaller isotopic effects than those of H_2. For a discrimination of heavy isotopic molecules on the basis of the effect of their total nuclear spins on the hyperfine structures requires high-resolution spectroscopy (Chap.5).

There are other weaker hyperfine effects existing in molecular energy-level structures, such as the recoil effect due to the molecular momentum conservation during the absorption of radiation is the order of magnitude of a few kilohertz, and the even weaker effects such as the isomeric shifts caused by the excitation of a nucleus to a long-lifetime state or the shifts due to the different frame structures of molecules. With high-resolution laser spectroscopy some effects have been demonstrated. A few examples are described in Chap.5.

2.11 Molecular Rydberg States

If one of the outer electrons in a molecule is in a highly excited state, whose orbtial motion is relatively far from the molecular core (consisting of the nuclei and the remaining electrons), it can be considered approximately equivalent to a single electron interacting with a charged core. In such a case the energies are approximately described by the term formula for a hydrogen atom. Such states are the so-called molecular Rydberg states, which have recently been studied spectrocopically in the optical domain.

Two models are usually used to describe molecular Rydberg sates. The first model describes an outer electron interacting with a molecular core in the ground state. The energy terms $T_e(n,\Lambda)$ tending to a constant deviation from those of the hydrogen atom

$$E(n) = E_\infty - \frac{R}{2n^2} , \qquad (2.101)$$

or being of the form of a hydrogenlike atomic energy level $E(n,\ell)$ as

$$T_e(n,\Lambda) = E(n,\ell) = E_\infty - \frac{R}{|n - \delta(n,\ell)|^2} , \qquad (2.102)$$

where E_∞ is the ionization potential of a series of Rydberg states; n and ℓ indicate the principle and the angular momentum quantum numbers, respectively; δ is the so-called quantum defect, determined by n and ℓ. As the electronic-energy part of the molecule, $T_e(n,\Lambda)$ corresponds to the minimum (the bottom) of the potential curve, with the relevant principle quantum number n and the total angular momentum quantum number ℓ.

A second useful model is the molecular orbital one. It describes a molecular state combined with the electronic configurations of the individual atoms in it. For example, the electronic configurations of a sodium dimer with two outer electrons of (3s+ns) gives rise to the molecular Rydberg series $(n)^1\Sigma_g^+$; (3s+nd) to the $(n)^1\Sigma_g^+$, $(n)^1\Pi_g$ and $(n)^1\Delta_g$ series; (3p+np) to the $(n)^1\Sigma_g^+$, $(n)^1\Sigma_g^-$, $(n)^1\Pi_g$, and $(n)^1\Delta_g$ series.

The first model is helpful for describing the characteristics of each molecular Rydberg series; the second model can be used to identify them. Usually at a higher energy than three quarters of the ionization potential the molecular Rydberg properties become evident, as summarized as follows:

(i) High excitation energy;

(ii) large orbital motion of the outer electron, whose distance from the molecular core may be much larger than the internuclear distance;

(iii) The influence of the core's electric potential on the outer electron motion is weak. The outer electron has only a weak effect on the molecular binding, so that even with such high inner energy a molecule in Rydberg states is not easily dissociated, although it is easily ionized or autoionized. On the other hand, however, the molecular frame vibration seriously influence the outer electron motion. Some new spectral phenomena have demonstrated the complete absence of Rydberg states with vibrationally excited cores [2.7];

(iv) ionized by collisions, the external electric field or the interactions between molecular Rydberg states with high efficiency;

(v) low fluorescence efficiency. The molecular lifetime in a Rydberg state, depending on its relaxation channels and environment, can be as much as 3 orders of magnitude larger or smaller than the spontaneous emission lifetime of low-lying electronic states. For example, lifetimes as long as one microsecond for the stable Rydberg states in molecular hydrogen [2.8,9] or as short as tens of femtoseconds in benzene and toluene have been measured by time-resolved ionization [2.7].

In view of these interesting specific properties, the related nuclear-physics mechanisms and processes, and the future scientific or practical applications, molecular Rydberg states have been extensively studied and have become an attractive topic during the last decade. The studies are about to measure or calculate molecular energy-level constants, to determine the values of precise ionization potentials, to examine or interpret the intramolecular or intermolecular interactions between Rydberg states and their interactions with external fields or incident particles, to determine the lifetimes and their influence, to explore various fast and efficient experimental techniques, and to develop a theory for the Rydberg states of various molecules.

Since the studies on the hydrogen molecule by *Herzberg* [2.10], the topic has been extended to many homonuclear diatomic molecules such as Li_2 [2.11-14], Na_2 [2.15-17], K_2 [2.18,19], Cs_2 [2.20], N_2 [2.21-26], and O_2 [2.26,27], in addition to the most detailed studies on H_2 [2.28-38,68] and its isotopic variants HD and D_2 [2.31]; heteronuclear diatomic molecules such as NO [2.26,39-41,63], BeH [2.42], CO [2.26], PO [2.45], SiF [2.43,44], etc.; linear or nonlinear polyatomic molecules such as CO_2 [2.26], N_2O [2.26], COS [2.46], H_2S, CS_2 [2.47,48], NH_3 [2.49], etc.; many Rydberg states of

34

the ground-state dissociated molecules and Van der Waals molecules such as ArH, H_3, NH_4 [2.50-52], He_2 [2.53], H_2F, H_3O, CH_5 [2.54], Kr_2 [2.55], Ar-CO_2 and Kr-CO_2 [2.56], etc; and even a lot of complicated organic molecules such as methyl iodide [2.57,58], benzene [2.59], trans-butadiene [2.60], trans-1,3,5-hexatriene [2.61], acetylene, acetonitrie and cyanamide [2.62], etc.. Among them those studies involving molecular hydrogen are fundamentally interesting for the simplicity of the structure of this molecule facilitates comparison with theoretical calculations. Compared to H_2, the advantage of the alkali dimers regarding Rydberg states studies is their relatively low ionization potential. Therefore, all of their "gerade" Rydberg states with even parity of their wave functions up to the ionization limit are accessible by two-step excitation with two visible lasers. As examples, some of the recent achievements by various methods in the study of Rydberg states in hydrogen and alkali dimers are listed in Table 2.2.

The new results on the quasidimer He_2 were obtained by using the conventional spectral analysis technique for the emission spectra consisting of almost the entire Rydberg series of bands [2.54]. The conventional spectrographic technique for single-photon absorption spectra was also used to investigate the series of molecular Rydberg states, such as in N_2, H_2S, D_2S, H_2Te and D_2Te [2.64]. Nevertheless, single-photon spectroscopy suffers many limitations. Most of the excited states of interest lie at energies that correspond to the vacuum ultraviolet region. Taking advantage of laser sources, most published results so far were performed by means of double-resonance, two-step, two-photon or multiphoton-excitation methods. Perhaps the most efficient method is the two-step polarization-labelling spectroscopy first used by *Carlson* et al. for exploring numerous unknown molecular Rydberg states [2.15,65,66]. They used a polarized pumping laser beam to populate unequally the degenerate angular momentum sublevels of the upper state. The induced optical anisotropy causes a change in the polarization properties of a second laser beam, which excited the molecules to still higher levels. It is notable that the method does not require a narrow-band probe laser to resolve the Rydberg states. By using a broad-band (30nm) probe laser it presents a simplified spectrum at a spectrograph behind the analyzer. Two dozen of new electronic states were found quickly and efficiently, and have been identified (Fig.2.18). The molecular constants were all determined with very good accuracy. The new states include Σ, Π and Δ states from the electron configurations of (3s+ns), (3s+np) and (3s+nd), etc., which can therefore be abbreviated as (n)Σ, (n)Π, (n)Δ, etc. Each of the energy sum values for a given electron configuration corresponds to the dissociation energy of a molecular electronic state (indicated in the figure is the atomic asymptote). The "bottom of the potential well" corresponds to the electron energy Te(n,Λ) in the corresponding excited state. The maximum n value of the observed states in the experiments was 14. Substituting the measured Te(n,Λ) values in (2.102), the deviations $(T_e - E_n)$ of the electronic excitation energies from the respective terms of the atomic hydrogen (shown in Fig.2.19) and the magnitudes of the quantum defects $\delta(n,\ell)$ (shown in Fig.2.20) were obtained. These two figures reveal that n > 8 states in the Δ and Π state series possess typical Rydberg

Table 2.2. Examples of experimental studies of molecular Rydberg states in hydrogen and alkali dimers

Molecules	Rydberg electron & states				Experimental arrangement
	Symbols	n	$(\ell_i)_{max}$	J	
H_2	Singlet np	30	1	0-2	Cooled gas; VUV(83.5-76.5nm) single-photon absorption
	Singlet np	25-90	1	0-3	DF; two-step, two-color
	Triplet nf	10-20 25-35	2		Electron impact and one-photon excitation
	Triplet nf	10-19	3		DF; electron bombard. 2 laser excitation
	Singlet nf,ng, nh,ni	10,27	3-6		Charge exchange & Stark or microwave ionization
Li_2	Singlet nd, np	4,5	2		DF; two-step, polarization
	Singlet	32			Stepwise, mol. beam
	Singlet	3-15	0,1		OODR
Na_2	$^1\Sigma_g^+$ $^1\Pi_g$ $^1\Delta_g$	3-14	0-2	19-41 70	Two-step polarization labelling
	Singlet	13-29			OODR
	$^1\Sigma_g,^1\Pi_g$ $^1\Delta_g$	15-90	2	4-45	OODR, mol. beam
	$^3\Sigma_g,^3\Pi_g$ $^3\Delta_g$			Small-J	OODR
	Triplet				Two-step polarization labelling
	Singlet Triplet	3	1	Large-J	DF, equal-frequency two-photon
	$ns\sigma\,^1\Sigma_g^+$ $nd\sigma\,^1\Sigma_g^-$ $nd\Pi\,^1\Pi_g$ $nd\delta\,^1\Delta_g$	28	0-2	0	

Results		Ref.
Mol. constants	Ionization potential [cm^{-1}]	
Perturbation & pressure shift constant for H_2^+	124417.2±0.4	2.10
Perturbation	124417.61±0.07 (para)	2.81
	124475.94±0.07 (ortho)	2.82
Perturbation		2.91
	124417.50 (para)	2.90
	124417.53 (ortho)	
FS & mag. HFS parameters for H_2^+		2.30
		2.83
		2.84
Dunham coeff. for one state; perturbation		2.14
Constants for Li_2	41236.4±2.5	2.12
		2.13
Constants for Li_2 & Li_2^+	41496±4	2.11
Dunham coefficient for 24 states of Na_2; consts. of Na_2^+ +	39490	2.15
		2.92
Coupling status of angular momentum		2.16
Check Hund's case		2.87
Dunham coefficient for 4 states		2.88
Dunham coefficient for 7 states		2.86
Dunham coefficient, mag., HFS consts.		2.89
iden. methods		2.69
Constants for Na_2^+		2.93
		2.17

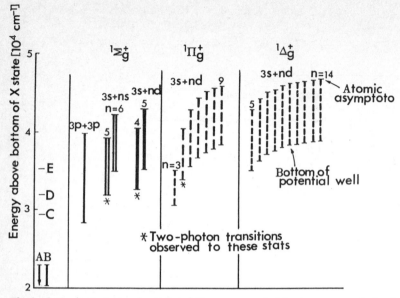

Fig.2.18 Excited electronic states of Na_2, as revealed by two-step polarization labelling [2.94]

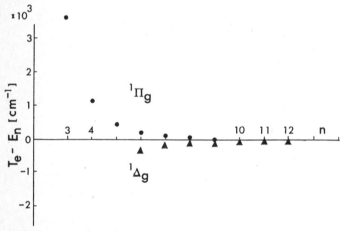

Fig.2.19. Deviations of the electronic energy for Na_2 molecular Rydberg states $(n)^1\Pi_g$ and $(n)^1\Delta_g$ from the energy term of atomic hydrogen [2.86]

characters. This means that the point-charge model is a fairly good approximate descriptions for this case. Figures 2.21 to 23 display other aspects of the series character of such molecular Rydberg states. Shown are the dissociation energies, vibrational constants and rotational constants as functions of $1/n^2$. They can all be extrapolated to obtain good values for the constants of the ground state of the Na_2^+ ion [2.67].

It is known that the density of Rydberg states increases as n^3 with increasing principal quantum number n. In addition, doubly excited states

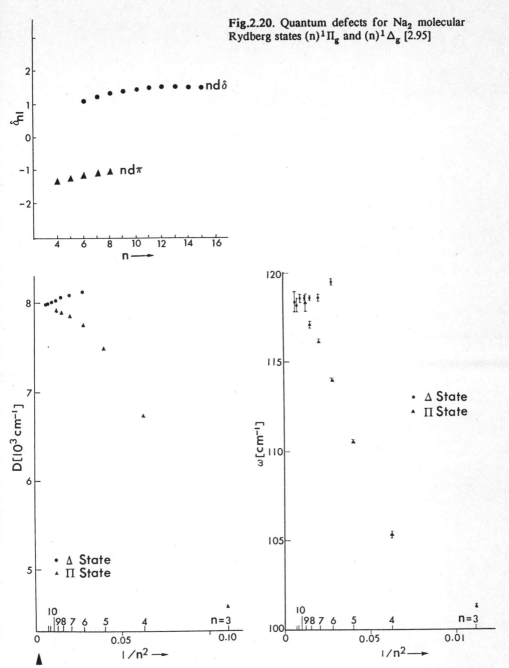

Fig.2.20. Quantum defects for Na_2 molecular Rydberg states $(n)^1\Pi_g$ and $(n)^1\Delta_g$ [2.95]

Fig.2.21. Dissociation energy for Na_2 molecular Rydberg states $(n)^1\Pi_g$ and $(n)^1\Delta_g$ as a function of $1/n^2$ [2.94]

Fig.2.22. Vibrational constant for Na_2 molecular Rydberg states $(n)^1\Pi_g$ and $(n)^1\Delta_g$ as a function of $1/n^2$ [2.96]

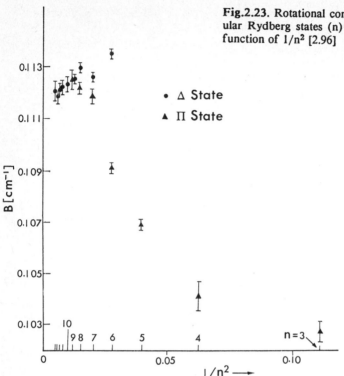

Fig.2.23. Rotational constant for Na_2 molecular Rydberg states $(n)^1\Pi_g$ and $(n)^1\Delta_g$ as a function of $1/n^2$ [2.96]

• Δ State

▲ Π State

may overlap with Rydberg states, causing various perturbations of the Rydberg series. The complicated coupling statuses between molecular angular momenta make the assignment of complex Rydberg spectra a difficult task. Therefore more precise techniques are used to reveal the spectral details with Doppler-free resolution. For example, techniques involving equal-frequency two-photon transitions allow one to distinguish singlet or triplet states and to measure directly the hyperfine constants of Rydberg states [2.69]. Also optical-microwave double-resonance [2.84], and molecular-beam techniques [2.85] etc. have been applied. A notable new technique has been developed, namely Doppler-free optical double-resonance polarization spectroscopy, in which different branch transitions show distinguishable signals according to differences in phases, line shapes, or intensity changes. This then gives unambiguous assignments to a perturbed spectrum [2.14].

The knowledge for understanding molecular Rydberg states is thus extensive. So far the highest principal quantum number n is 90 in molecular sodium [2.30]. The autoionization character for large n states is noticeable. The corresponding detection is therefore performed by ion-sensitive devices, often combined with a mass spectrometer [2.97,98]. Since different series with distinct ℓ values show great disparity in natures, the Rydberg series with large ℓ value is of interest. The angular momentum states as high as ℓ = 3-6 can be formed by charge-capture collisions with a fast beam and can then be further excited by a laser up to high-n and high-ℓ states [2.30], while in pure spectroscopic excitation the highest angular momentum quantum number is limited by the absorbed photons of a molecule. The character of single-electron excited states of H_2 changes markedly when the coupling of the excited electron's angular momentum to the orientation of the H_2^+ internuclear axis is much less than the free rotational energy of the ion core. For Rydberg states of low ℓ, this condition is satisfied only for sufficiently high n [2.68,69]. States with $\ell > 2$, however, are always near the uncoupled limit (Hund's case d) because of the rapid decrease in coupling for nonpenetrating orbitals. Figure 2.24b shows the uncoupling of the electronic angular momentum from the molecular axis to the axis of frame rotation, corresponding to the transfer from Hund's case (b) towards case (d). The treatments have been extended and completed by the recent observations of Rydberg states with $\ell > 2$ in H_2 [2.30,33], NO [2.42], and N_2 [2.22]. In this case, both the Rydberg electron's total orbital angular momentum (**L**) and the core rotational angular momentum (**O**) are approximately good quantum numbers which couple to form the total angular momentum **N = L+O** exclusive of spin. The radiative transitions between such states take place without change in core states ($\Delta v = \Delta J = 0$) and obey the usual atomic selection rules, $\Delta L = \pm 1$, $\Delta N = 0, \pm 1$.

The Multichannel Quantum Defect Theory (MQDT) has been applied to molecules [2.70]. In recent years it has emerged as a powerful theoretical tool for the calculation or analysis of the Rydberg states of molecular hydrogen [2.71,72], alkali dimers [2.73,74] and other diatomic or polyatomic molecules [2.54,75-79]. The readers are refered to the recent review by *Greene* and *Jungen* for a detailed discussion of molecular applications of quantum defect theory [2.80].

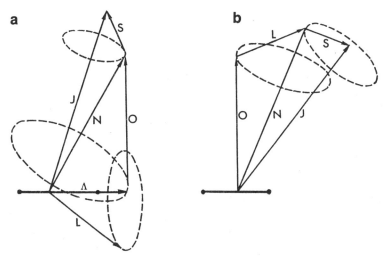

Fig.2.24 Angular momentum coupling schemes for molecular Rydberg states: (a) Hund's case (b) for low-n low-ℓ ($\ell < 2$) states; (b) Hund's case (d) for high-n high-ℓ states [2.2]

3. Linear Molecular Spectroscopy

On the basis of the energy-level discussion of Chap.2 we summarize here the various kinds of molecular spectra due to linear interactions of molecules with laser fields. For a detailed analysis of molecular-spectra regularities the reader is referred to [3.1-4].

3.1 Infrared Pure-Rotational Spectra

According to (2.44 and 47) for a diatomic molecule as a rigid rotor the allowed transition frequencies are

$$\nu = F(J + 1) - F(J) = 2B(J + 1) , \quad J = 0,1,2,3,... . \tag{3.1}$$

Therefore the rotational spectrum consists of a set of uniformly spaced lines with the spacing of 2B.

With the term formula (2.48) for the nonrigid rotor model, the transition frequencies are

$$\nu = F(J + 1) - F(J) = 2B(J + 1) - 4D(J + 1)^3 + ... \tag{3.2}$$

Obviously the lines are no longer exactly equidistant but their spacing decreases lightly with increasing J in agreement with experimental data.

For a linear polyatomic molecule with (2.52 and 53) the formula for the transition frequencies is the same as (3.1).

According to the term formula (2.55) and the selection rule (2.57) for an oblate symmetric top molecule in rigid rotor approximation the transition frequencies are

$$\nu = F(J + 1, K) - F(J, K) = 2B(J + 1) \tag{3.3}$$

as shown in Fig.3.1a. For a nonrigid rotor model, the following equation is obtained from (2.58):

$$\nu = F(J + 1, K) - F(J, K) = 2B(J + 1) - 4D_J(J + 1)^3 - 2D_{JK}(J + 1)K^2$$
$$= 2(J + 1)(B - D_{JK}K^2) - 4D_J(J + 1)^3 . \tag{3.4}$$

The rotational lines with the same J and different K values (Fig.2.7) can be resolved by a high-resolution spectrometer. Figure 3.1b exhibits the K-splitting for pure rotational lines.

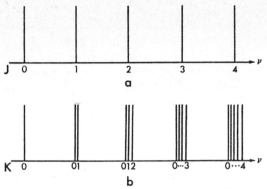

Fig.3.1. Pure-rotational transition frequencies for an oblate symmetric top molecule. (a: Rigid rotator approximation, b: K-splitting)

3.2 Infrared Vibrational Spectra

If the third and higher terms on the right-hand side of (2.29) are neglected, the vibrational transition frequencies of a diatomic molecule from the v'' to v' states are

$$\nu = (v' - v'')\omega_e - [v'(v' + 1) - v''(v'' + 1)]\omega_e x_e . \tag{3.5}$$

Taking $v'' = 0$, (3.5) becomes

$$\nu = \omega_e v' - v'(v' + 1)\omega_e x_e . \tag{3.6}$$

Then

$$\nu_1 = (1 - 2x_e)\omega_e , \quad \text{for } 0 \rightarrow 1 , \tag{3.7}$$
$$\nu_2 = 2(1 - 3x_e)\omega_e , \quad \text{for } 0 \rightarrow 2 , \tag{3.8}$$
$$\nu_3 = 3(1 - 4x_e)\omega_e , \quad \text{for } 0 \rightarrow 3 . \tag{3.9}$$

If, as mentioned in Chap.2, the anharmonic constant $\omega_e x_e$ is very small compared with ω_e, the ratios $\nu_1 : \nu_2 : \nu_3 : \nu_4$ are then roughly $1:2:3:4$.

Owing to the complex vibrational levels compared with diatomic molecules, the vibrational transitions of polyatomic molecules are complicated, too. For example, in the simplest case of a triatomic molecule the frequencies of the vibrational bands can be expressed approximately as

$$\nu = \Delta v_1 \omega_1 + \Delta v_2 \omega_2 + \Delta v_3 \omega_3 , \tag{3.10}$$

where Δv_1, Δv_2, and Δv_3 are small integers, including zero; ω_1, ω_2, and ω_3 are the three fundamental vibrational frequencies. When $\Delta v_1 = \Delta v_2 = 0$ and $\Delta v_3 = 2$, we have $v = 2\omega_3$, i.e., the first overtone; while for $\Delta v_1 = \Delta v_2 = \Delta v_3 = 1$, the combination band $(\omega_1 + \omega_2 + \omega_3)$ is obtained, etc.

Table 3.1. Infrared bands of H_2O vapor[a]

ν [cm^{-1}]	Upper state $v_1 v_2 v_3$	Lower state $v_1 v_2 v_3$	ν [cm^{-1}]	Upper state $v_1 v_2 v_3$	Lower state $v_1 v_2 v_3$
1595	0 1 0	0 0 0	12151	2 1 1	0 0 0
3152	0 2 0	0 0 0	12565	0 1 3	0 0 0
3657	1 0 0	0 0 0	13831	3 0 1	0 0 0
3756	0 0 1	0 0 0	14319	1 0 3	0 0 0
5331	0 1 1	0 0 0	15348	3 1 1	0 0 0
6872	0 2 1	0 0 0	15832	1 1 3	0 0 0
7252	1 0 1	0 0 0	16822	3 2 1	0 0 0
8807	1 1 1	0 0 0	16899	4 0 1	0 0 0
10613	2 0 1	0 0 0	17495	2 0 3	0 0 0
11032	0 0 3	0 0 0			

[a] After Ira N. Levine: *Molecular Spectroscopy* (Wiley-Interscience, New York 1975) p.263; references cited in W.S. Bebedict, N. Gailar, E.K. Plyler: J. Chem. Phys. 24, 1139 (1956)

Table 3.1 lists some infrared bands of H_2O vapor. The three fundamental bands occur at 1595, 3657 and 3756 cm^{-1}, corresponding to $(000) \rightarrow (010)$, $(000) \rightarrow (100)$ and $(000) \rightarrow (001)$ transitions, respectively.

Some of the infrared bands of gaseous CO_2 are listed in Table 3.2. The values of the quantum number ℓ for the vibrational angular momentum take

$$\ell = v, v - 2, v - 4, ..., 1 \text{ or } 0 \tag{3.11}$$

with the selection rule

$$\Delta \ell = \pm 1, 0 . \tag{3.12}$$

The vibrational frequencies $\nu_2 = 667$ cm^{-1} and $\nu_3 = 2349$ cm^{-1} refer to the infrared active transitions $(00^00) \rightarrow (01^00)$ and $(00^00) \rightarrow (00^01)$, respectively; whereas $\nu_1 = 1388$ cm^{-1}, corresponding to transition $(00^00) \rightarrow (10^00)$, is infrared inactive (Raman active). The well-known CO_2 laser output around 10.4 μm (961 cm^{-1}) and 9.4 μm (1064 cm^{-1}) is from the stimulated transitions $(00^01) \rightarrow (10^00)$ and $(00^01) \rightarrow (02^00)$, respectively.

3.3 Infrared Vibration-Rotational Spectra

For transitions of a diatomic molecule between the levels in (2.64) with $v'' \rightarrow v'$ and $J'' \rightarrow J'$ the frequency is given by

$$\nu = \nu_0 + B_v' J'(J' + 1) - D_v' J'^2(J' + 1)^2 - B_v'' J''(J'' + 1)$$
$$+ D_v'' J''^2(J'' + 1)^2 , \tag{3.13}$$

where ν_0 is the position of the band origin, i.e., the wave number of the pure vibrational transition ($J'' = J' = 0$). For the electronic ground state with

Table 3.2. Infrared bands of gaseous CO_2 [a]

ν [cm^{-1}]	Upper state $v_1 v_2^{\ell} v_3$	Lower state $v_1 v_2^{\ell} v_3$	ν [cm^{-1}]	Upper state $v_1 v_2^{\ell} v_3$	Lower state $v_1 v_2^{\ell} v_3$
667	$01^1 0$	$00^0 0$	618	$02^0 0$	$01^1 0$
1932	$03^1 0$	$00^0 0$	668	$02^2 0$	$01^1 0$
2077	$11^1 0$	$00^0 0$	721	$10^0 0$	$01^1 0$
2349	$00^0 1$	$00^0 0$	1881	$04^0 0$	$01^1 0$
3613	$02^0 1$	$00^0 0$	2093	$12^2 0$	$01^1 0$
3715	$10^0 1$	$00^0 0$	2130	$20^0 0$	$01^1 0$
4854	$04^0 1$	$00^0 0$	597	$03^1 0$	$02^2 0$
4978	$12^0 1$	$00^0 0$	647	$03^1 0$	$02^0 0$
5100	$20^0 1$	$00^0 0$	742	$11^1 0$	$02^2 0$
6076	$06^0 1$	$00^0 0$	961	$00^0 1$	$10^0 0$
6228	$14^0 1$	$00^0 0$	1064	$00^0 1$	$02^0 0$
6348	$22^0 1$	$00^0 0$			
6503	$30^0 1$	$00^0 0$			
6972	$00^0 3$	$00^0 0$			

[a] After Ira N. Levine: *Molecular Spectroscopy* (Wiley-Interscience, New York 1975) p.264; C.P. Courtoy: Ann. Soc. Sci. Bruxelles; Ser. I, 73, 5 (1959)

the quantum number of the resultant electronic orbital angular momentum about the internuclear axis equal to $\Lambda = 0$, the selection rule for rotational quantum number is

$$\Delta J = \pm 1 . \tag{3.14}$$

Since the transition with $\Delta J = 0$ is forbidden, a line at the band origin will not appear, but will have the P-branch ($\Delta J = -1$) and R-branch ($\Delta J = +1$) transitions on its opposite sides. Figure 3.2 shows the rotational fine structure of the vibration-rotational bands of a diatomic molecule. Due to vibration-rotation interaction ($D'_v, D''_v \neq 0$) the line spacing decreases in the R-branch and increases in the P-branch with increasing J value.

Neglecting D'_v and D''_v, which represent centrifugal effects, (3.13) is reduced to

$$\nu = \nu_0 + B'_v J'(J' + 1) - B''_v J''(J'' + 1) . \tag{3.15}$$

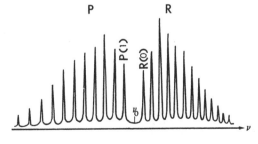

Fig.3.2. Diagrammatic representation of a vibration-rotational band of a diatomic molecule

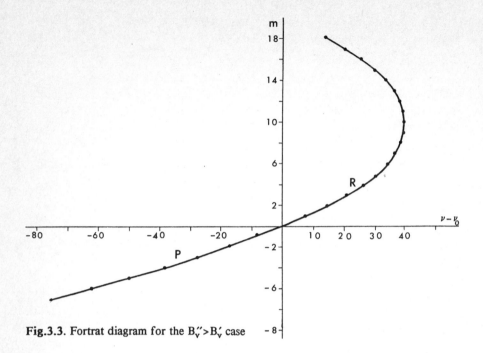

Fig.3.3. Fortrat diagram for the $B_v'' > B_v'$ case

For the P-branch $(\Delta J = J' - J'' = -1)$,

$$\nu_P = \nu_0 - (B_v' + B_v'')J + (B_v' - B_v'')J^2 \ , \quad J = 1,2,3,... \ , \tag{3.16}$$

where $J = J''$. For the R-branch $(\Delta J = J' - J'' = 1)$,

$$\nu_R = \nu_0 + 2B_v' + (3B_v' - B_v'')J + (B_v' - B_v'')J^2 \ , \quad J = 0,1,2,... \ . \tag{3.17}$$

Combining (3.16 and 17), the P- and R-branches can be represented by the following formula:

$$\nu = \nu_0 + (B_v' + B_v'')m + (B_v' - B_v'')m^2 \ , \tag{3.18}$$

where $m = -J$ refers to the P-branch, $m = J+1$ to the R-branch, and $m = 0$ to the band origin. Equation (3.18) is illustrated graphically by the Fortrat parabola shown in Fig.3.3. Owing to the quadratic term in (3.18), one of the two branches turns back to form a band head. When $B_v'' > B_v'$ the band head lies in the R-branch and the band is degraded toward longer wavelengths as the example in Fig.3.3. Conversely, when $B_v'' < B_v'$ the band head lies on the P-branch and the band is degraded toward shorter wavelengths.

If we consider the centrifugal distortion and assume the difference between D_v' and D_v'' is very small, the following formula usually gives an exact fit to the experimental observations:

$$\nu = \nu_0 + (B_v' + B_v'')m + (B_v' - B_v'')m^2 - 4D_v'' m^3 \ . \tag{3.19}$$

46

The exceptions are for molecules with ground states of $\Lambda'' \neq 0$. For parallel vibration of a linear polyatomic molecule the selection rule for the rotational quantum number is

$$\Delta J = \pm 1 , \qquad (3.20)$$

which corresponds to the R- and P-branches, respectively. For perpendicular vibration, the selection rule is

$$\Delta J = 0, \pm 1 , \qquad (3.21)$$

which gives the Q-, R-, and P-branches, respectively. The frequency equations for the P- and R-branches have the same forms as (3.16 and 17), resepctively, and the equation for the Q-branch is

$$\nu_Q = \nu_0 + (B_v' - B_v'')J + (B_v' - B_v'')J^2 , \quad J = 0,1,2,... . \qquad (3.22)$$

Since the value of $(B_v' - B_v'')$ is very small, the Q-branch occurs in the vicinity of the band origin. Figures 3.4a and b show the vibration-rotational spectra for a parallel band with R- and P-branches and a perpendicular band with Q-, R-, and P-branches, respectively.

For a prolate symmetric top molecule the selection rules for a parallel band are

$$\Delta J = \pm 1 , \qquad \Delta K = 0 , \quad (\text{for } K = 0) , \qquad (3.23)$$
$$\Delta J = 0, \pm 1 , \qquad \Delta K = 0 , \quad (\text{for } K \neq 0) . \qquad (3.24)$$

Then, according to (2.66), the transition frequencies for P-, Q-, and R-branches are given by

$$\nu_P = \nu_0 - 2BJ , \qquad (3.25)$$
$$\nu_Q = \nu_0 , \qquad (3.26)$$
$$\nu_R = \nu_0 + 2B(J + 1) , \qquad (3.27)$$

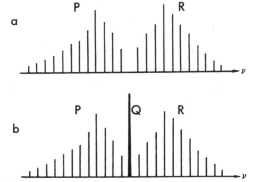

Fig.3.4. Schematic structure of the vibration bands for a linear molecule. a) A parallel band with R and P branches, b) a perpendicular band with R, P and Q branches

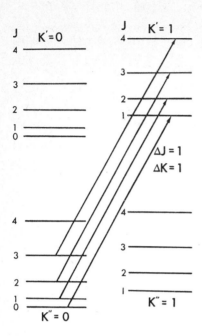

Fig.3.5. A few transitions for the R-branch of the first positive subsidiary band of a prolate symmetric top molecule

which have the same forms as those for a linear molecule, when $B'_v = B''_v$ and $D = 0$. Due to the vibration-rotation interaction the Q-branch frequency is, in fact, not limited to the one line indicated by (3.26). The selection rules for a perpendicular band are

$$\Delta J = 0, \pm 1 , \quad \Delta K = \pm 1 . \tag{3.28}$$

In the case of $\Delta K = +1$, i.e., $K' = K''+1$, the frequency equations are, from (2.66),

$$\nu_P = \nu_0 - 2BJ + (A-B)(2K''+1) , \qquad J = J'' = (K''+2), (K''+3), ... ; \tag{3.29}$$
$$\nu_Q = \nu_0 + (A-B)(2K''+1) , \qquad J = J'' = (K''+1), (K''+2) ... ; \tag{3.30}$$
$$\nu_R = \nu_0 + 2B(J+1) + (A-B)(2K''+1) , \quad J = J'' = K'', (K''+1), ... \tag{3.31}$$

Figure 3.5 illustrates a few transitions in (3.31) for the R-branch ($\Delta J = +1$) of the first positive subsidiary band ($\Delta K = +1$; $K''=0 \rightarrow K'=1$). The negative subsidiary bands with $\Delta K = -1$ also contain P-, Q-, and R-branches. The $K''=1 \rightarrow K'=0$ and $K''=2 \rightarrow K'=1$ are known as the first and second negative subsidiary bands, respectively.

3.4 Vibrational Band Systems of Diatomic Molecules

When the various interactions are neglected, the total term value T is given by (2.6). For the transitions between upper T' and lower T'' electronic states, the line frequencies are

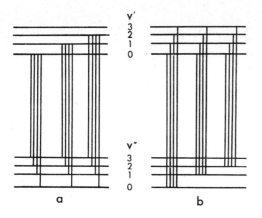

Fig.3.6. Some transitions indicating v″-progression (a) and v′-progression (b)

$$\nu = T' - T'' = (T'_e - T''_e) + [G'(v') - G''(v'')] + [F'_v(J') - F''_v(J'')]$$
$$= \nu_e + \nu_v + \nu_r. \tag{3.32}$$

In the vibrational analysis ν_r is temporarily neglected, and then the following formula is obtained:

$$\nu = \nu_e + [G'(v') - G''(v'')]$$
$$= \nu_e + (v' + \tfrac{1}{2})\omega'_e - (v' + \tfrac{1}{2})^2\omega'_e x'_e + (v' + \tfrac{1}{2})^3\omega'_e y'_e + ...$$
$$- [(v'' + \tfrac{1}{2})\omega''_e - (v'' + \tfrac{1}{2})^2\omega''_e x''_e + (v'' + \tfrac{1}{2})^3\omega''_e y''_e + ...]. \tag{3.33}$$

It is very useful for the vibrational analysis to pick out the band progressions and sequences from a large number of vibrational bands. Figure 3.6 illustrates some transitions corresponding to the v″- and v′-progressions. Obviously, the band separations in a v″-progression with the same v′ decrease toward shorter wavelengths. Figure 3.7 clearly shows the observed v″-progression with v′ = 4 in $C^1\Pi_u$ state, excited from $X^1\Sigma_g$ (v″=11) in

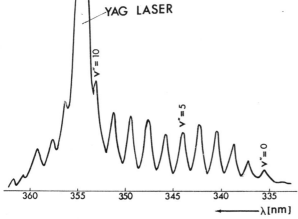

Fig.3.7. The observed v″-progression with v′ = 4 ($C^1\Pi_u$ state) excited by a 354.719 nm laser (initial level: v″ = 11 in $X^1\Sigma_g^+$ state) [3.5]

49

Na_2 by the third harmonic output (354.719nm) of a YAG laser system. On the other hand, the bands gathered with a constant value of $v'-v''$ are termed a sequence, where the bands are usually close together.

3.5 Rotational Spectra of Electronic Bands of Diatomic Molecules

Let us now consider the rotational structure of a particular band in an electronic transition with fixed values of v_e and v_v. Let $v_0 = v_e+v_v$ indicate the band origin. Then the frequencies of the rotational lines are given by

$$v = v_0 + v_r = v_0 + [F'_v(J') - F''_v (J'')] , \tag{3.34}$$

where $F'_v(J')$ and $F''_v (J'')$ are the rotational terms of the upper and lower electronic states, respectively. Thus

$$v = v_0 + [B'_v J'(J' + 1) - D'_v J'^2(J' + 1)^2 + ...]$$
$$- [B''_v J''(J'' + 1) - D''_v J''^2(J'' + 1)^2 + ...] . \tag{3.35}$$

If both Λ values of the upper and lower electronic states are zero (i.e., a Σ-Σ transition), the selection rule for J is

$$\Delta J = \pm 1 . \tag{3.36}$$

Provided that at least one of the electronic states of the transition has $\Lambda \neq 0$ (e.g., Π-Σ, Π-Π transitions, etc.) the selection rule is

$$\Delta J = 0, \pm 1 . \tag{3.37}$$

Neglecting the D_v term in the rotational expressions, the equations for P-, Q-, and R-branches are given by

$$v_P = v_0 - (B'_v + B''_v)J + (B'_v - B''_v)J^2 , \tag{3.38}$$
$$v_Q = v_0 + (B'_v - B''_v)J + (B'_v - B''_v)J^2 , \tag{3.39}$$
$$v_R = v_0 + 2B'_v + (3B'_v - B''_v)J + (B'_v - B''_v)J^2 . \tag{3.40}$$

It is to be noted that the rotational constants B'_v and B''_v now correspond to the different vibrational levels in two electronic states and that the difference between B'_v and B''_v may be larger than that in the same electronic state. As before, if $B'_v > B''_v$, the rotational lines of the band degrade toward shorter wavelengths, whereas in $B'_v < B''_v$ they degrade toward longer wavelengths. Since the J = 0 level in the Π electronic state is absent, the P(1) and Q(0) lines are also absent in each of the relevant bands.

3.6 Electric Quadrupole and Magnetic Dipole Hyperfine Spectra of Molecules

The spectral shift of a hyperfine component due either to electric quadrupole or to magnetic dipole interaction is determined by the difference between corresponding hyperfine splittings of two rotational levels. The selection rules among electric quadrupole hyperfine components for a one-photon molecular absorption transition are

$$\Delta J = 0, \pm 1 ; \quad \Delta F = 0, \pm 1 ; \quad \Delta I = 0 . \tag{3.41}$$

With increasing J the relative intensities of the hyperfine spectral components with different ΔF values tend towards greater disparity, and the components satisfying $\Delta F = \Delta J$ have the greatest intensities. The intensity ratios among $\Delta F = \Delta J$ components are proportional to individual F values; the larger the value of J, the less the differencies [3.4]. In fact, therefore, for J >> 0 lines in a molecular spectral band there are merely (2I+1) components which are observable.

The selection rules for magnetic hyperfine spectra are

$$\Delta J = 0, \pm 1 ; \quad \Delta F = \pm J ; \quad \Delta I = 0 . \tag{3.42}$$

For large J the components of spin I in direction of J are represented by M_I, which obey the selction rule

$$\Delta M_I = 0 . \tag{3.43}$$

These selection rules make the number of hyperfine components for an optical transition equal to the number of hyperfine splittings of the energy levels involved. The intervals between the hyperfine components are determined by the difference between the hyperfine constants of the levels involved and the J value of the initial level of the molecular transition. In fact, the hyperfine constants of excited vibrational levels in ground electronic states are essentially the same, which results in a spectral splitting order of magnitude smaller than the term splittings in the infrared region. In contrast, the hyperfine constants of excited electronic states may have great disparity from that of the ground state. The larger of them then dominates the hyperfine structure of the electronic vibration-rotational line. While the magnetic hyperfine constant in the ground state is orders smaller than the electric quadrupole constant, it might be dominant in excited electronic states, as observed in one-photon and two-photon absorption spectra in Na_2 [3.6, 7] (Fig. 5.18).

However, in either wavelength region of a molecular transition the hyperfine splittings are always obscured by molecular thermal motion broadening. The experimental measurement of the hyperfine structure of optical transitions of molecules can only be performed if the Doppler background of the spectral line is eliminated. Examples and references are given in Sect. 5.6.2.

3.7 The Goals for Experimental Studies of Molecular Spectroscopy

The important goals of molecular spectroscopy are to establish the relationship between molecular spectra and molecular structure, its intradynamic, external interactions and environment. Therefore the following pieces of information are essential to the analysis of molecular spectra:

1) electronic energy levels;
2) symmetry species;
3) vibrational energy levels;
4) rotational energy levels, corresponding to molecular moments of inertia;
5) bond lengths and bond angles;
6) dissociation energies;
7) ionization potentials;
8) electric dipole moments;
9) magnetic moments;
10) angular momentum couplings and energy level shifts;
11) oscillator strength of individual molecular lines;
12) line shape, linewidth and broadening;
13) life times of excited states;
14) molecular orientation, etc.

In its long history molecular spectroscopy has focused on extending the variety of molecular spectroscopic effects understood, and the variety of molecules investigated, with increasing accuracy and sensitivity. New results are emerging constantly, especially with lasers as light sources.

The features of molecular energy levels and the nature of molecular radiation has led to applications in many fields, such as

1) the identification of new materials;
2) the investigation of the dynamics of chemical reactions, including the effect of reagent rotational energy on the products state distribution [3.8] (e.g., Fig. 3.8);
3) isotopic enrichment, based on selective photoexcitation [3.9];
4) combustion diagnostics to extract accurate temperatures and concentrations with an accuracy strongly dependent on the accuracy of spectroscopic parameters (e.g., linewidth, cross-section, transition frequency) [3.10-12];
5) long distance propagation of laser beams in the atmosphere under various weather conditions [3.13];
6) air pollution monitoring [3.14];
7) production of optical frequency standards in metrology based on the precise stabilization of the laser frequency; this often begins with the accurate measurement of the parameters of a molecular resonance absorption [3.15-16], which in turn plays the role of the reference frequency for stabilizing the laser frequency as well as for checking the degree of stabilization.

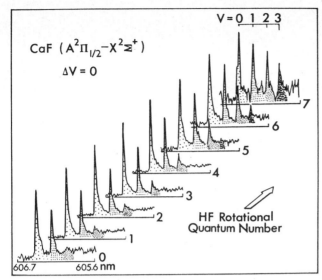

Fig.3.8. v = 0 band series in the laser-excitation spectra of CaF $(A^2\Pi_{1/2} - X^2\Sigma^+)$ formed in the Ca+HF $(X^1\Sigma^+, v = 1, J)$ reaction as a function of HF rotatioanl quantum number J. Bandheads corresponding to individual CaF vibrational levels are shaded differently to facilitate visual comparison between spectra [3.8]

A good example is the sodium dimer which was one of the first molecules observed and which has been constantly studied by various groups. It is still a unique molecule to test new effects, to verify new ideas, to monitor new phenomena, to obtain new results, and even to construct new lasers [3.17].

This wide range of applications with the aid of lasers has given rise to a breakthrough in molecular spectroscopy.

3.8 Advances of Molecular Spectroscopy
Through Linear Interaction of Molecules with Lasers

The fast development of various laser types has opened up a range of investigations of molecular spectra. Among them the wide-range tunable diode lasers in the infrared region, and CW and pulsed dye lasers in the visible regions are especially useful for systematic studies [3.18-20]. The optically pumped far-infrared lasers [3.21-23] and frequency mixing laser systems [3.23-25] are flexible complements in the research laboratory.

The use of lasers in the domain of linear interactions of a molecule with an electric-magnetic field has been marked by a new depth of investigation, and by accuracy and sensitivity at the limit spectral resolution given by the Doppler-broadening linewidth of molecular transitions and the time resolution limit of femtoseconds [3.26]. The noticeable achievements represented by recently published papers may be summarized by the following aspects:

1) Very weak transitions involving highly excited vibrational levels have been observed, revealing many previously unknown features of energy level structures and dynamic information. This was achieved by obtaining weak absorption spectra such as the high vibrational overtone spectroscopy performed by the sensitivity-enhanced intercavity method [3.28] or by noise-reduced low-temperature detection [3.29], by high-resolution absorption to the predissociation limit [3.30], or even by direct one-photon absorption to the photodissociation level to examine the velocity and angular distributions and internal states of the atomic fragments [3.31] or organic radicals [3.32]. The highly excited vibrational levels in a polyatomic molecule may reach a chaotic regime, where different vibrational modes are strongly coupled to provide a peculiar structure [3.33]. Such measurements allow one to determine the absolute absorption cross-section and its dependences on experimental conditions [3.30].

Another method was to use laser-induced fluorescence to observe the long series of spontaneous emission bands from a selected upper level, revealing the weak bound-continuum transitions [3.18]. Such data are an important complement to those from absorption spectra since one can accurately determine the dissociation energy and the molecular potential curve up to the very large internuclear separation region, where basic theoretical assumptions may not hold. When the selected upper level is a forbidden level from the ground state, the method may provide data for the determination of the potential curves of metastable electronic states [3.35, 36].

2) Various intramolecular perturbations have been examined. The importance of lasers in such precise studies is obvious. For example, the perturbations in the $A^1\Sigma_u^+$ state in Na_2 were systematically studied by *Kusch* and *Hessel* [3.37]. Even though they took the A-X absorption spectrum of Na_2 by using a conventional, long-focus grating spectrometer, the analyses for obtaining the perturbing shifts due to spin-orbital coupling between the singlet state $A^1\Sigma_u^+$ and the triplet states $b^3\Pi_{\Omega u}$ with $\Omega = 0, 1$, and 2 were confirmed by the cosntants of the $X^1\Sigma_g^+$ state, which were determined through the analysis of laser-induced fluorescence spectra of the $B^1\Pi_u - X^1\Sigma_g^+$ band system [3.38]. They were then able to obtain the beautiful curves for the frequency deviations of A-X bands from the regular Fortrat curves shown in Fig.3.9. The figure demonstrates that the rotational levels from J = 36 to 110 in the vibrational level of v' = 0 in $A^1\Sigma_u^+$ state are highly perturbed in three regions. Each of them has three perturbation maxima, which clearly means that the perturbing states are three sequential vibrational levels in a triplet state. The deviation signs changing at each of the perturbed center agree with the crossing-avoided perturbation version (Fig.4.16). In fact, for a diatomic molecule, the perturbed lines show only the repulsive shifts from the perturbing center in opposite directions rather than a J structure jumping.

The perturbations in polyatomic molecules can be surprisingly complicated. For example, by means of the stimulated emission pumping technique *Dai* et. al. were able to examine the energy-level structure of the electronic ground state of the asymmetric top molecule H_2CO at extremely highly excited vibrational levels [3.39]. They found that, while at low values

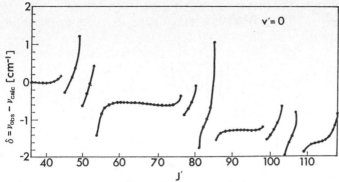

Fig.3.9. Differences between observed and calculated line frequencies for the transitions up to $A^1\Sigma_u^+(0,J')$ levels of Na_2. The data from $J' = 36$ down to $J' = 10$ omitted in the figure were all very nearly equal to zero. The differnces shown from $J' = 44,49,53,76,80,85,98,103,107$ to individual neighbors at larger J' changes sign, corresponding to individual perturbation centers. The distribution of these centers clearly shows that the lowest vibrational level $v' = 0$ in the $A^1\Sigma_u^+$ state of Na_2 are perturbed by three sequential vibrational levels in a triplet state overlaping with $A^1\Sigma_u^+$ of Na_2 [3.37]

of $J (\leq 3)$ the observed spectra were as expected, the spectra became complex with increasing J, and the level densities at $J = 10$ and $K_a = 2$ were several times larger than the known total density of the vibrational levels. Theses phenomena were attributed to the results of the interaction between rotation and vibration (Coriolis coupling). The study calls attention to the intramolecular vibrational randomization, which may limit the mode- or bond-selective excitation scheme to the chemically uninteresting low levels of vibrational energy.

For obtaining accurate nominal molecular constants, and for the practical uses of the idea of selective excitation for polyatomic molecules precise spectral measurements and a large amount of computation are required. *Zhao* et. al. have observed the J structure jumping in the ν_5 band of the symmetric top molecule CH_3F as shown in Fig.3.10, where the line with J

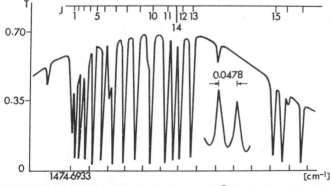

Fig.3.10. The high-resolution spectrum of $^RQ_0$ in the ν_5 band of CH_3F. The J structure jumping occurs at $J = 14$ [3.40]

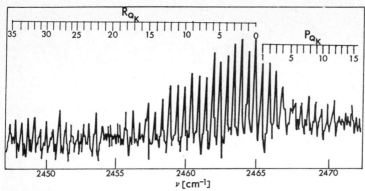

Fig.3.11. A well-resolved perpendicular $\nu_9+\nu_{10}$ combination band spectrum of cyclopropane. The most obvious perturbation causing both frequency and intensity irregularities appears in the region of $^PQ_5-^PQ_6$ [3.43]

= 14 is displaced to a point between the lines with J = 11 and 12 [3.40]. The experiment was carried out by using the diode-laser spectrometer with a spectral resolution of 0.002 cm^{-1}, which seems necessary for obtaining the improved 1(2,-1) type Coriolis resonance parameter of the ν_5 band since it has the small value of 2.3892·10^{-3} cm^{-1}. With the same apparatus they have found even more serious anomalies for higher J lines in the ν_5 band of the symmetric top molecule CH$_3$I [3.41]. They observed the J structure jumping of J = 45 in $^PQ_2$ to the spectral position close to J = 33 and the J progression of $^RQ_1$ compressed into one peak. These phenomena were mainly attributed to the Fermi resonance between the ν_5 and the $(\nu_3+\nu_6)$ combination bands and to xy Coriolis interaction between these bands as well. The further least-squares analysis had been carried out by diagonalizing the complete Hamiltonian matrix for the energy levels of the 10 observed sub Q-branches. A set of improved band constants and interaction parameters allows one to reproduce the observed spectra with a deviation of about 0.00083 cm^{-1} for J values up to 75 [3.42].

Indeed the spectroscopic perturbation implies the departure of transition intensities, in addition to the frequency locations, from an expected regular pattern. Figure 3.11 exhibits a part of the $(\nu_9+\nu_{10})$ perpendicular band of cyclopropane recorded using a Fourier spectrometer and a diode-laser spectrometer [3.43], in which the highly perturbed region of $^PQ_5-$ $^PQ_6$ was anomalously reduced. The relative changes in line shape, peak height and frequencies of a number of bands of acetamide and trioacetamide have been used to analyse the Fermi resonance bands. The change in line shape has been considered as the dynamic part of the Fermi interaction, which is affected through temperature control [3.44].

When the observations and analyses of molecular spectra are precise enough to establish very weak perturbations, they can also prove the absence of a certain sort of perturbation over the observed region as was done for the rotational spectra of several vibrational transitions of glyoxal [3.45].

There are publications on studies of molecular perturbations with even higher spectral resolution. One example is with microwave-optical and radio-frequency-optical double resonance spectroscopy to find the perturbation shifts as small as a few to a few hundred MHz in HNO [3.46]. Another example is with sub-Doppler laser-Stark and high-resolution Fourier-transform spectroscopy to analyse the Coriolis interaction between combined bands of HCOOH [3.47]. More examples with Doppler-free techniques will be given in Sect.5.6.

3) The sorts of molecules observable have been greatly extended. Firstly, for scientific and practical reasons molecular isotopic effects have been widely and precisely studied. The involved particles include stable or unstable molecules and molecular ions, such as CaOH and CaOD [3.48], H_2S and D_2S [3.49], $^{32}SF_6$ and $^{34}SF_6$ [3.50-52], $^{235}UF_6$ and $^{238}UF_6$ [3.53-55], H_2Te and D_2Te [3.56], $Si^{35}Cl$ and $Si^{37}Cl$ [3.57], $I^{35}Cl$ and $I^{37}Cl$ [3.58], $^{14}NH^-$ and $^{15}NH^-$ [3.59], etc. A few examples for the hyperfine structure due to the molecular isotopic effect are given in Sect.6.5. The advances in the study of molecular isotopic effects are certainly fundamental for finding an isolated absorption line for only one isotopic molecule and for achieving the desirable degree of selectivity for isotopic separations [3.9].

Other aspects, the very weak bonding force, molecular constants and vibration-excited dynamics for weakly bound Van der Waals molecules or molecular complexes, have been extensively studied recently with high spectral resolution. Examples include ArHCl [3.60,61], NeXe, ArXe, KrXe, and Xe_2 [3.62], CO_2HF [3.63], IClNe [3.64], and the further complexation of perylene-Benzene with Ar_n or N_2 in different sites [3.65], etc. For direct measurement of the vibrational motions in Van der Waals bonds supersensitive laser-spectroscopic techniques, such as the intracavity method [3.61], are required.

Moreover, laser spectroscopy provides the practical means of studying the spectra of short-lived molecules, such as transient molecules, free radicals and molecular ions, which were impossible to be studied by conventional spectroscopy because of low sensitivity and slow scanning speed. Different laser spectroscopic methods have successfully been used to probe short-lived molecules. Among them the fast scanning diode lasers (e.g., $1\,cm^{-1}\mu s^{-1}$) are used to obtain high-resolution kinetic absorption spectra of the vibration-rotational transitions for probing the nascent distribution of various photofragments (transient molecules) generated by another powerful laser [3.66], or the free radicals produced by chemical reaction in an electrical discharge [3.67]. Many other lasers are available for the spectral observation of various fragments or free radicals [3.26,68,69]. For enhancing the sensitivity of the direct-absorption detection multipass cells were used in these studies. A successful example for the application of the sensitive laser-induced fluorescence technique is the reported discovery of a large number of new organometallic free radicals, which were generated by the reaction of individual earth alkali metal vapor with a wide variety of organic molecules [3.70]. Some new properties of the photofragments were found, such as the non-thermal distribution of the translational energies of the nascent photofragments [3.66]; and the fact that the angular distributions of photofragments due to one-photon and two-photon absorption dissociation were isotropic and anisotropic, respectively [3.71,72]. The measurements of the angular distribution reveal in return the symmetry and lifetime of the final state and the photon excitation processes [3.72]; the study of relaxation processes in some photofragments has revealed the relative relaxation rates among the vibrational relaxation, rotational relaxation, fast chemical reaction rate involving free radicals, diffusion rate, and their different pressure dependences [3.73].

The kinetic studies of radicals and photofragments are also required for monitoring the combustion and atmospheric chemistry accompanying the propagation of a super-power laser beam in the atmosphere.

4) Quantitative studies of the characteristics of molecular spectral parameters have been experimentally developed. It is well known that Doppler broadening is dominant among line broadening mechanisms at low pressure (below a few Torr) and gives a Gaussian profile expressed by

$$K(\nu) = \frac{S}{\Delta\nu_D}\sqrt{\ln2/\pi}\exp\left[-\frac{\ln2}{\Delta\nu_D{}^2}(\nu - \nu_0)^2\right].$$ (3.44)

Collision broadening, however, is dominant at high pressure (e.g., beyond 20 Torr), where the absorption coefficient of a spectral line is described by a Lorentzian profile:

$$K(\nu) = \frac{S}{\pi}\frac{\Delta\nu_L}{(\nu - \nu_0)^2 + \Delta\nu_L{}^2}.$$ (3.45)

In the intermediate pressure range, when the two broadenings are effective simultaneously, the absorption coefficient at frequency ν is determined by the convolution of Gaussian and Lorentzian function in the form of the Voigt profile as follows

$$K(\nu) = \frac{\Delta\nu_L S\ln2}{\pi^{3/2}(\Delta\nu_D)^2}\int_{-\infty}^{\infty}\frac{\exp(-x^2)dx}{(\Delta\nu_L/\Delta\nu_D)^2\ln2 + [(\nu-\nu_0)\sqrt{\ln2}/\Delta\nu_D - x]^2}.$$ (3.46)

It follows that for a calculation of the magnitude of a molecular spectral line at a desired frequency ν the line strength S, halfwidth $\Delta\nu$ and line shift (corresponding to the variation of ν_0) are required, as well as a one-time measurement of its center frequency ν_0. The experimental determination of some parameter dependences by means of conventional spectroscopy, however, used to be poor for reasons of both low spectral accuracy and low detection sensitivity.

In fact, the pressure-broadening coefficient is on the order of one tenth wavenumber per atmosphere; the pressure shift of the line center may be two or three orders smaller. Nowadays with laser spectroscopy these parameters can be determined experimentally. In the low pressure region where the Doppler effect is dominant or takes a large part in the linewidth broadening and center frequency shift, various Doppler-free techniques have been applied to make direct measurements. A new nonlinear broaden-

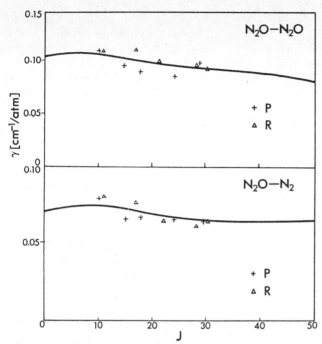

Fig.3.12. Self-broadened and N_2-broadened halfwidths of vibration-rotational lines in the 10.65 μm band of N_2O. (+: P-branch; Δ: R-branch; solid curves: calculated curves according to the ATC theory) [3.74]

ing zone in the extremely low-pressure region (millitorrs) was found (Sect.5.6, Table 5.10). In the high-pressure region, where collisional broadening is dominant., two methods have mainly been used. One is direct measurement with the use of tunable diode lasers or modern Fourier spectrometers. The other is an indirect one, which begins with the direct measurement of the absorption coefficient under the interesting pressure and temperature conditions, then with computer fitting to the theoretical expression to obtain the desired parameters and their variation factors. By means of the latter, for example, *Yu* et al. determined the absorption coefficients for 10 spectral lines in the ($\nu_3 - \nu_2$) difference-frequency bands of the N_2O molecule at a total pressure of above 200 Torr with a N_2O tunable laser [3.74]. Through least-square fitting the experimental data for the self-broadened and foreign gas (N_2) broadened half widths of N_2O lines shown in Fig.3.12 were obtained. The solid curves are the computed results, considering the perturbation potentials of the intermolecular collisions dominated by quadrupole-quadrupole interaction and by quadrupole-dipole interaciton for N_2O-N_2O and N_2O-N_2 collision boradening, respectively, according to the Anderson-Tsao-Curnutte (ATC) theory [3.75, 76]. The calculation was done according to the Quantum-Fourier-Transform (QFT) theory, which gave very close values with wandering deviations from the ATC curve for the self-broadening of N_2O, and a curve slightly bellow the ATC curve for the N_2 and O_2 broadening of N_2O [3.74].

As N_2O is an asymmetric linear molecule the first order pressure shift is predicted to be zero by both ATC and QFT theories. Investigators duly ignored the second-order effect for a long time. In fact, the second-order line shift of N_2O in the $(\nu_3-\nu_1)$ bands was nonzero, being a blue or red shift, depending on individual transitions, in the order of tenths of milli-wavenumber per atmosphere, as determined by *Yu* et. al. Most of the experimental data treated by least-square fitting were in agreement with the theoretical values calculated by the second-order ATC theory [3.74, 77].

Similar studies have been done for the planar molecule ethylene (C_2H_4) by *Shen* et. al. [3.78]. The data for the line strength S = 7.61 $cm^{-2}atm^{-1}$, self-broadening coefficient α_1 = 0.161 $cm^{-1}atm^{-1}$ and air broadening coefficient α_2 = 0.0693 $cm^{-1}atm^{-1}$ for the $28_{1,27} \leftarrow 28_{2,27}$ transition in ν_7 band of C_2H_4 have been obtained.

Temperature is another interesting external condition, causing an additional variation of the linewidth according to the well-known empirical formula [3.79]

$$\Delta\nu_T = \Delta\nu(T_0)(T_0/T)^\eta ,\qquad\qquad (3.47)$$

where $\Delta\nu(T_0)$ is the halfwidth of a line at temperature T_0, η is the so-called temperature factor, which was thought to be 0.5 in the classical theory. The experimental determinations have demonstrated that η is not a fixed value. For example, from the measured decreasing potential of the absorption coefficients of the above mentioned C_2H_4 lines as temperature increases at 1 atm, as shown in Fig.3.13, the factor η of 0.4 was obtained by least-square fitting of the measured absorption coefficients to (3.45 and 47) simultanesouly [3.80]. Detailed studies of the η factor for self-broadening of the CO_2 9.4μm and 10.4μm bands have been performed to obtain η values of 0.794 and 0.705, respectively, which approximate the theoretical value of 0.75 predicted by quadrupole-quadrupole collision interaction [3.81, 82]. It is interesting to find that the linewidths decrease rapidly with increasing J value and, in contrast, the η factor is a constant. Of course, the on-resonance absorption, e.g., molecule CO_2 responding to CO_2 laser lines, shows a temperature dependence inversely related to the off-resonance absorption of C_2H_4 shown in Fig.3.13. More information about the spectral parameters of CO_2 lines can be found in many recent publications [3.83-87].

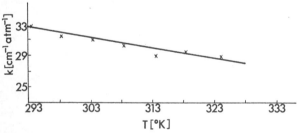

Fig.3.13. Variation of absorption coefficients of C_2H_4 at the CO_2-laser transition 10P(14) versus temperature [3.80]

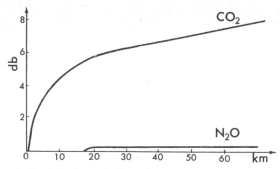

Fig.3.14. Estimated absorption attenuation of laser lines of CO_2 R(16) and N_2O R(10) propagating through atmosphere at the zenith angle of 20° [3.74]

An example of the application of the above studies is to estimate the attenuation of individual laser lines propagating in the atmosphere over a long distance, as displayed in Fig.3.14, which compares the R(16) line of the CO_2 laser with the R(10) line of the N_2O laser at a zenith angle of 20°.

In practice, atmospheric water vapor plays an important role. The above-mentioend line-by-line ATC theory for taking into account wing absorption of H_2O lines responding to individual CO_2 laser lines, in addition to the on-resonance absorption of CO_2 molecules, is terribly tedious and yields a poor accuracy. For solving these difficulties some available band models have been established based on weak absorption measurements with lasers [3.89-91].

Even though the models are successful for calculating the propagation attenuation of laser beams in a clear atmosphere, there still is the problem of estimating the attenuation in all weathers, which may cause strong scattering extinction, in addition to the absorption of radiation. Five infrared or visible laser lines together with thirty-six narrow infrared bands have been used for in-site measurements in all possible weather [3.13].

Moreover, the absorption attenuations of many other molecules, including various petrochemical organic components such as C_2H_4, C_2H_3CN, CH_3OH, $C_4H_9PO_4Cl_2$, C_3H_8, CH_3COOH, in addition to nonorganic components of H_2O, CO_2 and NH_3, have been measured [3.92]. For eliminating the influence of background attenuation the differential absorption method with individual, selected pairs for monitoring laser lines has been employed for the outdoor applications. Other supersensitive methods, including photoacoustic detection [3.93-96], intercavity detection [3.97,98] and heterodyne techniques [3.99], have been used for outdoor or indoor monitoring of air pollution components. Recently a review paper about the coherence and atmospheric attenuation aspects of atmospheric laser communications was published [3.100].

4. Spectral Characteristics of Molecular Two-Photon Transitions

The new level of molecular spectroscopy promoted by the development of laser spectroscopy is by no means limited to the advances summarized in the previous chapter, which were restricted to linear interaction processes between molecules and laser light. The high density of monochromatic energy laser light satisfied the requirements of nonlinear two-photon absorption, the strength of which is proportional to the square of the incident intensity. The process was first demonstrated by *Kaiser* et al. [4.1] and Doppler-free high-resolution two-photon spectroscopy of gases was first recognized by *Chebotayev* and co-workers [4.2], and has been achieved independently by *Cagnac* et. al. [4.3], by *Levenson* et al. [4.4] and by *Hänsch* et. al. [4.5]. Although the one-photon interaction continues to represent the mainstream in molecular spectroscopy, the follwing aspects have led to a rapid development in the field of two-photon absorption spectroscopy:

1) The final state is forbidden for one-photon absorption.

2) The energy level of the final state can be as high as twice that corresponding to the frequency of the incident laser light. The large and important energy level zone called Rydberg states, presently in need of much investigation, can to a considerable extent be studied by means of two-photon absorption with commercially available wide-range tunable dye lasers.

These two characteristics are common to investigations of both atoms and molecules. In parallel with atomic two-photon absorption studies, molecular two-photon excitations have extensively been studied for countless molecules, including H_2[4.6], Na_2[4.7-10], NO [4.11], IF [4.12], O_3 [4.13], NO_2 [4.14], CH_3F [4.15], CH_3 [4.16], SF_6 [4.17], C_6H_6 [4.18,19], XeCl [4.20], glyoxal [4.21], ammonia clusters [4.22], biological molecules [4.23], etc. The results obtained reveal new states and further excited states. These results have proved that molecular two-photon spectroscopy is an indispensable part of molecular spectroscopy as a whole.

Molecular two-photon transitions can take place between the types of molecular energy levels described in Chap.2, but with differences in selection rules. The fact that the induced two-photon transition rate (in the dipole approximation) depends on the inverse square of the offset of the intermediate enhancing level causes its spectral structure to be distorted beyond recognition in comparison with the conventional one of an induced one-photon transition. This can be seen in the following features:

1) The relative intensity distribution among observable molecular two-photon transition lines is no longer like that shown in Fig.3.2, which is dominated by the relative thermal population in their initial levels. Instead,

the reduced fine structure of a two-photon band is determined by a more or less accidental close coincidence with some allowed and therefore enhancing molecular lines.

2) There are reduced band progressions and band series in an observable coarse structure, compared with that shown in Fig.3.6.

3) The line shape of an isolated two-photon transition is observable in a combination of (4.41 and 43) with relevant parameters depending closely on the energy-level schemes in a three- or four-level system.

These features imply that a observed molecular two-photon absorption spectrum forms its own spectral regularity, which is not a regular, simplified spectrum as provided by other two-quantum excitations, e.g., double resonance or two-step excitation (Chap.7).

There are many excellent books on molecular spectroscopy, a few of which have been listed in previous chapters. In addition, there are excellent reviews on the topic of two-photon spectroscopy [4.24-27]. However, there is no text covering the spectral structure of an enhanced molecular two-photon absorption-band system and the particular line shapes encountered therein. It is the purpose of this chapter to develop such a discussion, starting from the experimental results observed in molecular sodium. Before doing this we will, in the next section, classify the observable signals. More about Doppler-free two-photon spectroscopic techniques will be given in Chap.5.

4.1 Classification of Equal-Frequency Molecular Two-Photon Transitions

The complicated structure of molecular energy levels and the actual environment of molecules encountered in practice (e.g., sodium dimers are enveloped in atoms, which outnumber them about a hundred to one in sodium vapor) suggest distinct approaches for near-resonant two-photon transitions, which may be classified as follows:

I) Directly excited two-photons transitions in sodium dimer:

$$Na_2(X) + 2h\nu \rightarrow Na_2^*(\Lambda) \tag{4.1}$$

including

$$Na_2(X^1\Sigma_g^+) + 2h\nu \xrightarrow{\quad A^1\Sigma_u^+ \quad} Na_2^*((n)^1\Lambda_g) . \tag{4.2}$$

$$Na_2(X^1\Sigma_g^+) + 2h\nu \xrightarrow{\quad b^3\Pi_u - A^1\Sigma_u^+ \quad} Na_2^*((n)^3\Lambda_g) , \tag{4.3}$$

where X means the ground state, and Λ represents a two-photon excited electronic state. The state symbols above the arrows indicate the intermediate enhancing states. The latter must be such that the transitions from it to

63

both the initial and final levels are electric-dipole allowed. The spectral structure of such processes consists mainly of separated lines. The following selection rules for molecular quantum numbers can be deduced from the combined selection rules for electric-dipole transitions, such as

$$\Delta\Lambda = 0, \pm1, \pm2, \dots ,$$
$$\Delta v = 0, \pm1, \pm2, \dots ,$$
$$\Delta J = \begin{cases} 0, \pm1, \pm2, & (\text{one of } \Lambda_i \neq 0) \\ 0, \pm2 & (\Lambda_i = 0, i = 1,2,3) \end{cases} \tag{4.4}$$
$$\Delta S = 0 ,$$
$$\Delta\Sigma = 0 ,$$
$$\Delta I = 0 ,$$
$$\Delta F = 0, \pm1, \pm2 .$$

For homonuclear molecules there are additional selection rules for the molecular symmetry:

$$\begin{array}{lll} g \longleftrightarrow g , & u \longleftrightarrow u , & g \leftarrow/\rightarrow u ; \\ + \longleftrightarrow + , & - \longleftrightarrow - , & + \leftarrow/\rightarrow - ; \\ s \longleftrightarrow s , & a \longleftrightarrow a , & s \leftarrow/\rightarrow a . \end{array} \tag{4.5}$$

A molecular two-photon transition from the singlet state $X^1\Sigma_g^+$ to the triplet state $(n)^3\Pi_g$ in (4.3) will be spin forbidden according to (4.4). Fortunately, with the aid of angular momentum coupling the singlet and triplet intermediate states can share their wave functions with each other; in other words, the triplet state can borrow excitation probability from the nearby singlet state. The resulting two-photon excitation process is then composed of two dipole transitions, one between singlet states and the other between triplet states. This means that the two-photon excitation mechanism in (4.3) is a four-level system (Fig.4.1b) while that in (4.2) is a three-level system (Fig.4.1a). The spectral differences between (4.2 and 3) are interesting, and are compared theoretically and experimentally later in this chapter.

II) Two-photon excitation with collision processes: Under certain experimental conditions a molecular two-photon excitation process can be performed by collisions. In sodium dimers, for example, the contribution of an effective collision process to the population of high-lying g-parity molecular states can be one of the following energy-transfer processes:

1) One-photon energy transfer between atoms and molecules (Fig. 4.1c):

$$\begin{array}{lll} Na(3S) + h\nu & \rightarrow Na^*(3P) + \Delta E , & (4.6) \\ Na^*(3P) + Na_2(X) & \rightarrow Na_2^*(A) + \Delta E + Na(3S) , & (4.7) \\ Na_2^*(A) + Na^*(3P) & \rightarrow Na_2^*(\Lambda) + \Delta E , & (4.8) \\ Na_2^*(A) + h\nu & \rightarrow Na_2^*(\Lambda) + \Delta E , & (4.9) \end{array}$$

where in (4.8) the atomic sodium plays a role in the energy transfer to an already excited molecule; their energies sum up to reach the high-lying

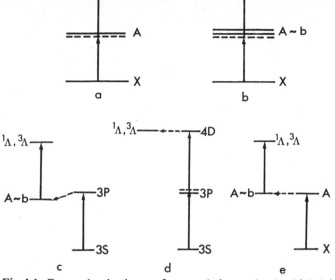

Fig.4.1. Energy-level schemes for populating molecular high-lying states with two-photon pumping. (A) Two-photon absorption without collision processes. a)Three-level system with one near-resonant enhancing level. b) Four-level system with near-resonant mixing levels as two enhancing levels. (B) Two-photon excitation aided by collision processes. c) Atomic one-photon excitation energy transfer to molecule. d) Atomic two-photon absorption energy transfer to molecule. e) Molecular energy pooling process

state Λ. The spectral character is a narrow band centered on the atomic one-photon resonant excitation; (4.9) represents a two-step process.

    2) Two-photon energy transfer between atoms and molecules (Fig. 4.1d):

$$Na(3S) + 2h\nu \quad\quad \rightarrow Na^*(4D) \, ,$$
$$Na^*(4D) + Na_2(X) \rightarrow Na(3S) + Na_2^*(^1\Lambda \text{ or } ^3\Lambda) + \Delta E \, , \quad\quad (4.10)$$

where Λ denotes high-lying states with the same parity of the ground state X. The spectral character is a specific, narrow band centered on the atomic two-photon excitation wavelength.

    3) Collision between molecules described by Fig.4.1e

$$2Na_2(X) + 2h\nu \rightarrow 2Na_2^*(A) \, , \quad\quad (4.11)$$

$$Na_2^*(A) + Na_2^*(A) \rightarrow Na_2^*(\Lambda) + \Delta E \, ; \quad\quad (4.12)$$

$$Na_2^*(\Lambda) + Na_2(X) \overset{(Ar)}{\longrightarrow} Na_2(X) + Na_2^*(\Lambda') + \Delta E \, . \quad\quad (4.13)$$

The spectral character of these processes is a "quasi-continuous" wide band similar to the molecular one-photon absorption spectra.

Of the processes listed above, the dominant one will depend upon experimental conditions, and their relative efficiency in populating particular high-lying states [4.28] will vary widely. These aspects are discussed in Chap.7; and their applications to the production of stimulated radiation are outlined in Chap.6. In order to explain molecular two-photon excitation spectra, the following sections are restricted to discussions on the near-resonantly enhanced direct two-photon excitation in (4.2 and 3), including the coarse (vibrational) structures of a band system, the fine (rotational) structures of a band, and the shape of a line of a molecular two-photon transition, where the semiclassical treatments for three-level and four-level systems are required for the case of (4.2 and 3), respectively.

4.2 Excitation Probability of a Two-Photon Transition with One Near-Resonant Enhancing Level

The enhancement of the transition probability by the presence of an intermediate level is one of the interesting features of two-photon spectroscopy. In general, the atomic or molecular two-photon absorption transitions have transition probabilities, depending on the degree of their intermediate levels on resonance [4.29]. Among atomic samples it is difficult to find a two-photon transition whose intermediate level is nearly coincident with the halfway point. In order to enhance two-photon transition probability, one has to use two-counterpropagating beams with different frequencies [4.30].

In contrast, a molecule has a great number of vibration-rotational levels which may form an almost equally spaced 3-level system. Therefore, equal-frequency two-photon transitions enhanced by an intermediate level give rise to considerable two-photon transition probabilities. For example, the $A^1\Sigma_u^+$ state in the sodium dimer possesses a series of near-resonant intermediate levels within the appropriate wavelength region for a dye laser with rhodamine 6G dye. Thus, a simple and accessible way to study the line shapes and the spectral band structure of molecular two-photon absorptions might become available [4.7-10]. To examine them in theory, we will begin with the semiclassical equations of motion to obtain the transition probability of two-photon excitation enhanced by one near-resonant intermediate level with the energy level scheme shown in Fig.4.2b.

A statistical ensemble of molecules with axial velocity v at position z and time t can properly be described by an ensemble-averaged density matrix (three-levels system case):

$$\rho(v,z,t) = \begin{pmatrix} \rho_{aa} & \rho_{ab} & \rho_{ac} \\ \rho_{ba} & \rho_{bb} & \rho_{bc} \\ \rho_{ca} & \rho_{cb} & \rho_{cc} \end{pmatrix}, \tag{4.14}$$

where a, b, and c represent three stationary levels. The normalization of the matrix is done by setting the diagonal elements of the matrix, ρ_{aa}, ρ_{bb}, and

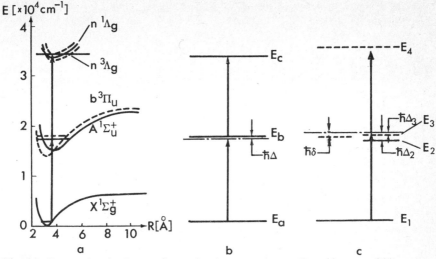

Fig.4.2. Energy-level schemes for molecular two-photon absorption transitions. a) A part of the electronic states in Na_2. Solid potential curves: singlet states. Dashed potential curves: triplet states. b) Energy-level scheme of a molecular two-photon transition between selection-rule-allowed states (three-level system). c) Energy-level scheme of a spin-forbidden molecular two-photon transition enhanced by a pair of mixing levels (four-level system) [4.44]

ρ_{cc}, equal to the population densities at the levels a, b, and c, respectively. The off-diagonal elements ρ_{ij} (i,j = a,b,c; i≠j) correspond to the oscillating dipole moments between i and j levels. In the laboratory frame the molecular density matrix obeys the following time-dependent Schrödinger equation:

$$\left(\frac{\partial}{\partial t} + v\frac{\partial}{\partial z}\right)\rho = \frac{1}{i\hbar}[\mathcal{H},\rho] + \Lambda \ . \tag{4.15}$$

The Hamiltonian

$$\mathcal{H} = \mathcal{H}^A + \mathcal{H}^I + \mathcal{H}^R \tag{4.16}$$

includes the interaction-energy operator $\mathcal{H}^I$ with the radiation fields. Here $\mathcal{H}^A$ is the Hamiltonian of the unperturbed molecule. In the electric dipole approximation we have μ_{ab}, $\mu_{bc} \neq 0$ and $\mu_{ac} = 0$. The interaction energy can be written as

$$\mathcal{H}^I = -\mu E(z,t) \tag{4.17}$$

with the radiation field E for the two traveling waves

$$\begin{aligned}
E(z,t) &= E_1(z,t) + E_2(z,t) \ , \\
E_1(z,t) &= \tfrac{1}{2}E_1\exp[-i(\omega_1 t - k_1 z)] + \text{c.c.} \ , \\
E_2(z,t) &= \tfrac{1}{2}E_2\exp[-i(\omega_2 t - k_2 z)] + \text{c.c.} \ .
\end{aligned} \tag{4.18}$$

Relaxation processes are introduced phenomenologically by the operator $\mathcal{H}^R$, which is defined by

$$[\mathcal{H}^R, \rho]/i\hbar = - \begin{pmatrix} \gamma_a \rho_{aa} & \gamma_{ab}\rho_{ab} & \gamma_{ac}\rho_{ac} \\ \gamma_{ba}\rho_{ba} & \gamma_b \rho_{bb} & \gamma_{bc}\rho_{bc} \\ \gamma_{ca}\rho_{ca} & \gamma_{cb}\rho_{cb} & \gamma_c \rho_{cc} \end{pmatrix}. \tag{4.19}$$

The term

$$\Lambda = \begin{pmatrix} \lambda_a & 0 & 0 \\ 0 & \lambda_b & 0 \\ 0 & 0 & \lambda_c \end{pmatrix} \tag{4.20}$$

in (4.15) introduces collisional excitation processes, i.e.,

$$\lambda_\alpha = n_\alpha \gamma_\alpha , \quad (\alpha = a,b,c) , \tag{4.21}$$

where

$$n_\alpha(v) = N_\alpha{}^0 f(v) \tag{4.22}$$

represents the initial population densities of the molecules on the respective level α; $f(v)$ is the Maxwellian velocity distribution. The Doppler effect due to the molecular thermal motion is taken into account by the spatial derivative on the left-hand side of (4.15).

The spectral characteristics of molecular two-photon transitions can be shown by the progressive perturbation solution of (4.15), provided the radiation fields are not too strong. The perturbation expansion of the density matrix is

$$\rho = \rho^{(0)} + \rho^{(1)} + \rho^{(2)} + \dots , \tag{4.23}$$

where the nth-order solution can be obtained by solving the following progressive equations [4.25]

$$i\hbar \left(\frac{\partial}{\partial t} + v\frac{\partial}{\partial z} \right) \rho^{(0)} - [\mathcal{H}^A + \mathcal{H}^R, \rho^{(0)}] = i\hbar\Lambda , \tag{4.24}$$

$$i\hbar \left(\frac{\partial}{\partial t} + v\frac{\partial}{\partial z} \right) \rho^{(n)} - [\mathcal{H}^A + \mathcal{H}^R, \rho^{(n)}] = [\mathcal{H}^I, \rho^{(n-1)}] . \tag{4.25}$$

For the sake of simplicity the following initial conditions are assumed:
1) The molecular sample is in thermal equilibrium, i.e., $(\partial\rho(0))/\partial t = 0$.
2) The spatial nonuniformity is negligible, i.e., $(\partial\rho(0))/(\partial z) = 0$.
3) The initial ensemble-averaged induced dipole moments of both the transitions a⟷b and b⟷c are zero, i.e., $\rho_{ij}^{(0)} = 0$ (i≠j).

4) For the transitions in the optical-wavelength region all molecules are initially in the lowest state, i.e., to set $N_b^0 = N_c^0 = 0$ is reasonable. Therefore, $\rho^{(0)}$ has the form of

$$\rho^{(0)} = \begin{bmatrix} \rho_{aa}^{(0)} & 0 & 0 \\ 0 & 0 & 0 \\ 0 & 0 & 0 \end{bmatrix} = \begin{bmatrix} n_a(v) & 0 & 0 \\ 0 & 0 & 0 \\ 0 & 0 & 0 \end{bmatrix}. \tag{4.26}$$

In the usual case of a three-level system, where $|\omega_{ab} - \omega_{bc}| >> \Delta\omega_D$ ($\Delta\omega_D$ being the Doppler width of the transition), we assume $\omega_1 \simeq \omega_{ab}$, $\omega_2 \simeq \omega_{bc}$. The interaction Hamiltonian is then

$$\mathcal{H}^1 = \begin{bmatrix} 0 & -\mu_{ab}E_1(z,t) & 0 \\ -\mu_{ba}E_1(z,t) & 0 & -\mu_{bc}E_2(z,t) \\ 0 & -\mu_{cb}E_2(z,t) & 0 \end{bmatrix}. \tag{4.27}$$

Substituting (4.26 and 27) into the right-hand side in (4.25), we obtain the driving term for the first-order solution:

$$\rho^{(1)} \sim [\mathcal{H}^1, \rho^{(0)}] = \begin{bmatrix} 0 & \mu_{ab}E_1(z,t)\rho_{aa}^{(0)} & 0 \\ -\mu_{ba}E_1(z,t)\rho_{aa}^{(0)} & 0 & 0 \\ 0 & 0 & 0 \end{bmatrix}. \tag{4.28}$$

For the first-order solution Eq.(4.25) becomes

$$i\hbar\left(\frac{\partial}{\partial t} + v\frac{\partial}{\partial z}\right)\rho_{ab}^{(1)} - (E_a - E_b)\rho_{ab}^{(1)} + i\hbar\gamma_{ab}\rho_{ab}^{(1)} = \mu_{ab}E_1(z,t)\rho_{aa}^{(0)}. \tag{4.29}$$

Other off-diagonal and diagonal terms of $\rho^{(1)}$ are zero on account the initial condition 4 above. Equation (4.29) means that $\rho_{ab}^{(1)}$ is related linearly to the radiation field $E_1(z,t)$. Thus we have

$$\rho_{ab}^{(1)} = \frac{V_{ab}}{\Delta_{ab}} \rho_{aa}^{(0)}. \tag{4.30}$$

where

$$V_{ab} = \mu_{ab}E_1/\hbar, \tag{4.31}$$

$$\Delta_{ab} = \omega_1 - \omega_{ab} - k_1 v + i\gamma_{ab}. \tag{4.32}$$

By analogy with the above steps from (4.28 to 32), the source term for the second-order solution is

$$\rho^{(2)} \sim [\mathcal{H}^{I}, \rho^{(1)}] = \begin{bmatrix} 2\mu_{ab}E_1(z,t)\rho_{ab}^{(1)} & 0 & \mu_{bc}E_2(z,t)\rho_{ab}^{(1)} \\ 0 & -2\mu_{ab}E_1(z,t)\rho_{ab}^{(1)} & 0 \\ -\mu_{cb}E_2\rho_{ba}^{(1)} & 0 & 0 \end{bmatrix}. \quad (4.33)$$

Note that the upper-right off-diagonal term is non-zero, corresponding to $\rho_{ac}^{(2)} \neq 0$. This means that the molecular charge is oscillating coherently between the levels a and c at the sum frequency of the fields. Substituting (4.33) into the right-hand side of (4.25), we have

$$\rho_{ac}^{(2)} = \frac{V_{bc}}{\Delta_{ac}}\rho_{ab}^{(1)} = \frac{V_{ab}V_{bc}}{\Delta_{ac}\Delta_{ab}}\rho_{aa}^{(0)} \qquad \text{where} \qquad (4.34)$$

$$V_{bc} = \mu_{bc}E_2/\hbar, \qquad (4.35)$$

$$\Delta_{ab} = (\omega_1 + \omega_2) - (\omega_{ab} + \omega_{bc}) - (k_1 + k_2)v + i\gamma_{ac}. \qquad (4.36)$$

The non-zero second-order diagonal terms $\rho_{aa}^{(2)}$ and $\rho_{bb}^{(2)}$ correspond to the well-known "hole burning" phenomena, when the corrections to the molecular-velocity distribution real population on the levels a and b, respectively, are substantial. In the rotating-wave approximation these diagonal terms are

$$\rho_{bb}^{(2)}(v) = -\rho_{aa}^{(2)} = \frac{|V_{ab}|^2}{\gamma_b}\rho_{aa}^{(0)}(v)\,\text{Im}\{1/(\Delta_{ab})\}$$

$$= \frac{|V_{ab}|^2}{\gamma_b}\frac{2\gamma_{ab}}{|\Delta_{ab}|^2}\rho_{aa}^{(0)}(v). \qquad (4.37)$$

Furthermore, according to the same procedures we can get the third solution using the second-order one as the driving source, i.e.,

$$i\hbar\left(\frac{\partial}{\partial t} + v\frac{\partial}{\partial z} + i\omega_{bc} + \gamma_{bc}\right)\rho_{bc}^{(3)} = \mu_{bc}E_2(z,t)\rho_{bb}^{(2)} - \mu_{ab}E_1(z,t)p_{ac}^{(2)}. \qquad (4.38)$$

This formula implies that there are two interaction paths which give rise to a nonlinear induced dipole moment of the molecules, corresponding to nonlinear absorption from level a to c. The two interaction paths can be illustrated by

$$1)\ \rho_{aa}^{(0)} \xrightarrow{E_1} \rho_{ab}^{(1)} \xrightarrow{E_1} \rho_{bb}^{(2)} \xrightarrow{E_2} \rho_{bc}^{(3)} \xrightarrow{E_2} \rho_{cc}^{(4)}; \qquad (4.39)$$

$$2)\ \rho_{aa}^{(0)} \xrightarrow{E_1} \rho_{ab}^{(1)} \xrightarrow{E_2} \rho_{ac}^{(2)} \xrightarrow{E_1} \rho_{bc}^{(3)} \xrightarrow{E_2} \rho_{cc}^{(4)}. \qquad (4.40)$$

70

The third-order solution via path 1 is a stepwise excitation process and is of the form

$$(\rho_{bc}^{(3)})_s = \frac{\rho_{bb}^{(2)} V_{bc}}{\Delta_{bc}} = \frac{|V_{ab}|^2 V_{bc}}{\gamma_{bc} \Delta_{bc}} \frac{2\gamma_{ab}}{\Delta_{ab} \Delta_{ba}} \rho_{aa}^{(0)} , \tag{4.41}$$

where $\rho_{bb}^{(2)}$ is given by (4.37), and

$$\Delta_{bc} = \omega_{bc} - \omega_2 + k_2 v_z + i\gamma_{bc} . \tag{4.42}$$

The third-order solution via path 2 is a two-photon excitation process, which does not really populate the intermediate level b. The result is of the form

$$(\rho_{bc}^{(3)})_T = - \frac{V_{ba}}{\Delta_{bc}} \rho_{ac}^{(2)} = - \frac{|V_{ab}|^2 V_{bc}}{\Delta_{ab} \Delta_{bc} \Delta_{ac}} \rho_{aa}^{(0)} , \tag{4.43}$$

where $\rho_{ac}^{(2)}$ is given by (4.34). The total value of $\rho_{bc}^{(3)}$ corresponds to the sum of (4.41 and 43) and is given by

$$\rho_{bc}^{(3)} = (\rho_{bc}^{(3)})_S + (\rho_{bc}^{(3)})_T$$

$$= |V_{ab}|^2 V_{bc} \rho_{aa}^{(0)} \left[\frac{2\gamma_{ab}}{\Delta_{ab} \Delta_{ba} \Delta_{bc} \gamma_b} - \frac{1}{\Delta_{ab} \Delta_{bc} \Delta_{ac}} \right] . \tag{4.44}$$

The expression for the lowest-order population of the final state c can be derived from (4.25) as

$$\rho_{cc}^{(4)} = - \text{Im}\{(V_{cb}\rho_{bc}^{(3)} - \rho_{cb}^{(3)} V_{bc})\}/\gamma_c = \rho_{cc}^{(4)}(v)_S + \rho_{cc}^{(4)}(v)_T , \tag{4.45}$$

where $\rho_{cc}^{(4)}(v)_S$ and $\rho_{cc}^{(4)}(v)_T$ are the partial populations of the final state by two-step and by two-photon processes, respectively. Substituting $\rho_{cb}^{(3)} = -\rho_{bc}^{(3)}$ in (4.45) because $\rho_{cc}^{(0)} = \rho_{cc}^{(2)} = 0$, we have

$$\rho_{cc}^{(4)}(v) = \frac{|V_{ab}|^2 |V_{bc}|^2}{\gamma_c} \rho_{aa}^{(0)}(v) \text{Im}\{S + T\} , \tag{4.46}$$

where

$$\text{Im}\{S\} = \text{Im}\left\{ \frac{-2\gamma_{ab}}{\gamma_b \Delta_{ab} \Delta_{ba} \Delta_{bc}} \right\}, \tag{4.47}$$

$$\text{Im}\{T\} = \text{Im}\left\{ \frac{1}{\Delta_{ab} \Delta_{bc} \Delta_{ac}} \right\}. \tag{4.48}$$

71

For the two-photon transition with the photon energy in the visible or ultraviolet wavelength regions the initial level a can be considered as the ground state. That means we can set $\gamma_a = 0$, giving $\gamma_{ab} = (\gamma_a+\gamma_b)/2 = \gamma_b/2$; $\gamma_{ac} = (\gamma_a+\gamma_c)/2 = \gamma_c/2$; $\gamma_{ab}+\gamma_{ac} = (\gamma_b+\gamma_c)/2 = \gamma_{bc}$. Therefore, the sum (S+T) can be simplified as

$$S + T = 1/\Delta_{ab}\Delta_{ba}\Delta_{bc} - 1/\Delta_{ab}\Delta_{bc}\Delta_{ac} = 1/\Delta_{ab}\Delta_{ba}\Delta_{ac} \; . \tag{4.49}$$

Substituting (4.49) into (4.46), we obtain the formula for the population rate of the final state

$$
\begin{aligned}
\rho_{cc}^{(4)}(v) &= |V_{ab}|^2|V_{bc}|^2\rho_{aa}^{(0)}(v)\, \mathrm{Im}\left\{ \frac{1}{\Delta_{ab}\Delta_{ba}\Delta_{ac}} \right\}/\gamma_c \\
&= |V_{ab}|^2|V_{bc}|^2\rho_{aa}^{(0)}(v)\, \frac{1}{[\omega_{ac} - (\omega_1+\omega_2) + (k_1+k_2)v_z]^2 + \gamma_{ac}^2} \\
&\quad \times \frac{1}{(\omega_{ab}-\omega_1+k_1 v_z)^2 + \gamma_{ab}^2} \; .
\end{aligned}
\tag{4.50}
$$

The last factor at the right-hand side is the so-called two-photon enhancing factor, which shows that an intermediate level b near-resonant with radiation frequency ω_1 will greatly enhance two-photon excitation processes. The second-last factor in (4.50) is the called two-photon line-shape factor, which shows that for the absorbtion of two counter-propagating photons the two-photon transition over molecular velocities will result in a sub-Doppler line shape with residual Doppler width $\Delta|k|v$, whereas absorbing two unidirectional photons will cause a Doppler-broadened background with linewidth $(|k_1|+|k_2|)v$.

Two-photon absorption can be induced by equal-frequency radiation fields, provided the radiation frequency ω satisfies

$$\omega_{ac} - 2\omega + (k_1 + k_2)v_z = 0 \; . \tag{4.51}$$

In this case the residual Doppler width could be cancelled if each of the molecules involved absorbed photons propagating in opposite directions. The full width at half maximum (FWHM) of the Doppler-free Lorentzian line shape is $2\gamma_{ac}$. However, for most of the allowed molecular two-photon transitions the offset value Δ of the intermediate level b from equal-frequency two-photon resonance

$$\Delta = \omega_{ab} - \tfrac{1}{2}\omega_{ac} \tag{4.52}$$

is much larger than the Doppler width $(\Delta\omega_D)$. In this case the Doppler shift kv_z and the relaxation rate γ_{ab} in the denominator of the enhancing factor in (4.50) are negligible and the equation describing the population rate on level c is simplified from (4.50) to become

$$\rho_{cc}^{(4)}(v) = |V_{ab}|^2|V_{bc}|^2\rho_{aa}^{(0)}(v)\, \frac{1}{|\Delta_{ac}|^2}\, \frac{1}{\Delta^2} \; . \tag{4.53}$$

The value of $\rho_{cc}^{(4)}(v)$ drops rapidly with increasing Δ. In fact, in the CW excitation case the offset value Δ cannot be more than about 1 cm^{-1} for a two-photon transition to be strong enough to be noticeable [4.9]. Comparing the pattern of $\rho_{cc}^{(4)}$ with that of $\rho_{bb}^{(2)}$, we find that, when a, c, and a, b vary through related successive rovibrational levels, the intensity distribution among the successive rovibrational lines within a molecular spectral band of two-photon absorption can vary greatly due to the sharply changing value of $1/\Delta^2$ in (4.53). This results in the particular spectral structure of molecualr two-photon bands so that it is uncomparable with the smooth variations of $\rho_{aa}^{(0)}(v)$ in an one-photon band corresponding to the thermal distribution on the rotational levels in (4.37).

Nevertheless, it is possible to find some of the molecular two-photon transitions with almost equally spaced three-level systems among numerous rovibrational levels. In fact, we have recently observed about 70 such enhanced CW two-photon transitions in the 16600–17600 cm^{-1} region [4.10] and dozens of CW two-photon transitions enhanced by a pair of near-resonant mixing levels in the 15000–16100 cm^{-1} region in molecular sodium [4.31, 32]. The measured offset range was from few GHz down to as little as 34 MHz. That is a large fraction of them are within the Doppler width of 1.5 GHz, and some are offset by even less than the lifetime width of about 40 MHz. For the particular case of molecular resonant two-photon transitions the frequency ω falls inside the Doppler widths of both the dipole transitions of a→b and b→c simultaneously. Therefore each dipole transition can be induced by either of the two equal-frequency opposite-travelling waves simultaneously. The two counter-propagating equal-frequency waves can be written as

$$E = E_+\exp[-i(\omega t + kz)] + E_-\exp[-i(\omega t - kz)] + c.c. .\qquad(4.54)$$

Instead of the interaction paths listed in (4.39,40), there are alternative paths in the interactions of the fields with molecules as follows:

I) Absorbing two photons from both of the beams:

$$1)\quad \rho_{aa}^{(0)} \underset{-1}{\overset{+1}{\to}} \rho_{ab}^{(1)} \underset{-1}{\overset{+1}{\to}} \rho_{bb}^{(2)} \underset{+1}{\overset{-1}{\to}} \rho_{bc}^{(3)} \underset{+1}{\overset{-1}{\to}} \rho_{cc}^{(4)} ,$$

$$2)\quad \rho_{aa}^{(0)} \underset{\substack{-1\\-1}}{\overset{\substack{+1\\+1}}{\to}} \rho_{ab}^{(1)} \underset{\substack{+1\\+1}}{\overset{\substack{-1\\-1}}{\to}} \rho_{ac}^{(2)} \underset{\substack{-1\\+1}}{\overset{\substack{+1\\-1}}{\to}} \rho_{bc}^{(3)} \underset{\substack{+1\\-1}}{\overset{\substack{-1\\+1}}{\to}} \rho_{cc}^{(4)} ;\qquad(4.55)$$

73

II) Absorbing two photons from one beam

3) $\rho_{aa}^{(0)} \overset{+1}{\underset{-1}{\longrightarrow}} \rho_{ab}^{(1)} \overset{+1}{\underset{-1}{\longrightarrow}} \rho_{bb}^{(2)} \overset{+1}{\underset{-1}{\longrightarrow}} \rho_{bc}^{(3)} \overset{+1}{\underset{-1}{\longrightarrow}} \rho_{cc}^{(4)}$;

4) $\rho_{aa}^{(0)} \overset{+1}{\underset{-1}{\longrightarrow}} \rho_{ab}^{(1)} \overset{+1}{\underset{-1}{\longrightarrow}} \rho_{ac}^{(2)} \overset{+1}{\underset{-1}{\longrightarrow}} \rho_{bc}^{(3)} \overset{+1}{\underset{-1}{\longrightarrow}} \rho_{cc}^{(4)}$, (4.56)

where the notation +1 or -1 above the arrows denote waves propagating in the +z or -z direction, respectively. Then, instead of (4.50), the population rate of the final state c for the near-resonant equal-frequency two-photon transition is given by

$$
\rho_{cc}^{(4)}(v) = \frac{\rho_{aa}^{(0)}}{\hbar^4} \left[\frac{1}{(\omega_{ac}-2\omega)^2 + \gamma_{ac}^2} \left| \frac{(\mu_{ab}E_1)(\mu_{bc}E_2) + (\mu_{ab}E_1)(\mu_{bc}E_2)}{(\omega_{ab}-\omega+kv_z) + i\gamma_{ab}} \right|^2 \right.
$$

$$
+ \frac{1}{(\omega_{ac}-2\omega+2kv_z)^2 + \gamma_{ac}^2} \left| \frac{(\mu_{ab}E_1)(\mu_{bc}E_1)}{(\omega_{ab}-\omega+kv_z) + i\gamma_{ab}} \right|^2
$$

$$
\left. + \frac{1}{(\omega_{ac}-2\omega-2kv_z)^2 + \gamma_{ac}^2} \left| \frac{(\mu_{ab}E_2)(\mu_{bc}E_2)}{(\omega_{ab}-\omega-kv_z) + i\gamma_{ab}} \right|^2 \right] . (4.57)
$$

The first term on the right-hand side, corresponding to the interaction paths *1* and *2* in (4.55), will contribute to the Doppler-Free (DF) peak of the two-photon transition signal, whereas the second and third terms in (4.57), corresponding to +1 and -1 paths in (4.56), respectively, will result in a Doppler-broadened (DB) background. Comparing the numbers of the paths in (4.55) and (4.56), the integral area of the DF Lorentzian profile with respect to v_z is larger than that of DB one, so that the peak ratio of DF to DB is larger than the width ratio $\Delta\omega_D/\gamma_{ac}$. This conclusion is correct if the radiation field is weak and the higher-order solutions of the density matrix equation are negligible.

4.3 Coarse Structure of Near-Resonantly Enhanced Molecular Two-Photon Absorption Spectra

Let us first consider the spectral structure due to vibrations of the observable molecular two-photon transitions that can take place between two electronic states, each with a series of vibration-rotational levels denoted by (v″,J″) and (v, J), respectively, enhanced by a series of levels denoted by (v′,J′) in an intermediate electronic state. The energy level scheme for one

74

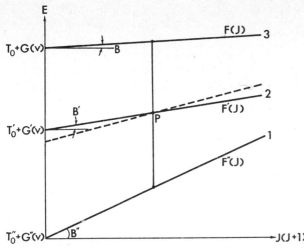

Fig.4.3. Rotational term curves for a molecular two-photon absorption band. The dashed curve indicates the individual middle positions for every two-photon transition in the spectral band. The intersection P corresponds to the resonantly-enhanced two-photon transition. The separations of the middle solid curve from the dashed curve show the offset values for individual transitions [4.10]

band of such a band system can be seen in Fig.4.3. The energies of each series of vibration-rotational levels with a given vibrational quantum number v is in the form of

$$T(v, J) = T_0 + G(v) + F(J) = T_0 + \omega_e(v+\tfrac{1}{2}) + B_v J(J+1) , \qquad (4.58)$$

which can be plotted approximately as a line with a particular slope corresponding to the rotational constant B_v. The dashed line with the average slope of lines *1* and *3* represents the midpositions between the individual upper and lower rotational levels. The intersection P corresponds to the two-photon resonance, provided P is at an integral J_m-value, with a maximum observed signal among the spectral lines of the two-photon absorption band. The separations of the individual values on the middle solid curve from the dashed curve with respect to each integral J-value in Fig.4.3 correspond to the offset values of the relative intermediate levels. Thus, while the offset Δ at P (i.e., at J_m) is zero, it increases rapidly away from $J = J_m$ with the following relation

$$\Delta = \left[\frac{B''+B}{2} - B' \right] [J(J+1) - J_m(J_m+1)] . \qquad (4.59)$$

Obviously, only if the rotational constant B′ of the intermediate state is approximately equal to the average value $\tfrac{1}{2}(B''+B)$ of the initial and final states would all the allowed two-photon transitions be greatly enhanced. Only this circumstance would result in a two-photon absorption spectrum with a structure similar to a one-photon absorption band. In contrast, the rotational constant of the intermediate state does not satisfy this condition. Therefore the situation is as in Fig.4.3 that there is only one intersection P between the dashed line and the line 2. If the intersection P were located at half integral value of J, there would be two lines in a two-photon band with comparable strengths, and the neighbouring lines would only have one nineth of the transition strength according to (4.53,59). For the case of the intersection P being very close to an integral value of J the strength ratio between the adjacent molecular two-photon lines will have great disparity. For example, an offset as small as 34 MHz has been observed in Na_2 [4.9]. For a medium J_m-value the line next to the strongest-enhanced two-photon transition in the same vibrational band ought to be weaker than 10^{-3} of the strongest two-photon transition signal. The further are the two-photon lines from the enhancing center J_m, the weaker are the transitions. This means that with the practical signal-to-noise limit of an experimental detection apparatus in a vibration-rotational spectral band of molecular two-photon transitions only a very small number of lines are observable, and this results in a simplified molecular spectral structure. As is the case in conventional molecular spectroscopy, varying one of the vibrational quantum numbers is the way to form the band progression in a molecular two-photon absorption spectrum. Figure 4.4. illustrates the rotational-term curves for two successive two-photon bands of a v-progression from a given initial vibrational level v'' enhanced by the same intermediate level v'. The dashed lines represent the mid-positions between curves 1 and 3, and between 1 and 4, and give the intersections at P_1 and P_2 on curve 2, corresponding to the near-resonant enhancements of these two bands from term T_1'' to T_1 and from T_2'' to T_2, respectively. Therefore we can write down the term relations as follows

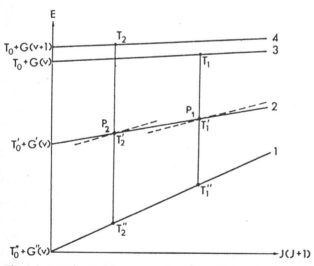

Fig.4.4. Rotational-term curves for two neighbouring two-photon absorption bands in v-progression. Dashed lines represent the mid-positions between curves *1* and *3* and between *1* and *4* giving the intersections P_1 and P_2 on curve *2*, respectively

$$T_1 - T_1' = T_1' - T_1'' \, ,$$

or

$$T_1'' + T_1 = 2T_1' \quad (\text{at } P_1) \, ; \tag{4.60a}$$

$$T_2'' + T_2 = 2T_2' \quad (\text{at } P_2) \, , \tag{4.60b}$$

where each term has the form given in (4.58). Subtracting (4.60a) from (4.60b), we obtain an equation for calculating the frequency spacing in a band progression of a molecular two-photon transition spectrum:

$$T_2'' - T_1'' + T_2 - T_1 = 2(T_2' - T_1') \, . \tag{4.61}$$

Let J_1 and J_2 correspond to the rotational quantum numbers at P_1 and P_2; respectively. Let the number 2 refer to the higher v band. $\Delta[J]$ denotes the projected difference between P_2 and P_1 on the transverse axis, i.e.,

$$\Delta[J] \equiv J_2(J_2 + 1) - J_1(J_1 + 1) \, . \tag{4.62}$$

For v-progression (Fig.4.4) the vibrational quantum number of the term T_2 is one greater than that of T_1. Substituting (4.58) into (4.61) for every term T, we obtain

$$\Delta[J] (B''+B) + \omega_e = 2B' \Delta[J]. \tag{4.63}$$

Therefore, for successive bands in a v-progression with a given v'' enhanced by a certain v' level we have the following relations between their enhancing centers

$$\Delta[J] = \frac{-\omega_e}{B'' + B - 2B'} \quad (\text{J relation}) \, , \tag{4.64}$$

$$\Delta\nu = - \Delta[J] (B'' - B') = \frac{\omega_e(B'' - B')}{B'' + B - 2B'} \quad (\text{frequency spacing}) \, , \tag{4.65}$$

where $\Delta\nu$ corresponds to the frequency spacing of the v-progression in a two-photon absorption spectrum. By analogy we can obtain the formula for the v''-progression in Fig.4.5

$$\Delta[J]'(B''+B) + \omega_e'' = 2B' \Delta[J]'' \, ,$$

so that

$$\Delta[J]'' = \frac{-\omega''_e}{B'' + B - 2B'} \quad (\text{J relation}) \, , \tag{4.66}$$

$$\Delta\nu'' = - \Delta[J] (B' - B) = \frac{\omega_e''(B' - B)}{B'' + B - 2B'} \quad (\text{frequency spacing}) \, . \tag{4.67}$$

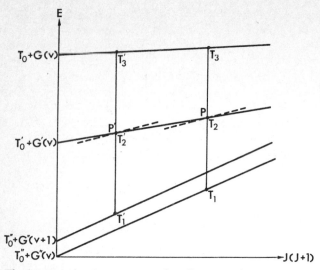

Fig.4.5. Rotational term curves for v″-progression

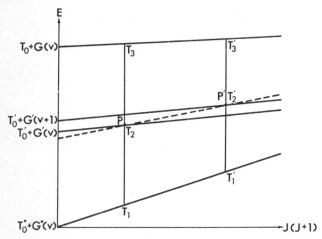

Fig.4.6. Rotational term curves for the extra v′-progression occuring in the molecular two-photon absorption spectrum

For the v′-progression in Fig.4.6, we have

$$\Delta[J]' = \frac{2\omega_e'}{B'' + B - 2B'} \quad \text{(J relation)} , \tag{4.68}$$

$$\Delta\nu' = -\tfrac{1}{2}\Delta[J](B'' - B) = \frac{-\omega_e'(B'' - B)}{B'' + B - 2B'} \quad \text{(frequency spacing)} . \tag{4.69}$$

Summarizing the results from (4.64 to 69), we can describe the characteristics of the coarse structure of equal-frequency molecular two-photon transitions as follows:

1) A molecular electronic spectrum of two-photon transitions near-resonantly enhanced by an intermediate level can be decomposed into three classes of band-progressions rather than two as in the conventional molecular one-photon absorption spectrum.

2) The three progressions have distinct frequency spacings and rotational quantum-number relations between maximally enhanced lines. Their absolute values are determined by their respective vibrational constants and by the differences between the rotational constants of the three related electronic states. Because the equilibrium internuclear distance in the ground state is usually smaller than that in excited electronic states, i.e.,, $B'' > B'$ or B, the v-progression is usually a blue shift and the v'-progression a red shift, but the shift direction of the v''-progression depends on the sign of $(B' - B)$ (Tables 4.1,2).

3) The coarse structure of molecular two-photon absorption spectra contains the information needed to estimate the rotational constants of the molecular high-lying electronic states of interest. From the ratio of (4.67) to (4.69) we have

$$B = \frac{\Delta v'' \omega_e' B'' + \Delta v' \omega_e'' B'}{\Delta v' \omega_e'' + \Delta v'' \omega_e'}, \qquad (4.70)$$

where the vibrational constants ω_e'' and ω_e' are usually known, and $\Delta v''$, $\Delta v'$ are determined experimentally. One cannot, in conventional one-photon absorption spectroscopy, expect to obtain a molecular rotational constant from a coarse (vibrational) structure. Substituting (4.70) into the ratio of (4.65 to 67), we obtain a formula for estimating the vibrational constants of the molecular high-lying electronic states of interest

$$\omega_e = \frac{\Delta v \omega_e''(B' - B)}{\Delta v''(B'' - B')} = \frac{-\Delta v \omega_e'' \omega_e'}{\Delta v' \omega_e'' + \Delta v'' \omega_e'}, \qquad (4.71)$$

ω_e being positive since the denominator is negative. Equations (4.70,71) show that we can easily determine the molecular constants from a simplified molecular two-photon spectrum, even at low resolution. This is especially useful for identifications of two-photon transitions.

4) In contrast with conventional molecular one-photon electronic spectra, the frequency spacing between the successive bands in a band pro-

Table 4.1. Some of the molecular constants in Na$_2$

States	ω_e [cm^{-1}]	B_e [cm^{-1}]	Ref.
Initial: X$^1\Sigma_g^+$	159.10449	0.1547337	4.33
Intermediate: A$^1\Sigma_u^+$	117.32322	0.1107818	4.34
Final: $\begin{cases} (3)^1\Pi_g \\ (3S+5S)^1\Sigma_g^+ \end{cases}$	102.579 109.1	0.10305 0.1078	4.35 4.35

Table 4.2. Expected frequency spacings and rotational quantum number spacings for respective band progressions in enhanced two-photon spectra in Na_2

Transitions	Progressions	$\Delta\nu$ [cm^{-1}]	$\Delta[J(J+1)]$
	v″-progression	33.96	−4393
$X^1\Sigma_g^+ \xrightarrow{A} (3)^1\Pi_g$	v′-progression	−167.41	6478
	v-progression	124.48	−2832
	v″-progression	11.57	−3883
$X^1\Sigma_g^+ \xrightarrow{A} (3S+5S)^1\Sigma_g^+$	v′-progression	−134.53	5727
	v-progression	117.04	−2663

gression in molecular two-photon electronic spectra does not correspond to the relative vibrational constant. Whereas ω_e'' is the largest value in the second column in Table 4.1, $\Delta\nu''$ is the smallest value in the third column in Table 4.2. It is remarkable that, although the coarse structure still corresponds to changes in the vibrational quantum number as in the conventional molecular electronic spectrum, the observed two-photon vibrational structure is no longer determined by one series of vibrational quantum-number changes and is no longer approximately equal to the vibrational constant as it is in molecular one-photon absorption spectrum.

5) The J relations in (4.64, 66, 68) imply that all three of the band progressions or band sequences will be very short. This is because the lower limit of the J-values is J_1, $J_2 \geq 0$; the upper limit of the J-values is restricted by the experimental conditions. That is to say, the population distribution among the vibration-rotational levels obeys the Maxwell-Boltzmann distribution. In the case of molecular sodium, for example, contained in an oven heated at about 400° C, the rotational quantum number J with maximum population is about 40. The population at J = 80 is less than half of that at J = 40. The larger the J value the lower is the population. In addition, unfortunately, the ΔJ values between the observable lines in the adjacent bands of the band progressions are rather large. Taking a practical example, by using the molecular constants listed in Table 4.1, the $\Delta[J]$ values for v″-, v′- and v-band progressions can be calculated according to (4.64 to 68), as listed in Table 4.2. The sign + or − of $\Delta\nu$ in Table 4.2 means blue or red shift in the band progressions with the respective vibrational quantum number increasing by one unit. The sign in the last column indicates the direction in which the rotational quantum number changes when the vibrational quantum number increases by one unit in the relative progression. The large ΔJ values listed in Table 4.2 mean that molecular two-photon band progression will rapidly cut-off at J limits. For example, the near-resonance enhanced two-photon transition in Na_2 at 16601.803 cm^{-1}, which has been well identified by complementary experiments in addition to

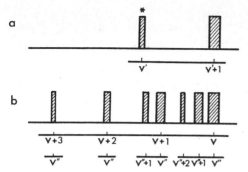

Fig.4.7. Illustration of the coarse structures of near-resonantly enhanced two-photon absorption spectra. a) v'-progression. b) v- and v''-progressions. The block marked with a star indicates the band with the enhancing centre of the near-resonant two-photon transition line at 16601.803 cm^{-1}. The widths of the blocks are proportional to the J values of the individual calculated enhancing centres for the three progressions for the two-photon transitions with B">B'>B

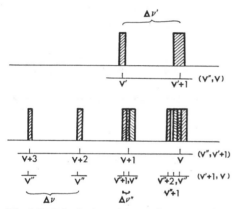

Fig.4.8. Principal positions of a part of the decomposed coarse structures of two-photon transitions with B">B', B'~B. The block marked with a star is the same band as in Fig.4.7. The v''-progression has been partially overlapped

numerical calculations [4.8, 10], is a Q_1 branch two-photon transition from the ground state $X^1\Sigma_g^+$ (2,50) to the upper state $(3S+5S)^1\Sigma_g^+$ (18,50) enhanced by the intermediate state $A^1\Sigma_u^+$ (22,49). By using the data listed in Table 4.2 to estimate the frequency positions of the enhancing centers of its neighbouring bands a section of the calculated vibrational structures round 16601.803 cm^{-1} has been drawn in Fig.4.7,8, corresponding to from $x^1\Sigma_g^+$ to $(3)^1\Pi_g$ with the conditions of B" > B' > B and to $(3S+5S)^1\Sigma_g^+$ with B" > B' ~ B enhanced by $A^1\Sigma_u^+$, respectively. The widths of the blocks correspond to the J values of the enhancing center of each band, see (4.77 and 81). A molecular two-photon spectrum was here decomposed into a v'-progression containing the two bands shown in Fig.4.7a; Fig.4.7b shows a v-progression containing four bands, and four v''-progressions containing different numbers of bands, the largest being three. The block marked with an asterisk represents the band involving the two-photon transition signal at

16601.803 cm^{-1} with a J value of 50. Referring to the $\Delta[J]$ values in Table 4.2, we see that there is no way to have a successive band in v- or v''-progression from the band marked with an asterisk by increasing the value of v or v'' by one unit on account of the minimum J = 0 limit. It is possible to have a neighbouring band with (v'+1) in v'-progression as shown by the right-side block in Fig.4.6a, but the J value might be too large for observation (J~95). However, from this principal position we can estimate the wave-number positions for v- and v''-band progressions with the given value (v'+1) of the intermediate level for the two-photon transitions of $X^1\Sigma_g^+ \rightarrow (3)^1\Pi_g$ enhanced by $A^1\Sigma_u^+$. One can then see that within either of them the bands on the higher frequency side have successively smaller J values. This results in the longest progression in the two-photon spectrum in Na$_2$ containing only 3 bands. The spacing ratios among the three progressions depend on the related energy-level constants. If two of the three electronic states for the two-photon transitions had similar rotational constants the band progressions of the third state might overlap each other, as shown in Fig.4.8. Comparing with Fig.4.7, it can be seen that the closer the values of B' and B the smaller the frequency spacing in the v''-progression. The entire coarse structure of two-photon molecular absorption spectra can be calculated by iteratively expanding from left to right along the bands in (a) followed by expanding from right to left in (b) of Figs.4.7 and 8 for every block. The expansion will contain the band progressions with v'', v' or v given and the band sequences with the Δv values given. A more accurate calculation would involve the higher-order corrections of the molecular constants for the three related electronic states. The vibrational structure calculated for the observed molecular two-photon electronic transitions in Na$_2$ from the $X^1\Sigma_g^+$ to the $(3S+5S)^1\Sigma_g^+$ state with the published constants is shown in Fig.4.9 [4.10]. It illustrates the fact that the observable two-photon band progressions are short (the J region was taken for the calculation from 0 to 80).

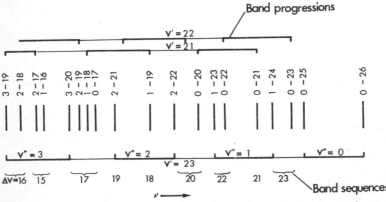

Fig.4.9. The vibrational structure of the observed two-photon electronic transitions in Na$_2$, assigned by calculations with the known molecular constants. Horizontal lines represent the v-progressions with given v'' or v' values. The individual sequences of bands are indicated by brackets [4.10]

Note that the vibrational structure of the two-photon absorption spectrum discussed above reveals the imaginary band centers, corresponding to the intersections of P_i (i = 1,2) in Figs.4.3-6, rather than the real signal frequencies. In fact, only if an intersection P_i corresponded to an integral value of J within its limitations would it have physical meaning as an observable two-photon transition line belonging to the Q-branch and enhanced by a Q-branch one-photon transition. Sometimes it could even be forbidden if the two-photon transition dealing with more than two Σ electronic states. Nevertheless, the blocks do mark the region of the near-resonantly enhanced two-photon lines belonging to various two-photon branches.

4.4 Fine Structure of Near-Resonantly Enhanced Molecular Two-Photon Transitions

The equation for the population rate on the final state of a molecular two-photon transition can be utilized to discuss the fine structure (i.e., the rotatioanl structure) of such spectra. As mentioned above, the individual rotational transitions in a two-photon absorption band from the term curves *1* to *3* in Fig.4.3 possess gradually changed offset values (Δ) from the respective enhancing levels on the term curve *2*. The largest enhancement is gained at the intersection P where the Δ value is approximately zero. One might then consider predicting the observable two-photon rotational structures by calculating those transitions from the intermediate to the upper levels which coincide with the transitions from the ground to the intermediate level for every allowed rotational branch. An example of Na_2 is illustrated in Fig.4.10. Curves *1* and *3*, *5*, and *2* and *4*, *6* represent R and P branches of the $X^1\Sigma_g^+$ (v'' = 2) $\rightarrow$ $(A^1)\Sigma_u^+$ (v' = 22) and $A^1\Sigma_u^+$ (v' = 22) $\rightarrow$ $(3S+5S)^1\Sigma_g^+$ (v = 19, 18) bands, respectively. The insert at the upper-right corner shows the relative rotational quantum numbers of the curves. The intervals between the curves , for example between *2* and *3* or *5*, for every J are equal to twice the Δ values, respectively (represented by vertical bars). So they give information about the intensity of all Q_1-branch lines (corresponding to $\Delta J = 0$ from the ground state to the higher excited state). This is true also for the O-branch ($\Delta J = -2$) between lines *2* and *4* (or *6*), and for the S ($\Delta J = 2$) and Q_2 branches ($\Delta J = 0$) between lines *1* and *3* (or *5*) and between lines *1* and *4* (or *6*), respectively. If any intersection is located very close to an integral J value (let us, say, Δ less than the Doppler width) [4.9], there may appear a strong two-photon line of this branch (if the Franck-Condon factors, thermal distribution and the statistical weight are also favourable). However, the neighbouring transition will have a much larger Δ value; for Na_2 the magnitude typically is 0.04J cm^{-1} according to (4.59), which may be as great as several wave numbers for large J. Thus the intensity of the adjacent two-photon transitions around the enhancing center line would be reduced by several orders of magnitude and be difficult to observe. This means that under typical experimental condi-

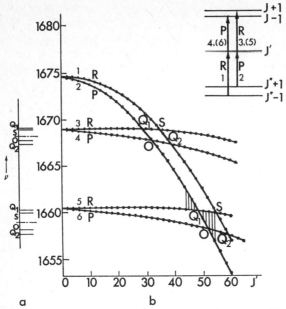

Fig.4.10. An example of how to predict observable two-photon transitions from the $X^1\Sigma_g^+$ (v″ = 2) to the $(3S+5S)^1\Sigma_g^+$ (v = 18, 19) vibrational-rotation levels enhanced by $A^1\Sigma_u^+$ (v′ = 22). Only the individual frequencies with even J are plotted. The insert shows the relative rotational quantum numbers. On the left side of the vertical axis is shown the resulting spectrum; a dashed line represents the centre frequency of each band [4.10]

tions (e.g., CW dye laser) we may observe only one line for one branch in this case. When the crossover point is located far from an integral J, we may even miss the whole branch. This contrasts with the large number of lines and smoothly changing intensities in an one-photon absorption band in Na_2, where the intensity varies from J = 3 to J = 100 by only a factor of 5. Among the very limited number of observable lines there are new relative frequency arrangements for the branches in a two-photon absorption band. In the case of B″>B′~B, the enhancing centers of the two-photon branches show the same frequency distribution as the spectrum at the left side in Fig.4.10, where the enhanced Q1 and Q_2 branches are at the outer sides of the band. However, in the case where B″,B>B′ (e.g., B″=0.154 cm^{-1}, B=0.124cm^{-1} > B′=0.1107cm^{-1} in Na_2 [4.36]) the enhancing centers of the Q_1 and Q_2 branches may be close to the middle of the band, as illustrated at the left side in Fig.4.11a, instead of near the edges. This is because the curve of the transition frequencies from the intermediate to the upper state as a function of J is curved upward (Fig.4.11b) and the projection of these crossover points onto the vertical (frequency axis) will show a different pattern from Fig.4.10a. The comparison of the patterns different reveals that the correct identification of individual lines in a two-photon transition spectrum requires the auxiliary spectral method (Sect.7.8).

Instead of four branches in a molecular two-photon absorption band of Σ-Σ transitions enhanced by a Σ state, there are spectra having as many as

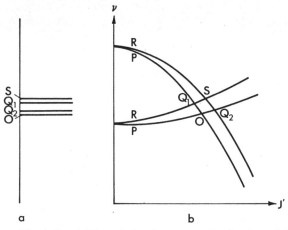

Fig.4.11. a) The relative frequency distribution of the enhancing centres as given by the projections of the intersections in (b). **b)** The transition frequencies for forming a band of two-photon transitions between Σ states enhanced by a Σ state with rotational constants B'', $B > B'$

six or nine branches in a near-resonantly enhanced two-photon absorption band, provided that one of the states has a nonzero quantum number of the electronic angular momentum on the molecular axis, $\Lambda_i \neq 0$. Figure 4.12 shows, for example, a) $\Lambda'' \neq 0$, $\Lambda, \Lambda' = 0$; b) $\Lambda \neq 0$, $\Lambda'', \Lambda' = 0$; c) at least $\Lambda' \neq 0$, in which cases the Q_i-branch centers for the respective transitions appear. The notations $Q_i (i=1,2,3)$, R_j and $P_j (j=1,2)$, S and O correspond to the rotational quantum number differences between the final and initial levels being $\Delta J_{gf} = 0, +1, -1, +2,$ and -2, respectively.

Obviously, the spectral extension range of the near-resonantly enhanced fine structure in a molecular two-photon spectrum depend on the rotational quantum numbers represented by the line in the Q_3-branch enhanced by a Q-branch of the relative one-photon transition. The determination of the rotational constant B for the final state, according to the coarse structures in a molecular two-photon absorption spectrum with (4.70), will help us to select an upper electronic state from all the candidates. The integral value J_c, corresponding to the enhancing center of the Q_3 branch in Fig.4.12c, can be evaluated by (4.60), where J_1 is replaced by J_c, as

$$J_c = -\tfrac{1}{2} + \sqrt{\tfrac{1}{4} + \Delta T_{ev}/\Delta B} \;|_{integral} , \qquad (4.72)$$

where the notation "$|_{integral}$" means taking the integral value closest to he calculated one for J_c; $\Delta T_{ev} = \overline{T_0}' + G'(v) - \overline{T_0} - \overline{G}(\nu)$; $\overline{T_0} = (T_0'' + T_0)/2$; $\Delta B = B' - \overline{B}$; $\overline{G} = [G''(v) + G(v)]/2$; $\overline{B} = (B'' + B)/2$. Note that J_c does not always means an observed two-photon line. In the cases of Figs.4.10 and 12a,b the Q_3-ranch is not allowed. A better way to present the extension range is to derive some formulas which allow comparison with the observed signals. In Fig.4.12a we see that the Q branch from initial to intermediate states was

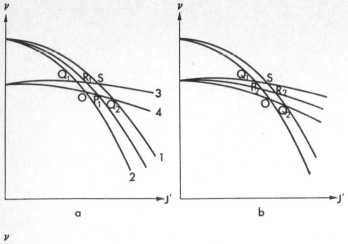

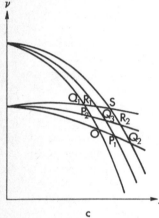

Fig.4.12. Illustration for forming more enhanced two-photon branches in the cases of a) $\Lambda'' \neq 0$, $\Lambda', \Lambda = 0$; b) $\Lambda \neq 0$, Λ'', $\Lambda' = 0$; c) at least $\Lambda' \neq 0$

inserted between the curves *1* and *2*. It intersects curves *3* and *4*, and gives the frequency positions corresponding to the enhancing centers of the R and P branches in the two-photon absorption band. They take the positions between Q_1 and S, and between O and Q_2, respectively. By analogy the enhancing centers of the P_2 and R_2 branches in Fig.4.12b and the R_1, R_2, P_1, P_2 and Q_3 lines in Fig.4.12c have frequency positions within the region from Q_1 to Q_2 in the case of B''>B'>B; or within the region from S to O in the case of B'', B>B'. The expected extension range of the enhanced fine structure in a molecular two-photon band are then characterized by the frequency intervals from Q_1 to Q_2 or from S to O for almost all of the two-photon absorption bands. The frequency separations between these enhancing centers of the branches can be derived as follows:

1) $Q_1 - Q_2$ separation

According to the definitions for the molecular two-photon absorption branches [4.10], the Q_1 branch is enhanced by the P branch of the one-photon transition from the initial to the intermediate state. Thus the enhancing center of Q_1 corresponds in the first order approximation to the following energy-term relation

$$\overline{T_0} + \overline{G}(v) + \overline{B}J_1(J_1+1) = T_0' + G'(v) + B'(J_1-1)J_1 . \qquad (4.73)$$

Therefore, the integral J_1-value near the enhancing center of the Q_1 branch is

$$J_1 = -\tfrac{1}{2}b + \sqrt{(\tfrac{1}{2}b)^2 + \Delta T_{ev}/\Delta B} \mid_{integral} , \qquad (4.74)$$

where $b = (\overline{B}+B')/(\overline{B}-B')$. On the other side, the Q_2 branch of the two-photon transitions is enhanced by the R branch of the one-photon transitions from the initial to the intermediate levels, corresponding to the energy-term relation

$$\overline{T_0} + \overline{G}(v) + \overline{B}J_2(J_2+1) = T_0' + G'(v) + B'(J_2+1)(J_2+2) . \qquad (4.75)$$

Subtracting (4.75) from (4.73) we get the relation between the rotational quantum numbers of the two enhanced Q_1 lines

$$J_2 - J_1 = 2B'/(\overline{B} - B') . \qquad (4.76)$$

Therefore the Q_1-Q_2 frequency separation $\Delta\nu_{Q_1 Q_2}$, corresponding to the extension range of the enhanced fine structure of a molecular two-photon absorption band in the case of $(B'' > B' > B)$ is obtained as follows

$$\Delta\nu_{Q_1 Q_2} = \tfrac{1}{2}(B'' - B)[J_2(J_2+1) - J_1(J_1+1)]$$

$$= \frac{B'' - B}{B - B'} B' \left[2J_1 + \frac{B'' + B}{2(\overline{B} - B')} \right]. \qquad (4.77)$$

2) S - O separation

As shown in Figs.4.10-12, an S- or O-branch of a molecular two-photon absorption band is enhanced by the R- or P-branch of the corresponding one-photon transitions from the initial to the intermediate state, respectively. Denote as J_3 and J_4 the integrals closest to the enhancing centers of the S- and O-branch, respectively. The energy term relation for J_3 is of the form

$$2\overline{T_0} + 2\overline{G}(v) + B(J_3+2)(J_3+3) + B''J_3(J_3+1)$$
$$= 2[T'_0 + G'(v) + B'(J_3+1)(J_3+2)] , \qquad (4.78)$$

and for J_4 the form

$$2\overline{T_0} + 2\overline{G}(v) + B(J_4-2)(J_4-1) + B''J_4(J_4+1)$$
$$= 2[T'_0 + G'(v) + B'(J_4-1)J_4] . \qquad (4.79)$$

By analogy with the above steps for Q_1-Q_2 separations, we can get the relation between J_3 and J_4

$$J_3 - J_4 = 2(B' - B)/(\overline{B} - B') , \qquad (4.80)$$

and the S - O frequency separation $\Delta \nu_{SO}$ is

$$\Delta \nu_{SO} = 2 \frac{BB'' - B'\overline{B}}{\overline{B} - B'} \left[2J_3 + \frac{5B+B''-6B'}{2(\overline{B}-B')} \right] , \qquad (4.81)$$

where J_3 was determined by (4.78) and the related electronic states were recognized by comparing the data from (4.70, 71) with the candidate states.

The above formulae for the Q_1-Q_2 and S-0 separations in the first-order approximation are illustrated by the widths of the blocks in Figs. 4.8, 9.

The direct proportionality between the extension range of the enhanced fine structure and the rotational quantum number, as shown by (4.77 and 81), has been confirmed by applying the $\Delta|J|$ relations (e.g., the data listed in Table 4.2) to the assignment of the coarse structure for an experimentally recorded molecular two-photon absorption spectrum. The further line-by-line identification should be performed by means of numerical-data fitting with the aid of the comprehensive identification methods, as described in Chap. 7. Finally the two-photon spectrum analyses would result in a set of improved molecular constants for the high-lying electronic states.

According to the previous discussion about the two-photon excitation probability and the conditions required for a near-resonant two-photon transition, we may conclude that in practical measurements the observable number of lines in the fine structure in a two-photon molecular band will not be definite but will depend on the experimental conditions (e.g., light source, optical arrangement and detection sensitivity) and on the relative molecular energy states. While under some conditions (e.g., by using pulsed laser light or when one is dealing with mixing states as the enhancing states) several lines can be observed in one branch of a two-photon spectral band, in another cases one may miss the whole fine structure. For example, with careful scanning of the single-mode CW dye laser covering 1000 cm^{-1} around 600 nm, we have observed in Na_2 only 79 two-photon lines with widely varying intervals. It is interesting to note that the 34 two-photon transitions are composed of 24 different bands, with 15 of them appearing alone [4.10]. In such an incomplete spectral structure, the missing branches are often due to the non-integral J values of the intersections in Figs. 4.10-12 in the region of large J. The resulting, incomplete two-photon spectrum is difficult to assign. Fortunately, various ways have been demonstrated to be useful for discriminating molecular two-photon branches and for examing the J values (Chap. 7). Nevertheless, even with these incomplete data one is able, using the above mentioned evaluations together with the two-photon energy (determined by the scanning-laser wave numbers), to recognize a final state from the tens of candidates of high-lying states predicted by the theoretical calculations, such as the RKR approximation, or as measured by two-step excitations [4.36-39].

4.5 Line Shapes and Higher-Order Corrections for Near-Resonant Two-Photon Transitions in Three-Level Systems

Now let us turn to an analysis of the spectral characteristics of a single two-photon transition. The enhancement by an intermediate level is one of the interesting features of two-photon spectroscopy.

A general expression for the two-photon population rate into the final level of a molecule with velocity v irradiated by two equal-frequency counterpropagating light fields has been derived, see (4.57). Adding the thermal distribution of molecules to (4.57) and integrating over all the velocities from $-\infty$ to ∞, the total transition-rate expression (4.57) for one molecule has been transformed into the line profile of a particle ensemble. It consists of a Doppler-Broadened (DB) profile from the last two terms of the equation (corresponding to the absorption of two photons out of the first or second beam, respectively), upon which the Doppler-Free (DF) spectral part from the first term in (4.57) is superimposed (corresponding to the absorption of one photon from each of the beams), as shown in Fig.4.13.

By means of the error function with a complex argument, an analytical solution of the line shape, depending on the detuning of the light fields ($\delta = \omega - \frac{1}{2}\omega_{ac}$) and the offset Δ of the effective intermediate level, was obtained. Furthermore, for $\delta = 0$ we obtain both the DF peak (the first term) and DB peak (the later term vlaues), and can consequently find the peak ratio of the DF peak to the highest of its DB pedestals. Figure 4.14 shows these forms as functions of the offset Δ (normalized to the Doppler width $\Delta\nu_D$). The calculation reveals that the peak ratio approaches the constant value of

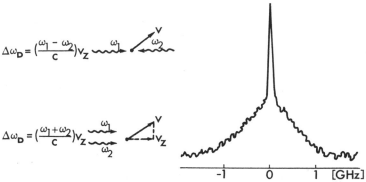

$$\Delta\omega_D = \left(\frac{\omega_1 - \omega_2}{c}\right)v_z$$

$$\Delta\omega_D = \left(\frac{\omega_1 + \omega_2}{c}\right)v_z$$

Fig.4.13. Typical line shape of a near-resonantly enhanced two-photon transition consisting of a Doppler-free peak and Doppler-broadened pedestal

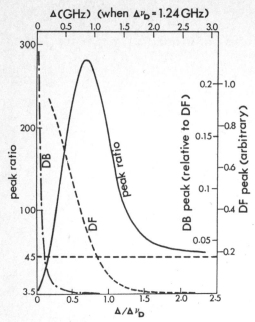

Fig.4.14. Theoretical intensity curves of the two-photon transition. DF: the intensity of the Doppler-free signal (arbitrary scale); DB: the center intensity of the Doppler-broadened signal relative to the DF curve; peak ratio: DF signal/DB signal [4.9]

$1.35 \Delta \nu_D / \Delta \nu_k$ with Δ larger than twice Doppler width, which is the approximation for the far-off-resonance case. Moreover with decreasing Δ values the peak ratio increases up to a maximum and then drops down to a minimum which is about one-tenth of the constant value for the far-off-resonant case [4.9].

In fact, the observed peak ratio of the DF to the DB parts of a near-resonantly enhanced two-photon absorption line is smaller than that expected by the theoretical results shown in Fig.4.14 (solid curve). This is because the higher-order corrections are particularly important in this kind of energy level scheme, as can be explained as follows. In a molecular two-photon transition with a near-resonant intermediate level satisfying $\Delta \leq \Delta \omega_D$, a near-equally spaced 3-level system is formed. In this case the velocity distribution of the "on resonance" molecules, satisfying

$$kv_z = \pm \Delta = \pm (\tfrac{1}{2}\omega_{ac} - \omega_{ab}) \ , \tag{4.82}$$

is no longer located on the far-off wings of the Gaussian function and has considerable values. These molecules absorb two counter-propagating photons not only simultaneously, but also stepwise at large rates since the resonant transition of a→b and b→c shown in Fig.4.2b is also satisfied. Note that the presence of the two-step transition is based on the hole-burning of the thermal distribution at the initial level a and the actual population of the intermediate level b. As the light intensity increases, the one-photon

90

process of the "on resonance" molecules may greatly deplete their number in the lower level a and populate the intermediate level b. Instead of the interaction paths of the radiation fields with molecules listed in (4.55, 56), the lowest-order population of the fianl state, based on the hole-burning distribution at the initial state, is then given by the sixth-order diagonal element $p_{cc}^{(6)}$ formed as follows [4.40]

$$1)\quad \rho_{aa}^{(0)} \xrightarrow{t} \rho_{ab}^{(1)} \xrightarrow{t} \rho_{aa}^{(2)} - \rho_{bb}^{(2)} \xrightarrow{p} \rho_{ab}^{(3)} \xrightarrow{q} \rho_{ac}^{(4)} \xrightarrow{r} \rho_{bc}^{(5)} \xrightarrow{s} \rho_{cc}^{(6)},$$

$$\text{(4.83)}$$

$$2)\quad \rho_{aa}^{(0)} \xrightarrow{t} \rho_{ab}^{(1)} \xrightarrow{t} \rho_{aa}^{(2)} - \rho_{bb}^{(2)} \xrightarrow{p} \rho_{ab}^{(3)} \xrightarrow{p} \rho_{bb}(4) \xrightarrow{q} \rho_{bc}^{(5)} \xrightarrow{q} \rho_{cc}^{(6)}.$$

Simultaneously there are some other interaction paths, which give the population corrections as the additional sixth order element $\rho_{cc}(6)$ for the final state c

$$3)\quad \rho_{aa}^{(0)} \xrightarrow{t} \rho_{ab}^{(1)} \xrightarrow{t} \rho_{bb}^{(2)} \xrightarrow{p} \rho_{bc}^{(3)} \xrightarrow{q} \rho_{ac}^{(4)} \xrightarrow{r} \rho_{bc}^{(5)} \xrightarrow{s} \rho_{cc}^{(6)},$$

$$4)\quad \rho_{aa}^{(0)} \xrightarrow{p} \rho_{ab}^{(1)} \xrightarrow{q} \rho_{ac}^{(2)} \xrightarrow{r} \rho_{bc}^{(3)} \xrightarrow{s} \rho_{cc}^{(4)} \xrightarrow{t} \rho_{bc}^{(5)} \xrightarrow{r} \rho_{cc}^{(6)},\qquad \text{(4.84)}$$

$$5)\quad \rho_{aa}^{(0)} \xrightarrow{p} \rho_{ab}^{(1)} \xrightarrow{p} \rho_{bb}^{(2)} \xrightarrow{q} \rho_{bc}^{(3)} \xrightarrow{q} \rho_{bb}^{(4)} \xrightarrow{t} \rho_{bc}^{(5)} \xrightarrow{t} \rho_{cc}^{(6)},$$

where we assumed that $\rho_{bb}^{(0)} = \rho_{cc}^{(0)} = 0$. The letters p, q, r, s, t above the arrows denote distinct propagating waves. We use the notation +1 or -1 in the calculation in order to label the direction of propagation of the waves as +z or -z. The paths 1 and 2 represent the influences of the hole-burning at the lower level and the actual population at the intermediate level for the two-photon transition and the two-step transition, respectively. Path 3 describes the dynamic Stark effect since the middle level is populated. The last two paths represent the reemission processes from the final and the intermediate level, respectively.

The "on resonance" molecules with v_z in (4.82) are the main components of the population in the final state and the major contributors to the Lorentzian peak, which might be seriously reduced by hole-burning at path 1 in (4.83) especially with intense fields. But the main contribution to the background center is still the packet of molecules with $v_z = 0$, which should not deplete much under the same experimental conditions. Therefore the correction will result in the reduction of the height ratio with increasing laser intensity. Moreover, the population differnce $\rho_{bb} - \rho_{cc} \neq 0$ in path 3 in (4.84), the dynamic Stark effect, should give a decrement in the terminal population with $v_z = \pm\Delta/k$, corresponding to a negative peak without Doppler broadening at the spectral location of $\omega = \omega_{ac}/2$. The reemission processes in path 4 and 5 in (4.84) should deplete the molecules with exactly $v_z = \pm\Delta/k$ rather than with $v_z = 0$.

Thus, if the laser intensity is low enough, the lowest term $\rho_{cc}^{(4)}$ of $\rho_{cc}(v)$ in (4.57) and Fig.4.14 will be good enough to describe the line shapes. The higher-order corrections have to be taken into account when the field intensity increases. The total population rate on level c is therefore given by the sum

$$\rho_{cc}(v) = \rho_{cc}^{(4)}(v) + \sum_{i=1}^{5} \rho_{cc}^{(6)}(v)_i \, , \tag{4.85}$$

where

$$\rho_{cc}^{(6)}(v)_1 = - h_1 \gamma_{ab}^2 \sum_t [\rho_{cc}^{(4)}(v)_T / \Delta_{ab}(t) \Delta_{ba}(t)] \, ,$$

$$\rho_{cc}^{(6)}(v)_2 = - h_2 \gamma_{ab}^2 \sum_t [\rho_{cc}^{(4)}(v)_S / \Delta_{ab}(t) \Delta_{ba}(t)] \, , \tag{4.86}$$

$$\rho_{cc}^{(6)}(v)_3 = - \frac{\gamma_{ab} \gamma_a}{\gamma_c} h_1 \sum_t \frac{1}{|\Delta_{ab}(t)|^2} \mathrm{Im} \left\{ \frac{|V_{ab}|^2 |V_{bc}|^2}{\Delta_{bc}^2(-t) \Delta_{ac}(0)} \right\} \rho_{aa}^{(0)}(v) \, ,$$

$$\rho_{cc}^{(6)}(v)_4 = - h_2 \gamma_{bc}^2 \sum_t [\rho_{cc}^{(4)}(v)_T / \Delta_{bc}(t) \Delta_{cb}(t)] \, ,$$

and

$$\rho_{cc}^{(6)}(v)_5 = - h_2 \gamma_{bc}^2 (1 + \gamma_c / \gamma_b) \sum_t [\rho_{cc}^{(4)}(v)_S / \Delta_{bc}(t) \Delta_{ab}(t)]$$

were derived according to (4.25). All of the resonant denominators in (4.86) show that the smaller the offset Δ the bigger the correction $\rho_{cc}^{(6)}(v)_i$. The quantities h_1 and h_2 are given by

$$h_1 = 2(1/\gamma_a + 1/\gamma_b) |V_{ab}|^2 / \gamma_{ab} \, ,$$
$$h_2 = 2 |V_{bc}|^2 / \gamma_c \gamma_{bc} \, . \tag{4.87}$$

Their physical meanings are the signal dips in the background profile at the frequencies $\omega_1 = \frac{1}{2}(\omega_{ab} + \frac{1}{2}\omega_{ac})$ and $\omega_2 = \frac{1}{2}(\omega_{bc} + \frac{1}{2}\omega_{ac})$ due to the cross-over resonances between one- and two-photon transitions, respectively. Figure 4.15 shows a comparison between the calculated line shapes and the experimental results for one of the two-photon transitions with the smallest offset value ($\Delta = 34$MHz) among the measured lines for Na_2 within the operational range of a R6G dye laser. In calculating the profiles in Fig.4.15b the parameters, such as $\Delta\nu_D$, Δ, γ_{ab} and γ_{ac}, were determined by the two-

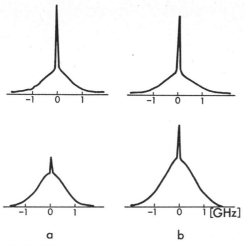

Fig.4.15. Line shape comparisons between experimental traces and theoretical calculations for a near-resonant two-photon transition at 16815.571 cm^{-1} in Na$_2$. (a) Experimental traces: lower: with focused beams (about 150 μm focusing waist); upper: with unfocused beams (3mm diameter, 30mW). (b) Theoretical line shapes in the Δ = 34 MHz case: upper: computation for $\rho_{cc}^{(4)}$ only; lower: computation including $\rho_{cc}^{(6)}$ [4.40]

photon transition line at the laser frequency of 16815.8713 cm^{-1}. The upper profile shows the computed shape, taking into account only the lowest term in (4.85) without the sixth-order corrections; the lower profile shows the theoretical shape involving the sixth-order corrections. Correspondingly, in (30mW CW unfocused laser beams of 3mm diameter), and the lower trace was recorded by using focused beams with 150 μm beam waists. Evidently the reduced peak ratio of DF signal to DB signal can be deduced from the higher-order corrections under the conditions of strong fields. The ratio was measured to be inversely proportional to the laser intensity [4.40].

When a two-photon transition involves a molecular multiplet electronic state, its line profile might present the superposition of several Doppler-free lines. The spectral analysis reveals the hyperfine splittings, which can be used to obtain the molecular hyperfine constants for the as yet not well-known molecular states.

4.6 Molecular Two-Photon Transitions Enhanced by Mixing Levels

According to the selection rules for the dipole approximation, the changes of the electronic spin quantum numbers should obey $\Delta S = 0$. However, due to the perturbation mechanisms there are molecular energy levels from distinct multiplet states for which the eigenfunctions mix, as described in Chap.2. Provided a pair of mixing levels is right at the middle position between the forbidden two-photon transition levels, the selection rule $\Delta S = 0$ will be not a restriction.

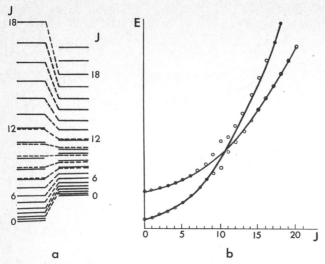

Fig.4.16. Illustration of the mutually perturbing rotational term series. (a) Two series of unperturbed levels. (b) Transition frequencies between unperturbed (solid curve) and perturbed (circles) levels [4.46]

For example, Fig.4.2a shows a part of the electronic states in Na_2, where solid and dashed lines represent the potential curves of singlet and triplet states, respectively. Because the lowest triplet state $a^3\Sigma_u^+$ (Fig.2.1) has an energy of more than 6000 cm^{-1}, it is not thermally populated under common experimental conditions. Only the ground singlet state $X^1\Sigma_g^+$ is normally populated. This presents a problem for the direct study of excited triplet states by allowed absorption transitions in Na_2. Fortunately, there is a wide overlapping area between the first excited singlet and the triplet states of $A_1\Sigma_u^+$ and $b^3\Pi_u$ at an energy between 15000 to 23000 cm^{-1}.

According to general perturbation theory, as was mentioned previously in Chap.2, the magnitude of the perturbation shift depends not only on the smallness of the energy difference between the nominal levels but also on the magnitude of the matrix element W_{12} of the perturbation function. The conditions for nonvanishing W_{12} is that both states must have the same total angular momentum J, as shown in Fig.4.16, in addition to the other selection rules. The perturbation center then can be determined by the related energy level terms. Assume that the mutual perturbation occurs between state A and b. The maximum perturbation at J_p satisfies the following total term relation

$$T_A(v_A, J_p) = T_b(v_b, J_p) . \qquad (4.88)$$

Substituting (4.58) into (4.88), we obtain the J_p position in first-order approximation from the nominal levels:

$$J_p(J_p+1) = (\Delta T_0 + \Delta G)/\Delta B , \qquad \text{where} \qquad (4.89)$$

$$\Delta T_0 = (T_0)_A - (T_0)_b \, ,$$
$$\Delta G = (\omega_e)_A (v_A + \tfrac{1}{2}) - (\omega_e)_b (v_b + \tfrac{1}{2}) \, , \qquad\qquad (4.90)$$
$$\Delta B = B_b - B_A \, .$$

For a constant value ΔT_0 of the perturbations between two electronic states ΔB varies slowly with vibrational quantum numbers. The variations of the vibrational quantum numbers v_A and v_b give a series of perturbation centers corresponding to distinct J_p values.

The absolute magnitude of the perturbation effect depends on practical perturbation mechanisms. *Atkinson* et. al. have systematically studied the perturbation coefficients for the mutually perturbing states $A^1\Sigma_u^+$ and $b^3\Pi_u$ in Na_2 [4.41], where the most important perturbation is spin-orbit interaction with the negligible spin-spin and orbit-rotation interactions.

For a certain electronic configuration (S and Λ values defined) the interaction Hamiltanian between the electronic spin and orbit motions is of the form

$$\mathscr{H}_{SO} = A\hat{L}\hat{S} \, , \qquad\qquad (4.91)$$

where A is the coupling constant. The $b^3\Pi_u$ state in Na_2 has A ~ 7.85 cm^{-1} [4.42]. Its fine electronic states can be denoted by subscripts derived from quantum numbers of the total angular momentum projected on the internuclear axis, $\Omega = \Lambda+\Sigma = 0,1,2$, as $^3\Pi_{0u}$, $^3\Pi_{1u}$ and $^3\Pi_{2u}$. Another kind of L and S interaction occurs between different multiplet electronic states, such as $b^3\Pi_{0u}$ and $A^1\Sigma_u^+$ in Na_2, both with $\Omega = 0$. Combining the two kinds of spin-orbital interactions means that the couplings between $A^1\Sigma_u^+$ and each of the triplet states are nonvanishing. The relative mixing coefficients for these mutually perturbing states depend on the smallness of the energy term difference and also on the degree of overlap of the vibrational wavefunctions. The mixing coefficients $|S_{\Omega\Sigma}|^2$ between the particular series of rotational levels in $A^1\Sigma_u^+$ ($v' = 17$) and in $b^3\Pi_{0u}$ (with v-given) in Na_2 have been determined, as shown in Fig.4.17. The maximum mixing coefficients correspond to the perturbation centers, each possessing a relative maximum in the repulsive perturbation shift. Therefore one of the vibration-rotational term curves for the $A^1\Sigma_u^+$ state, as a function of $J(J+1)$, will exhibit three perturbed regions with unequal distortions, as shown in the middle part of Fig.4.18. The solid and dashed lines represent the rotational term curves for singlet and triplet states with fixed vibrational quantum numbers. The regionally paired heavy curves display avoided crossing effects at the three perturbation centers. The lengths of these heavy paired curves correspond to the widths of the mixing coefficient curves in Fig.4.17.

Now, owing to the wave-function mixing, the transitions from the mutually perturbing levels to either the singlet ground or to the upper triplet states are allowed by the selection rules for dipole transitions. Thus the originally spin forbidden two-photon transition can take place.

If any one of the mixing regions is half way between $X^1\Sigma_g^+$ and a high-lying triplet state, it will near-resonantly enhance a molecular spin forbidden two-photon transition. Comparing Figs.4.18 with 4.3, one can

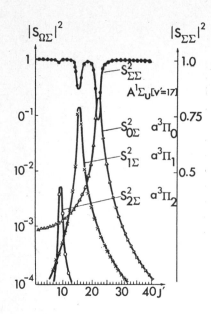

Fig.4.17. Mixing coefficients for the mutually perturbing levels $(v' = 17, J')$ in the $A^1\Sigma_u^+$ state and (v_x+4, J') in the $b^3\Pi_{\Omega u}$ states as a function of the rotational quantum number J' [4.41]

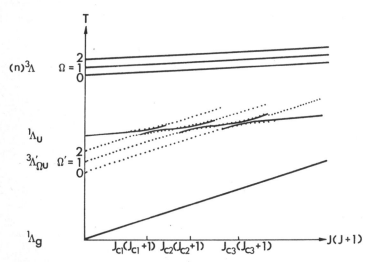

Fig.4.18. The rotational term curves for spin forbidden two-photon transition band, near-resonantly enhanced by the regionally mixing levels with vibrational quantum numbers given. J_{ci} means the ith mixing center, which is near the enhancing centre for two-photon resonances

predict the following different spectral characteristics of the $\Delta S \neq 0$ two-photon transitions from that of $\Delta S = 0$ in molecules:

1) On the one hand, the spin-forbidden two-photon transitions are similar to spin-allowed transitions in giving rise to a simplified system of molecular spectral bands.

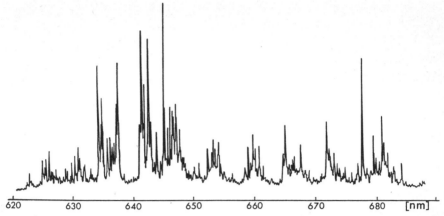

Fig.4.19. Recording of the band distributions of spin-forbidden two-photon transitions in Na$_2$ with a dye laser (CW broad band) in the DCM dye operation region

2) On the other hand, the repulsive perturbation shifts cause a widening of the enhancing level widths. As illustrated in Fig.4.18, the slightly separated heavy curves, instead of a straight line as in Fig.4.3, meet the requirements of near-resonant enhancement condition more often. In this way we were able to observe an even more regular, simplified two-photon transition spectrum for the $\Delta S = 1$ (Fig.4.19) than for the $\Delta S = 0$ case.

3) Moreover, within each of the enhancing frequency regions the spectral density of the observable lines for $\Delta S = 1$ is greatly increased. This is because the rotational constant at a mutual perturbation center takes on the average value of the mixing levels concerned. Instead of (4.59), the offset value increases for the adjacent lines near the perturbating center, having the form

$$\Delta_i = \left| \frac{B_a + B_c}{2} - \frac{B_{b1} + B_{b2}}{2} \right| 2J_{ci} |\Delta J| , \tag{4.92}$$

where J_{ci} is the rotational quantum number at the ith perturbtion center; ΔJ is the J difference of the line under consideration from J_{ci}. B_a and B_c indicate the rotational constants of the initial and the final states, respectively; B_{b1} and B_{b2} correspond to those of the mixing intermediate states. As an example, the rotational constants for several states in Na$_2$ are listed in Table 4.3. For the sake of comparison, the rotational constant differences and the relative enhancing factors for the ajacent two-photon transitions next to the individual enhancing centers for $X \rightarrow {}^1\Lambda_g$ amd $X \rightarrow {}^3\Lambda_g$ are listed in Table 4.4. The two orders of magnitude larger enhancing factor indicates, according to (4.53), a much slower changing of the signal intensities from an enhancing center J_{ci} to its neighboring line with rotational quantum number J in a spectral band for $\Delta S = 1$ than for $\Delta S = 0$. Figure 4.20 illustrates this comparison for one branch of two-photon absorptions [4.44].

97

Table 4.3. Relative rotational constants for two-photon transitions from the singlet to one of the triplet electronic states in Na_2

States		Rotational Constants [cm^{-1}]	References
initial state: $X^1\Sigma_g^+$		B_a : 0.15473	4.33
final state: $(3)^3\Pi_g$		B_c : 0.10305	4.43
mixing states $\begin{cases} A^1\Sigma_u^+ \\ b^3\Pi_u \end{cases}$		B_{b_1} : 0.11078	4.33
		B_{b_2} : 0.1434	4.42

Table 4.4. Comparison of the enhancing factors between the dipole approximation allowed and the forbidden two-photon transitions in Na_2

Transition	Enhancing level	Rel. rotational constants Term	Value [cm^{-1}]	Enhancing ratio for J near J_{ci}[a]
Singlet-singlet	$A^1\Sigma_u^+$	$\dfrac{B_a+B_c}{2} - B_{b1}$	0.0181	1
Singlet-triplet	$A^1\Sigma_u^+ - b^3\Pi_u$	$\dfrac{B_a+B_c}{2} - \dfrac{B_{b1}+B_{b2}}{2}$	0.0018	100

[a] J_{ci} denotes the rotational quantum number at the ith enhancing center

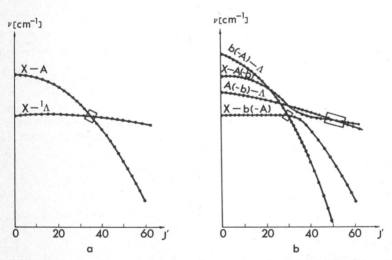

Fig.4.20. Comparison of the observable numbers for one branch of two-photon transitions between a) $\Delta S = 0$ and b) $\Delta S = 1$ cases. Enhancing centres are boxed [4.44]

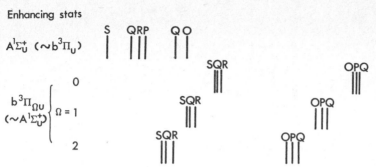

Fig.4.21. Illustration of the branch distributions for spin-forbidden two-photon transitions near-resonantly enhanced by mixing levels with given vibrational quantum numbers in Na_2 (mixing states: $A^1\Sigma_u^+$ with $b^3\Pi_{\Omega u}$). Upper: enhanced by A (~b) levels; lowers: that by b (~A) levels with $\Omega = 0,1,2$, respectively

4) Furthermore, the great increase in the number of branches is another reason for the much higher spectral density of the two-photon absorption lines in the $\Delta S = 1$ case. As shown in Fig.4.18, there are three fine electronic states now as the upper states with small but nearly constant separations (e.g., the fine splitting constant for $(3)^3\Pi_g$ state is $3\,\text{cm}^{-1}$ [4.43]). To each of them six branches are allowed from the ground state $X^1\Sigma_g^+$ enhanced by $A^1\Sigma_u^+ \sim b^3\Pi_u$ mixing as is illustrated in Fig.4.12b, and each branch has four enhancing centers, corresponding to the intersections of the rotational term curves with the vibrational quantum numbers v_A, v_0, v_1, v_2 in states $A^1\Sigma_u^+$, $b^3\Pi_{0u}$, $b^3\Pi_{1u}$ and $b^3\Pi_{2u}$, respectively. As an example, the predicted six branches of resonantly enhanced two-photon transitions from $X^1\Sigma_g^+$ to one of the high-lying triplet states, $(3)^3\Pi_{0g}$, with $v = 1$ in Na_2 have been divided into four parts as shown in Fig.4.21. For the sake of clarity another twelve branches up to the remaining two fine electronic states $(3)^3\Pi_{1g}$ and $(3)^3\Pi_{2g}$ have been omitted. Taking note of the above-mentioned remarks about the rotational-constant differences, one may consider that there are several lines appearing in a recorded trace accompanying each of the enhancing centers shown by the vertical lines in Fig.4.21. In short, within a few tens of Angstroms one can observe more than a hundred of molecular spin-forbidden two-photon transition lines.

By comparison with the only about 70 Doppler-free two-photon lines observed with relatively uniform distribution in a scanning range of 1000 cm^{-1} for two-photon transitions from the $X^1\Sigma_g^+$ to either of two excited g-parity singlet states $^1\Sigma_g^+$ or $^1\Pi_g$ at 33000 cm^{-1}, enhanced by the singlet state $A^1\Sigma_u^+$, we observed hundreds of lines between 620 and 680 nm with much more complicated fine structures (compare Fig.4.21 with Fig.4.11a) and with orders of magnitude higher spectral density in some wavelength regions. A section of the transitions observed using a pulsed laser with an output linewidth of 0.1 cm^{-1} around 646 nm is exhibited in Fig.4.22, corresponding to the fine structure in Fig.4.21 and the omitted parts.

5) In principle, the coarse structure for spin-forbidden two-photon transitions is predicted by searching the common intersections among three sets of rotational term curves (Fig.4.20b), where there are two parameters

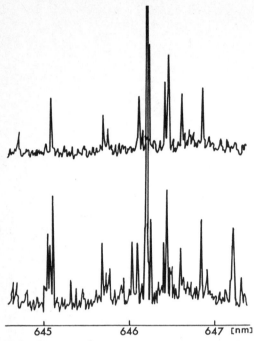

645   646   647 [nm]

Fig.4.22. A section of the trace showing the fine structure of an observed incomplete band of spin-forbidden two-photon transitions at 646 nm in Na_2. Interference filters were used at 436 nm for the lower trace and at 360 nm for the upper trace, corresponding to monitoring triplet and singlet final states, respectively

influencing the excitation rate of the spin-forbidden two-photon transition rather than one as for $\Delta S = 0$. One is the energy gap between the coupling levels, which influences the degree of mixing; the other is the smallness of the offset for an intermediate level from two-photon resonance. In general, the common intersection between three curves should occur less often by chance than that between two. The experiments have demonstrated, however, that the two-photon resonance is more critical than the mixing resonance. Therefore the above-mentioned width of the enhancing region gives an even more regular coarse structure. This makes it easier to evaluate the molecular constants B and ω_e according to (4.70, 71) for the upper electronic triplet states of these spin-forbidden two-photon transitions. In contrast, the complicated fine structures with high spectral density makes it even more difficult to perform line-by-line identifications than for a spin-allowed two-photon molecular absorption spectrum. Fortunately, the observed hyperfine structures within a two-photon vibration-rotational line have provided conclusive evidence for distinguishing between the respective upper fine electronic states (Chap.5). The comprehensive method can then be applied to identify the complex structure of the spin-forbidden two-photon absorption spectra (Chap.7)

4.7 Semiclassical Theory for a Two-Photon Transition in a Four-Level System

A two-photon transition enhanced by a pair of near-resonant mixing levels forms a four-level system (Fig.4.2c). The density-matrix motion equation describing the interaction of a four-level system with two travelling-wave fields is in principle the same as (4.15) in the form

$$i\hbar\left(\frac{\partial}{\partial t} + v_z\frac{\partial}{\partial z}\right)\rho_{ij} - \hbar(\omega_i - \omega_j)\rho_{ij} + i\hbar\gamma_{ij}\rho_{ij} = i\hbar\Lambda_{ij} - E\sum_k(\mu_{ik}\rho_{kj} - \rho_{ik}\mu_{kj}) ,$$

$$(4.93)$$

where subscripts $i,j = 1,4$ correspond to the initial and final levels, respectively, and $k = 2,3$ to the lower and higher intermediate levels. Assume that the levels 2 and 3 are nominally in singlet and in triplet states under the de-perturbation consideration and that levels 1 and 4 are in singlet and triplet states, respectively. $\Lambda_i = \Lambda_{ij}$ ($i=j$) is the thermal excitation rate of level i, $\gamma_i = \gamma_{ij}$ ($i=j$) is the relaxation rate of level i. Ignoring collisional dephasing, $\gamma_{ij} = \frac{1}{2}(\gamma_i+\gamma_j)$. Before perturbation and with the dipole-transition approximation we have $\mu_{12}, \mu_{34} \neq 0$ and $\mu_{13}, \mu_{24}, \mu_{14}, \mu_{23} = 0$. Once mixing conditions are satisfied between the two close levels 2 and 3, their wave functions become mixed, which is equivalent to having $\mu_{13}, \mu_{24} \neq 0$, in addition to our original $\mu_{12}, \mu_{34} \neq 0$. Taking the initial condition of thermal excitation as $\rho_{22}^{(0)} = \rho_{33}^{(0)} = \rho_{44}^{(0)} = 0$ and $\rho_{11}^{(0)} = N$ (density of the interested molecules), the interaction of such a system with the light field (4.18) may take the following interaction paths for the lowest-order population of the final level 4 [4.45]:

$$1)\quad \rho_{11}^{(0)} \xrightarrow{p} \begin{matrix}\rho_{12}^{(1)}\\[2pt]\rho_{13}^{(1)}\end{matrix} \xrightarrow{p} \rho_{14}^{(2)} \xrightarrow{p} \begin{matrix}\rho_{24}^{(3)}\\[2pt]\rho_{34}^{(3)}\end{matrix} \xrightarrow{p} \rho_{44}^{(4)} ,$$

$$\rho_{11}^{(0)} \xrightarrow{p} \begin{matrix}\rho_{12}^{(1)}\\[2pt]\rho_{13}^{(1)}\end{matrix} \xrightarrow{p} \begin{matrix}\rho_{22}^{(2)}\\[2pt]\rho_{33}^{(2)}\end{matrix} \xrightarrow{p} \begin{matrix}\rho_{24}^{(3)}\\[2pt]\rho_{34}^{(3)}\end{matrix} \xrightarrow{p} \rho_{44}^{(4)} ,$$

$$2)\quad \rho_{11}^{(0)} \xrightarrow{p} \begin{matrix}\rho_{12}^{(2)}\\[2pt]\rho_{13}^{(1)}\end{matrix} \xrightarrow{q} \rho_{14}^{(2)} \xrightarrow{p} \begin{matrix}\rho_{24}^{(3)}\\[2pt]\rho_{34}^{(3)}\end{matrix} \xrightarrow{q} \rho_{44}^{(4)} ,$$

$$\rho_{11}^{(0)} \xrightarrow{p} \begin{matrix}\rho_{12}^{(1)}\\[2pt]\rho_{13}^{(1)}\end{matrix} \xrightarrow{p} \begin{matrix}\rho_{22}^{(2)}\\[2pt]\rho_{33}^{(2)}\end{matrix} \xrightarrow{q} \begin{matrix}\rho_{24}^{(3)}\\[2pt]\rho_{34}^{(3)}\end{matrix} \xrightarrow{q} \rho_{44}^{(4)} ,$$

$$3)\quad \rho_{11}^{(0)} \xrightarrow{p} \begin{matrix}\rho_{12}^{(1)}\\[2pt]\rho_{13}^{(1)}\end{matrix} \xrightarrow{q} \rho_{14}^{(2)} \xrightarrow{q} \begin{matrix}\rho_{24}^{(3)}\\[2pt]\rho_{34}^{(3)}\end{matrix} \xrightarrow{p} \rho_{44}^{(4)} ,$$

$$\rho_{11}^{(0)} \xrightarrow{p,q} \begin{matrix}\rho_{12}^{(1)}\\[2pt]\rho_{13}^{(1)}\end{matrix} \xrightarrow{q,p} \begin{matrix}\rho_{22}^{(2)}\\[2pt]\rho_{33}^{(2)}\end{matrix} \xrightarrow{q} \begin{matrix}\rho_{24}^{(3)}\\[2pt]\rho_{34}^{(3)}\end{matrix} \xrightarrow{p} \rho_{44}^{(4)} ,$$

4) $\rho_{11}^{(0)} \xrightarrow{p} \begin{array}{c}\rho_{12}^{(1)}\\[2pt]\rho_{13}^{(1)}\end{array} \xrightarrow{p} \rho_{14}^{(2)} \xrightarrow{p} \begin{array}{c}\rho_{34}^{(3)}\\[2pt]\rho_{24}^{(3)}\end{array} \xrightarrow{p} \rho_{44}^{(4)},$ (4.94)

$\rho_{11}^{(0)} \xrightarrow{p} \rho_{12}^{(1)},\rho_{13}^{(1)} \xrightarrow{p} \rho_{23}^{(2)} \xrightarrow{p} \begin{array}{c}\rho_{34}^{(3)}\\[2pt]\rho_{24}^{(3)}\end{array} \xrightarrow{p} \rho_{44}^{(4)},$

5) $\rho_{11}^{(0)} \xrightarrow{p} \begin{array}{c}\rho_{12}^{(1)}\\[2pt]\rho_{13}^{(1)}\end{array} \xrightarrow{q} \rho_{14}^{(2)} \xrightarrow{p} \begin{array}{c}\rho_{34}^{(3)}\\[2pt]\rho_{24}^{(3)}\end{array} \xrightarrow{q} \rho_{44}^{(4)},$

$\rho_{11}^{(0)} \xrightarrow{p} \rho_{12}^{(1)},\rho_{13}^{(1)} \xrightarrow{p} \rho_{23}^{(2)} \xrightarrow{q} \begin{array}{c}\rho_{34}^{(3)}\\[2pt]\rho_{24}^{(3)}\end{array} \xrightarrow{q} \rho_{44}^{(4)},$

6) $\rho_{11}^{(0)} \xrightarrow{p} \begin{array}{c}\rho_{12}^{(1)}\\[2pt]\rho_{13}^{(1)}\end{array} \xrightarrow{q} \rho_{14}^{(2)} \xrightarrow{q} \begin{array}{c}\rho_{24}^{(3)}\\[2pt]\rho_{34}^{(3)}\end{array} \xrightarrow{p} \rho_{44}^{(4)},$

$\rho_{11}^{(0)} \xrightarrow{p,q} \rho_{12}^{(1)},\rho_{13}^{(1)} \xrightarrow{q,p} \rho_{23}^{(2)} \xrightarrow{q} \begin{array}{c}\rho_{34}^{(3)}\\[2pt]\rho_{24}^{(3)}\end{array} \xrightarrow{p} \rho_{44}^{(4)},$

where the letters p and q above the arrows denote the travelling waves for every interaction with $E_p \neq E_q$. In the rotating-wave approximation for the perturbation expansion of the density matrix the solutions of the above listed interaction paths have been derived as

$$\rho_{44}^{(4)}(\omega, v_z)_1 = N \sum_p \frac{1}{|\Delta_{14}(2p)|^2} \left[\frac{(V_{12}^p)^2 (V_{24}^p)^2}{|\Delta_{12}(p)|^2} + \frac{(V_{13}^p)^2 (V_{34}^p)^2}{|\Delta_{13}(p)|^2} \right],$$

$$\rho_{44}^{(4)}(\omega, v_z)_2 = N \sum_{p \neq q} \frac{1}{|\Delta_{14}(0)|^2} \left[\frac{(V_{12}^p)^2 (V_{24}^q)^2}{|\Delta_{12}(p)|^2} + \frac{(V_{13}^p)^2 (V_{34}^q)^2}{|\Delta_{13}(p)|^2} \right],$$

$$\rho_{44}^{(4)}(\omega, v_z)_3 = N \sum_m^{2,3} \frac{V_{1m}^p V_{1m}^q V_{m4}^p V_{m4}^q}{|\Delta_{14}(0)|^2}$$

$$\times \left[\frac{1}{|\Delta_{1m}(p)|^2} + \frac{1}{|\Delta_{1m}(q)|^2} - \frac{(2kV_z)}{|\Delta_{1m}(p)|^2 |\Delta_{1m}(q)|^2} \right], \quad (4.95)$$

$$\rho_{44}^{(4)}(\omega, v_z)_4 = N \sum_p \frac{V_{12}^p V_{13}^p V_{24}^p V_{34}^p}{|\Delta_{14}(2p)|^2}$$

$$\times \left[\frac{1}{|\Delta_{12}(p)|^2} + \frac{1}{|\Delta_{13}(p)|^2} - \frac{(\Delta_2 - \Delta_3)^2 + (\gamma_{12} - \gamma_{13})^2}{|\Delta_{12}(p)|^2 |\Delta_{13}(p)|^2} \right]$$

$$\rho_{44}^{(4)}(\omega, v_z)_5 = N \sum_{p \neq q} \frac{V_{12}^p V_{13}^p V_{24}^q V_{34}^q}{|\Delta_{14}(0)|^2}$$

$$\times \left[\frac{1}{|\Delta_{12}(p)|^2} + \frac{1}{|\Delta_{13}(p)|^2} - \frac{(\Delta_2 - \Delta_3)^2 + (\gamma_{12} - \gamma_{13})^2}{|\Delta_{12}(p)|^2 |\Delta_{13}(p)|^2} \right]$$

$$\rho_{44}^{(4)}(\omega, v_z)_6 = N \sum_{p \neq q} \frac{V_{12}^p V_{13}^q V_{24}^q V_{34}^p}{|\Delta_{14}(0)|^2}$$

$$\times \left[\frac{1}{|\Delta_{12}(p)|^2} + \frac{1}{|\Delta_{13}(q)|^2} - \frac{(\Delta_2 - \Delta_3 - 2kV_z)^2 + (\gamma_{12} - \gamma_{13})^2}{|\Delta_{12}(p)|^2 |\Delta_{13}(q)|^2} \right]$$

where $\Delta_2 = \omega_{21} - (\omega_{41}/2)$, $\Delta_3 = \omega_{31} - (\omega_{41}/2)$ are the offsets of the intermediate levels 2 and 3 from two-photon resonance, respectively. The total population rate of the final level 4 is

$$\rho_{44}^{(4)}(\omega, v_z) = \sum_{i=1}^{6} \rho_{44}^{(4)}(\omega, v_z)_i . \tag{4.96}$$

The above-listed interaction paths can be classified into three different excitation mechanisms:

1) Combined three-level two-photon excitation processes. The interaction paths 1)–3) in (4.94) belong to this class, where each of the two near-resonant intermediate levels independently (i.e., taking a full role) enhances the two-photon transition from level 1 to 4, both involving two-photon excitation process (with the characteristic term $\rho_{14}^{(2)}$) and two-step excitation processes (with the characteristic term $\rho_{mm}^{(2)}$, m = 2 or 3). The total population rate of the final level by this mechanism can be easily obtained by summing up the rates, corresponding to (4.57) in both m = 2 and 3 cases.

2) Four-level sum-frequency coherent excitation. The characteristic term of this kind of two-photon transition is still $\rho_{14}^{(2)}$ as usual. However, a remarkable feature of this excitation mechanism is that the first-order element and the third-order one of the density matrix deal with non-coincident intermediate levels as listed by the upper lines of the paths 4)–6) in (4.94), rather than with the same intermediate level as for paths 1)–3). It appears that each of the two intermediate levels plays a part in the enhancement for a two-photon transition.

3) Four-level coherent enhancing excitation. The characteristic term for this mechanism is a new nonzero off-diagonal second-order element of the density matrix, $\rho_{23}^{(2)}$, which governs the effective population in the coherent superimposed state of the intermediate levels for the two-step excitation processes, listed by the lower lines of the paths 4)–6) in (4.94),

103

rather than the diagonal elements $\rho_{22}^{(2)}$ or $\rho_{33}^{(2)}$ in the lower lines of paths 1)–3). The wave function of the coherent superimposed state is

$$\Psi_c = C_2 \Psi_2 + C_3 \Psi_3 , \qquad (4.97)$$

where Ψ_2 and Ψ_3 are the eigenfunctions of levels 2 and 3, respectively. The normalization condition is $|C_2|^2 + |C_3|^2 = 1$. According to the quantum mechanics principle, provided the interaction time of the system in the superimposed state is short enough, the photon energy will cover the energy gap between the two intermediate levels. It should be possible to populate the two superimposing levels simultaneously.

Since all of the perturbation interactions are effective simultaneously, there is still time for the four-level system to be excited to the final state or to radiate back to the initial state during the very short interaction time with the intermediate superimposed state. Moreover, the coherent irradiating fields result in a definite phase factor, which forms a coherent transfer process between the population distributions on level 2 and 3. The ensemble-statistical average of the effective population on a superimposed state for the two-photon excitation transition will exhibit quantum interference effects, which might cause a considerable constructive or destructive contribution, especially when its offset from a two-photon resonance is comparable with the Doppler width.

Checking the interacting fields in (4.94) we see that paths 1) and 4) correspond to the absorption of two photons from unidirectional fields, whereas the paths 2), 3), 5), and 6) relate to the absorption of two photons from oppositely-propagating fields. They are therefore effective on the Doppler-broadened background and the Doppler-free peak, respectively.

4.8 Coherent Effects on the Line Shape of a Near-Resonant Two-Photon Transition in a Four-Level System

The two-photon absorption line shape can be calculated by integrating $\rho_{44}^{(4)}(\omega, v_z)$ in (4.96) over all velocities of the molecules as follows:

$$L(\omega) \sim \int \rho_{44}^{(4)}(\omega, v_z) dv_z . \qquad (4.98)$$

Of course, the individual population terms for the final level, $\rho_{44}^{(4)}(\omega, v_z)_i$ in (4.95), have different dependences on energy level schemes and, on the other hand, make distinguishable contributions to the whole line shape for a given scheme. For the first class from the three mechanisms, i.e., the so-called combined three-level two-photon excitation processes, the discussions in Sect. 4.6 stand still, i.e., the offset values Δ_2 and Δ_3 of the intermediate levels 2 and 3 from the two-photon resonance independently determine the signal ratio of Doppler-free peak to Doppler-broadened background. The particular coherent terms appearing for the second and the third classes of the mechanisms result in the line shapes depending not only on the offset vlaues Δ_2 and Δ_3 themselves, but also on their difference $(\Delta_2 - \Delta_3)$, corresponding to the appearance of the following enhancement factor in the last three equations in (4.95):

$$K_c = \frac{1}{|\Delta_{12}(p)|^2} + \frac{1}{|\Delta_{13}(p)|^2} - \frac{(\Delta_2 - \Delta_3)^2 + (\gamma_{12} - \gamma_{13})^2}{|\Delta_{12}(p)|^2 \, |\Delta_{13}(p)|^2}. \qquad (4.99)$$

This means that the total enhancing effect of the two intermediate levels depends on their relative parameters. The numerical computation of (4.99) shows the following results:

1) If one of the intermediate levels is near-resonant with another but with a wide gap, the four-level system is equivalent to a three-level system with large enhancing factor K_c, containing a negligible third term (coherent enhancement) on the right-hand side of (4.99).

2) In the case of the four-level system with either $|\Delta_2|$ and $|\Delta_3|$ having comparable magnitudes or having the same sign, the coherent enhancing factor represented by the third term on the right-hand side of (4.99) is small, irrespective of whether the magnitude of the total enhancing factor K_c is large or small.

3) When both Δ_2 nad Δ_3 are comparable in magnitude to the Doppler width but have different signs, they might have a considerable difference value $|\Delta_2 - \Delta_3|$. In this case the total enhancing factor K_c may be negative around the center of the two-photon resonance, i.e., the contribution of the quantum interference effects to the total excitation probability of this two-photon transition might in practice be a reducing factor rather than an enhancing factor at the Doppler-free peak, while in the wing region of the Doppler-broadened background it still is active (positive).

Finally, since the various interaction paths listed in (4.94) are effective simultaneously during the interaction time of the light fields with the molecules, different parts in a line shape for such a two-photon transition may be enhanced by greatly differing amounts to produce unusual line shapes. As an exmaple, Fig.4.23 shows a comparison between computed line shapes with the given offsets of the intermediate levels in the four-level system in which the solid and dashed curves refer to the case where the last term in (4.99) is or is not taken into account, respectively. The solid curve shows a remarkable line shape for a molecular two-photon transition with an almost vanished Doppler-free peak at the two-photon resonance frequency. Referring back to the regular line shape, we would expect that the Doppler-free signal should be at least one order of magnitude larger than Doppler-broadened background for three-level two-photon transitions, whenever the offset Δ of the intermediate level is comparable or larger than Doppler width (Fig.4.14).

Varying the Δ_2 and Δ_3 values in (4.99), one can obtain a variety of theoretically predicted line shapes. In practice, we have observed various line shapes for four-level two-photon transitions in the study of the high--lying triplet states enhanced by mixing levels in Na_2 by means of Doppler-free two-photon techniques. Some of the recorded lines shapes are shown in Fig.4.24, where we can clearly see that the DF/DB signal ratios are chang-

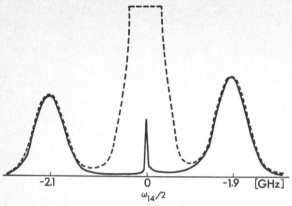

Fig.4.23. Computed noval line shape of a two-photon transition in a four-level system. Solid cure: including coherent enhancing factor; dashed curve: excluding coherent enhancing factor [4.45]

ing from smaller to larger to smaller again and that the DF component finally vanishes; we can also see that the Doppler-free peak shifts away from the center of its background. The explanation of the ratio altering is easy. When one of the intermediate levels is almost on two-photon resonance, the enhancing factor is dominated by the contribution of one out of two mixing levels, and the four-level system is, in fact degraded to a three-level system. It then should obey the curves in Fig.4.14, and show a reduced DF/DB signal ratio with decreasing offset value for the effective enhancing level in the region of about $\Delta < \Delta\nu_D$ corresponding to Fig.4.24 from d to a. As it is effectively a three-level system the line shapes with reduced peak ratio are quite symmetric. With the potential of the Δ_2 and Δ_3 values approaching equality, the effective four-level systems cause the two-photon line shape distortions to become gradual by more serious as shown in Fig.4.24 from (d) to (f). The peak ratio reductions from (d) to (a) are then expected on account of the increasingly important contributions from the higher order correction, $\rho^{(6)}$, following the decreasing offset of the efficient intermediate level as discussed in Sect.4.6. The peak ratio reductions from (d) to (f) are considered to be due to the increasing quantum interference on the two-photon enhancement as the two intermediate levels about the two-photon resonance approach more and mroe balanced positions. Therefore the line shape shows a gradual shift of the Doppler-free peak from the center of one background to the center between the two backgrounds from (a) to (f) in Fig.4.24 [4.46].

The multiple peak profile in Fig.4.24(b) shows the recently observed hyperfine structure for a molecular two-photon absorption line at 645 nm in Na_2. It can be easily recognized by its peculiar appearance. The hyperfine structures are useful for identifying molecular two-photon transitions (Chap.7) and for obtaining the hyperfine constants of the individual upper levels (Chap.5). The line shape dependence on some of the experimental conditions is also useful for obtaining molecular parameters (Chap.5).

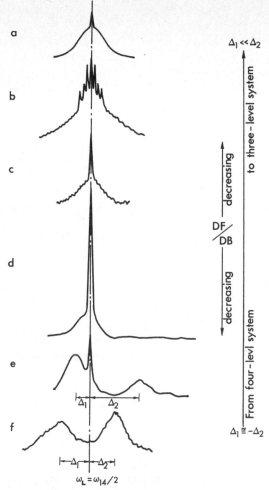

Fig.4.24. Observed two-photon line shape variations characterized by the DF/DB peak ratio, and the position of DF peak relative to the DB background [4.47]

5. Molecular Nonlinear Uncoupling Spectra with Doppler-Free Spectroscopy

Since the introduction of various methods of Doppler-free nonlinear high-resolution laser spectroscopy during the past two decades, they have been applied to atoms and to molecules, from the most simple diatomic to even large polyatomic molecules. It has been possible to combine techniques to advantage and to allow the study of increasingly complicated molecular spectra. The particular advantages of nonlinear high-resolution spectral techniques have become increasingly important in molecular spectroscopy [5.1-5].

For studies of intermolecular or intramolecular interactions it is often necessary to resolve components with a separation less than the normal molecular line width due to Doppler broadening. One way to reduce Doppler broadening is to cool the gas to a low temperature. This produces only rather small improvements because the line width decreases only as the square root of the absolute temperature. Moreover, most gases cannot be cooled very far before they condense. Somewhat greater reductions can be obtained by generating a beam of molecules and observing them in a direction perpendicular to their motion. However, if the beam is highly collimated, so as to produce a large reduction in the Doppler effect, it is likely to contain only a small number of molecules, and thus to be difficult to observe. Fortunately, as we will describe in this chapter, the laser's high directionality, intensity and monochromaticity allow many ways to eliminate the Doppler width with convenient, external sample cells and thus to observe very fine spectral details. The well-developed Doppler-free spectroscopic techniques can be performed with a commercial or home-made single-mode scanning tunable laser to obtain molecular lines with a spectral width of a few thousandths of the Doppler width (i.e., a few MHz). This is sufficient, and such techniques are in wide use. Some spectroscopic measurements require various combinations of Doppler-free techniques. The so-called super-high resolution techniques, which are based on the improved frequency-stabilized tunable laser, have been used to reach kHz resolution to observe the super-hyperfine structures of molecular spectra [5.5].

Being in pursuit of high resolution with Doppler-free spectroscopic techniques, the power broadening effect on the interested molecular lines should be minimized, which is assured by a low power density of the light and a small transition dipole moment of the matter. Therefore, on one side, the perturbation method for solving the equation of motion is available, on the other side, the induced polarization as the reemission source in the propagation equation of the field is negligible. The former provides the theoretical base of Chapters 4 and 5; the latter gives the reason for the title of the chapter. This is because the uncoupling aspect of nonlinear spectroscopy, distinguishing from the topic of nonlinear-coupling spectroscopy in the following chapter where the spectral phenomena arise from the reemission of the molecular polarization induced by optical pumping.

5.1 Doppler-Free Saturation Spectroscopy and Its Development

Doppler broadening is an example of inhomogeneous broadening. Molecules in a gas necessarily undergo thermal motions, typically having velocities of the order of 10^5 centimeter per second. Different molecules in the gas thus have different absorption frequencies, depending on their velocities with respect to the propagation direction of the interacting field, as shwon in Fig.5.1, where the arrows indicate the velocity projections of molecules on the propagating axis of the travelling light field. Each of these arrows corresponds to a package of molecules with homogeneously broadened absorption linewidth, as indicated by the shadowed region in the figure. If an individual molecule at rest has fine structure arising, for instance, from the interaction of the nucleus with the electrons, this would be observable as several absorption lines, closely spaced in frequency. When the molecule is moving, all of these components are shifted by nearly the same amount, while another molecule with a different velocity has the same components all shifted by a different amount. That is, the thermal motions merely canceal the fine structure, they do not obliterate it. If one can somehow pick out and observe just those molecules with a particular velocity, then the fine structure will be observable at least if the component spacing is larger than the homogeneous width.

The intensity of the single-mode laser is typically so large that it excites large numbers of molecules to another state, resulting in a distortion of the population of the relevant levels away from a thermal distribution. Such a population distortion can be probed by another beam passing through the same sample cell to generate a Doppler-free saturated, narrow resonance signal. Both the saturating and the probing beams can interact with the same transition (a two-level system) or with different transitions (a three-level system). Usually the theoretical treatment for the latter neglects the

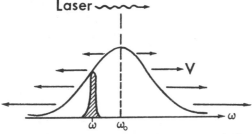

Fig.5.1. Illustration of Doppler-broadened line shape caused by Doppler-shifted absorption. Arrows mean the velocity projection of the molecules absorbing at the indicated frequencies

coherent contribution of the two laser fields on the population corrections of the levels when each of the two photons are far off resonance, but concerns the coherent effect at the sum freqency between the initial and final level, which is the so-called two-photon transition, as described in Chap.4. Coherent effects involving a single pair of levels may be measured by changes in the magnitude of the absorbed signal or by detection of some additional signals. Consiedering the more complicated coherent effects in a two-level system, we shall discuss two-level systems in this chapter following the detailed discussion of three- and four-level systems in Chap.4.

Let us assume that there is a transition from a lower level a to an upper level b with the initial population $N_a^0 \neq 0$ and $N_b^0 = 0$. Writing down a two-level density matrix ρ in the expansion form of (4.23) and substituting it to the motion equations (4.24 and 25) with the interacting fields described by (4.18), where $E_1(z, t)$ refers the field of the intense saturating beam, and $E_2(z, t)$ to that of the weak probe beam, one can derive the first-order non-diagonal element of the density matrix, $\rho_{ab}^{(1)}$, corresponding to the linear response of the molecular assemble to the fields E_1 and E_2

$$\rho_{ab}^{(1)} = \sum_{i=1}^{2} \rho_{aa}^{(0)} \frac{V_i}{\Delta_i} \exp[-i(\omega_i t - k_i z)] , \tag{5.1}$$

where $V_i = \mu_{ab} E_i / \hbar$, $\Delta_i = \omega_i - \omega_{ab} - k_i v_z + i\gamma_{ab}$, $i = 1, 2$.

Substituting this into (4.25), the second order diagonal element of the matrix in the rotating-wave-approximation can be obtained.

$$\rho_{aa}^{(2)} = \sum_{i=1}^{2} \left| \frac{V_i}{\Delta_i} \right|^2 \frac{2\gamma_{ab}}{\gamma_a} \rho_{aa}^{(0)}$$

$$+ \sum_{\substack{i,j=1 \\ i \neq j}}^{2} \frac{V_i V_j^*}{\Delta_i \Delta_{i,j}} \rho_{aa}^{(0)} \exp\{-i[(\omega_i - \omega_j)t - (k_i - k_j)z]\} , \tag{5.2}$$

where $\Delta_{i,j} = (\omega_i - \omega_j) - (k_i - k_j)v + i\gamma_{ab}$. The first term on the right is the sum of the hole-burning effects of the two beams; the second term represents the coherent effects introduced by the waves as the time modulation of the population with frequency $|\omega_1 - \omega_2|$ and the spatial modulation of the population with a modulation length of one half of the wavelength, corresponding to unidirectional and opposite propagation of the two waves, respectively.

Whereas the probe-beam absorption intensity in the absence of a saturating beam is proportional to the total population difference between the levels a and b, $\rho_{aa} - \rho_{bb} = \rho_{aa}^{(0)}$, the probe absorption is now proportional to the corrected population difference which is given by

$$\rho_{aa} - \rho_{bb} = (\rho_{aa}^{(0)} + \rho_{aa}^{(2)}) - (\rho_{bb}^{(0)} + \rho_{bb}^{(2)}) = \rho_{aa}^{(0)} - 2|\rho_{aa}^{(2)}| . \tag{5.3}$$

As was pointed out in Chap.4, the lowest-order nonlinear absorption correction should be the third-order off-diagonal element of the density matrix. In the weak probe-beam case the signal should correspond to the element $\rho_{ab}^{(3)}$, including all of the contributions with the amplitude $|E_1|^2 E_2$.

This brief theoretical discussion suggests that the saturating beam may label the sample in different ways during its passage through and interaction with the sample molecules. Such labels can be sensed by the probe beam, when the latter interacts with the same molecules. They are

I) the intensity label, corresponding to the term with $|V_1|^2$ in (5.2). In the simplest case of a two-level resonance one can find an additional driving term involving the factor $|V_1|^2$ when one substitutes (5.2) into (4.25) to obtain the formula for the third-order correction of the probe signal, responding to the molecular dipoles at the probe frequency. While the above-mentioend first part of the intensity labelling of the saturating beam corresponds to the term with $i = 1$ in the first summation for the population "hole burning" in (5.2), the second part of the intensity labelling corresponds to the term with $j = 1$ and $i = 2$ in the second summation in (5.2), which is caused by the interference excitation of the two fields, and which causes additional frequency or phase features for the probe signal [5.1,2].

II) The frequency label, left by the saturating beam, occurs in the exponent of the second summation [$(\omega_1 - \omega_2)t$ in (5.2)] in the case of two independent travelling waves with unequal frequencies. It allows one to detect a signal at the beat frequency between the probe and saturating waves from the probe beam.

III) The phase label left by the saturating beam is another factor of the exponent in the second summation [$(k_1 - k_2)z$ in (5.2)] and can be due either to different frequencies or to different directions between the two interacting waves. It can be observed with phase sensitive techniques at the difference frequency.

IV) The effective probed volume, i.e., the common interacting volume of the saturating and probe beams, which determines the total number of the interacting molecules by its multiplying with the population density of the sample molecules.

5.1.1 Intensity Modulation Labelling

The Doppler-free saturation spectroscopic techniques developed so far are summarized in Table 5.1 with typical applications. First of all, *Hänsch* et al. [5.6,7] and *Borde* [5.8] independently introduced the arrangement shown in Fig.5.2 with an external sample cell to give Doppler-free spectra of gases. The change in absorption due to saturation is detected by a weak probe beam which is split off from the same laser beam before the chopper and made to pass through the absorbing gas in the opposite direction. When the saturating beam is chopped, the probe-beam absorption signal contains the chopping frequency of the saturating beam.

This modulation, however, occurs only if both beams interact with the same atoms or molecules. Molecules having zero longitudinal velocity with respect to the laser beams, interact with both beams, and the transmitted

Table 5.1 Doppler-free saturation spectroscopy, its development and application examples for study of molecular spectra and structures. Classification I-IV refer to laser intensity modulation, frequency modulation, polarization modulation and beam overlaping modulation with their modifications, respectively.

Class	Name	Modulation style	Feature	Application example			Ref.
				Excitation wavelength	Molecules	Results*	
I	Doppler-free saturation spectroscopy	intensity modulated at ω for pump beam	intensity modulation transfered to probe beam	visible	isotope I_2	EHF and MHF consts.	5.6-8
				visible	Na_2	Dunham Consts.; Franck-Condon factors; perturbation shifts; pressure effects	5.9 5.10,11 5.20
				infrared	isotope SF_6	MHF consts.	5.12
				infrared	CH_4, CH_3F, CH_3I	MHF consts.	5.13,14
				infrared	NH_3	dipole moment	5.15
				RF disch. & visible	N_2	Mol. const. of Rydberg states	5.16
					LaO	EHF; spin-rotation interaction	5.26
				infrared	HCOOH	Dunham const.	5.41
				visible	$H_2C_2N_4$ (S-tetranzne)	rotational constants pressure broadening	5.42
	intermodulation spectroscopy	intensity modulated at ω1 & ω2 for pump & probe beams	fluorescence monitored at (ω1+ω2)	visible	I_2	hyperfine structure	5.17

	Technique	Modulation	Background / signal	Spectral region	Molecule	Application	Ref.
	saturation interference spectroscopy	intensity modulation of pump beam	background balanced out	visible	Na_2		5.18,19
II	FM saturation spectroscopy	frequency modulation for probe beam	unbalanced beats of sideband absorption	visible	I_2	hyperfine structure	5.21 / 5.22-24
				infrared	HF	precise EHF const. & v- & J- dependences	5.23,24
				visible	H_2O	pressure broadening	5.25
	modulation transfer spectroscopy	frequency modulation for pump beam	modulation transfer to probe beam	visible	I_2	hyperfine structure four-wave mixing	5.27,28
|t|	polarization modulation spectroscopy		see Table 5.2, POLINEX				5.34
IV	beam overlap modulation spectroscopy	space modulation for pump beam	modulation of common interacting volume	visible	I_2	hyperfine structure	5.29

* EHF-electric quadrupole hyperfine structure; MHF-magnetic dipole hyperfine structure

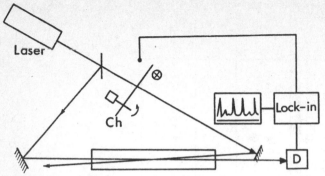

Fig.5.2. Experimental setup for Doppler-free saturation spectroscopy with external sample cell [5.4]. See text

probe beam out of the cell is modulated when the laser is tuned to the center of the absorption line of any one of the spectral components. Longitudinally moving molecules see the two beams as Doppler shifted to different frequencies, so that they cannot be resonant with both beams at the same time. Thus, this method picks out just those molecules that have a zero velocity component along the laser-beam propagating axis, and ignores the others.

The intensity-modulated Doppler-free nonlinear laser spectroscopies all provide the intensity label by the saturating beam as classified above. A summary of their properties and some examples from the extensive range of applications are listed in Part I of Table 5.1 (detailed discussions see Sect. 5.6).

In order to compare these methods with other kinds of Doppler-free techniques we have to first solve the motion equation of the density matrix to third order and then to derive the formula for the probe signal. The off-diagonal matrix element of the molecular dipoles, responding to the probe frequency, is described by

$$\rho_{ab}{}^{(3)} = \frac{V_2}{\Delta_2} (\rho_{aa}^{(2)} - \rho_{bb}^{(2)}) , \tag{5.4}$$

which is obtained by substituting (5.2) into (4.25). The absorption coefficient of the probe beam can be obtained from

$$\alpha = \int_{-\infty}^{\infty} 4\pi\omega_2 c^{-1} \text{Im}\{\chi\} dv_z , \tag{5.5}$$

where the susceptibility

$$\chi = \chi^{(1)} + \chi^{(3)} E_1{}^2 .$$

114

Therefore the absorption coefficient α in (5.5) can be divided into two terms as

$$\alpha = \alpha_0 + \Delta\alpha = \int_{-\infty}^{\infty} 4\pi\omega_2 c^{-1} \text{Im}\{\chi^{(1)}\} dv_z$$

$$+ \int_{-\infty}^{\infty} 4\pi\omega_2 c^{-1} \text{Im}\{\chi^{(3)}\} |E_1|^2 dv_z \ . \tag{5.6}$$

Substituting

$$\chi^{(1)} = \mu_{ab}\rho_{ab}^{(1)}/E_2 \quad \text{and} \quad \chi^{(3)} = \mu_{ab}\rho_{ab}^{(3)}/E_1^2 E_2 \tag{5.7}$$

into (5.6), one obtains the linear absorption coefficient of the probe beam [5.1]

$$\alpha_0 = 4(\pi)^{1/2} \mu_{ab}^2 (N_a^0 - N_b^0)(\hbar\bar{v})^{-1} \exp\left[-\left(\frac{\omega_2 - \omega_{ab}}{k_2\bar{v}}\right)^2\right] ; \tag{5.8}$$

and the nonlinear absorption coefficient of the probe beam, which is proportional to the intensity of the saturating beam

$$\Delta\alpha = \frac{64\pi^{5/2}\mu_{ab}^4\gamma_{ab}}{\hbar^3 c\bar{v}} \left[\frac{1}{\gamma_a} + \frac{1}{\gamma_b}\right](N_a^0 - N_b^0)I_1$$

$$\times \frac{\exp[-(\omega_2 - \omega_{ab})^2/(\Delta\omega_D)^2]}{[(\omega_1 - \omega_{ab}) \pm (\omega_2 - \omega_{ab})]^2 + (2\gamma_{ab})^2} , \tag{5.9}$$

where the signs − and + refers to the two co- and counter-propagating fields, respectively. The Doppler width corresponds to $k_2\bar{v}$.

From the ratio of $\Delta\alpha$ in (5.9) to α_0 in (5.8) one can obviously find a narrow spectral peak, corresponding to a resonance in the denominator in (5.9). The resultant Doppler-free signal therefore possesses the following spectral features:

1) Line shape: the nonlinear absorption part (5.9) has a narrow Lorentzian profile, whereas the linear absorption part (5.8) provides a wide Doppler-broadened background.
2) Linewidth: the full width at half maximum (FWHM) of the Doppler-free peak is $4\gamma_{ab}$, which is two to three orders narrower than its Doppler-broadened background.
3) Line center: the center of the Lorentzian profile for the counter-propagating probe beam, corresponding to the sign + in (5.9), is at $\omega_2 = 2\omega_{ab} - \omega_1$, which is on resonance ($\omega_2 = \omega_{ab}$) in the case of equal frequency beams; whereas for the unidirectional probe beam, corresponding to the sign − in (5.9), the center is at $\omega_2 = \omega_1$. That means that the

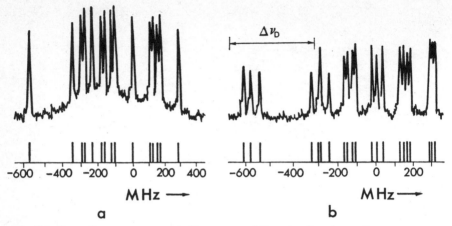

Fig.5.3. Hyperfine spectra recorded by means of Doppler-free saturation spectroscopy in $^{127}I_2$ with (a) even and (b) odd rotational quantum numbers in initial levels respectively [5.9]

    molecules are resonant with $v_z = 0$ for the former, and with v_z determined by the saturating beam for the latter.

4) Intensity: The nonlinear Doppler-free signal is proportional to the intensity I_1 of the saturating beam. However, as I_1 increases the linewidth of the Doppler-free peak will exhibit power broadening. The half width $2\gamma_{ab}$ in (5.9) should be concerned about it. Thereore, increasing I_1 will be accompanied by decreasing spectral resolution, so that the nonlinear signal to linear background ratio is limited (much less than 1).

    The above analysis concerning the line center indicates that the counter-propagating probe beam will provide a narrow spectral peak right at the resonance frequency ω_{ab}. This implies that when a narrow-band laser beam scans over the hyperfine structure of a molecular rotational line, with the arrangement as shown in Fig.5.2 for example, one can obtain a spectrum free of blurring of the hyperfine components due to the Doppler-broadening envelope. A good example is displayed in Fig.5.3 for the iodine molecule $^{127}I_2$, recorded with two counter-propagating waves split from one beam of a scanning single-mode argon ion laser [5.9]. The frequency distributions and the relative intensities of the hyperfine components allow one to obtain many interesting constants and other information on the molecules, as discussed in detail in the following section.

    For the above-mentioned Doppler-free saturation spectroscopy the detector is responding directly to the intensity of the transmitted probe beam. The sensitivity is then limited by the intense background (zero absorption) illuminance on the detector.

    To avoid direct exposure of the detector to the probe beam, the so-called intermodulated fluorescence spectroscopy was introduced (Fig.5.4). This is characterized by monitoring the fluorescence in the perpendicular direction and at the sum of the different chopping frequencies for the two counter-propagating beams by using two rings with different numbers of

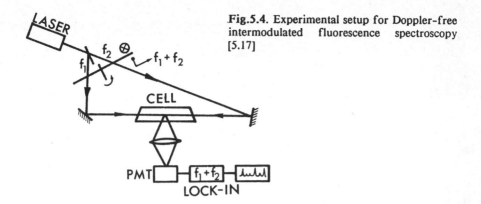

Fig.5.4. Experimental setup for Doppler-free intermodulated fluorescence spectroscopy [5.17]

holes on the chopping wheel. This technique greatly enhances the detection sensitivity and picks up only the signals corresponding to both of the beams. It eliminates the influence of the scattered light from the windows of the cell and greatly reduces the Doppler-broadened background [5.17]. Another technique with the detector working under dark background conditions is based on the destructive interference of the probe beam with the partially-split reference probe beam and is called Doppler-free saturated interference spectroscopy [5.18, 19]. In this technique the signal is picked up directly from the differential absorption of the probe beam with the reference beam, whereas the previously mentioned technique measures the part of the spontaneous emission effected by the two beams. There are more indirect measurement methods utilizing different effects of the absorption for various applications, being known collectively as supersensitive spectroscopy [5.4]. Either of the methods with the background eliminated exhibits a Doppler-free narrow resonance signal to background ratio much larger than one.

5.1.2 Frequency Modulation Labelling

One of the improvement steps is to introduce frequency modulation, a well-known microwave technique to laser spectroscopy. Frequency modulated (FM) saturation spectroscopy possesses the advantage of having an extremely low level of noise arising from dye-laser amplitude fluctuations, especially with radio frequency modulation (MHz) and with the background noise effectively eliminated by using the heterodyne beat and double balanced devices.

One way to perform frequency modulation is to set an Electro-Optical Modulator (EOM) in the path of the probe beam, as shown, for example, in Fig.5.5. The saturating beam with the frequency shifted and the intensity chopped by the Acousto-Optic Modulator (AOM) is collimated and aligned coaxially and antiparallel to the probe beam in the sample (I_2) cell. The signal-bearing probe beam is detected by a fast photodiode (D), whose output is filterd for the EOM modulation frequency and applied to the signal port of a radio-frequency Double-Balanced Mixer (DBM). The reference signal is phase-shifted by an adjustable delay line. The output of the

117

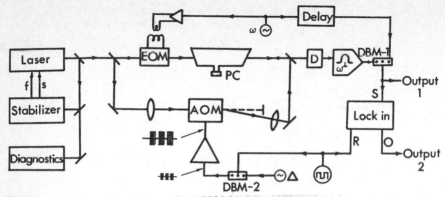

Fig.5.5. Experimental setup for FM saturation spectroscopy with probe beam frequency-modulated. EOM: electro-optic modulator (for frequency modulation of the probe beam); AOM: acousto-optic modulator (for frequency off-set and square wave amplitude modulation of the saturating beam); PC: pressure control cell; Delay: delay line (i.e phase shifter); D: photodetector, which is followed by a narrow band amplifier; DBM: doubly-balanced mixer. [5.23]

double-balanced mixer *1* may be further processed by a lock-in amplifier to recover the signal (output *2*) synchronously with the chopping of the saturation beam.

In the apparatus described the EOM puts two sideband frequencies $\omega_0 \pm \Delta\omega$ with opposite phases on the probe beam, in addition to the carrier frequency ω_0 which is equal to that of the saturating beam [5.22,23]. With the modulation frequency larger than the homogeneous width of a spectral component the hole burning effect of the saturation beam is on one of the sidebands of the probe beam. The spectral features which are revealed are:

1) **Line center:** This is the point at which there is no signal out of the double-balanced mixer on account of the balanced status of the two beat signals of the two sidebands, when the carrier is on resonance with the same molecules for the saturating beam. Furthermore, due to the carrier frequency shift of the saturation beam the zero-signal line center does not occur at the resonance frequency of a transition but interacts with the molecules having a Doppler shift of one half of the AOM modulation frequency to show a line center shift with this amount.

2) **Offset resonance:** However, if one of the sideband absorption is under the influence of the saturating beam, the mixer will provide an unbalanced beat signal. Thus, during the scanning of the laser output over a spectral line, there are two cases for the saturating beam to have common interacting molecular packages with the probe beam. These correspond to $kv_z = \pm\Delta\omega/2$ ($\Delta\omega = \omega_m + \omega_{sh}$ is the sum of the modulation frequency of the probe beam and the frequency shift of the saturation beam) with the positive and the negative sidebands of the probe beam, respectively. Therefore, for each spectral line or hyperfine component there are two signal peaks, where the positive-going peak occurs when the low-frequency sideband is resonant with the saturation hole produced by the counter-propagating saturating beam. The whole spectral pattern of positive-going peaks, due to

118

the low-frequency sideband, reproduces the saturation spectrum except that the resonant condition is $\omega_{abs} = \omega_{ab} - \Delta\omega/2$. The negative peaks result from absorption of the high-frequency sideband and occur when $\omega_{abs} = \omega_{ab} + \Delta\omega/2$. Finally the recorded trace is in the form of a frequency shifted mirror image spectrum (around the zero level) if each of the signals is in phase.

3) **Linewidth**: The half width is $2\gamma_{ab}$ as in the original Doppler-free saturation spectroscopy.

4) **Line shape**: If the phase-shifter (i.e. a delay line) is adjusted in phase, either positive-going or negative-going peaks correspond to absorption and exhibit Lorentzian profiles. If the phase is shifted by 90°, each of the positive or negative signals will provide an apparent dispersion line-shape.

5) **Intensity**: The relative value of Doppler-free peak to Doppler-broadenend background is much larger than 1. But the absolute magnitude of the Doppler-free signal is related to the modulation frequency $\Delta\omega$ value compared to the Doppler width $\Delta\omega_D$. Obviously, the modulation frequency is proper to leave only one of the sidebands of the probe beam in an absorption component. In contrast with the original Doppler-free saturation spectroscopy where the illuminance of the probe beam on the detector is adverse for sensitivity but produces a large DC load, the carrier radiation is important in the FM spectroscopy for producing a linearly increasing amplitude of the beat signal.

A second way to realize Doppler-free frequency modulation spectroscopy is to set an electrooptical modulator in the path of the saturating beam with the proper modulation frequency, which is not too much larger than the homogeneous width of a spectral component but large enough to avoid serious amplitude fluctuation of the laser output. This technique imposes saturating effects onto the single probe frequency not only from the carrier but also from the two opposite-phase sidebands of the saturating beam in every aspects, involving the hole burning effects and coherent excitation terms. We may simply treat the saturating beam as three unidirectional coincident coherent traveling waves. Thus the following expression applies to the second-order correction of the level population instead of (5.2)

$$\rho_{aa}^{(2)} = d_1 |E_1|^2 + 2d_2 |E_1'|^2 + d_4[E_1 E_1' \exp[-i(\Delta\omega t - \Delta kz)) + c.c] . \quad (5.10)$$

Refering to (5.2), d_i ($i = 1, 2, 4$) are velocity- and frequency-dependent amplitude factors; E_1 and E_1' are the amplitudes of the carrier and sideband fields, respectively. The conjugate term in the last term above is introduced as a result of the opposite phases between the sidebands. Substituting (5.10) into (5.4), we get the third order element $\rho_{ab}^{(3)}$ of the density matrix

$$\rho_{ab}^{(3)} = f_1 |E_1|^2 E_2 \exp[-i(\omega_2 t - k_2 z)] + f_2 |E_1'|^2 E_2 \exp[-i(\omega_2 t - k_2 z)]$$
$$+ f_4 (E_1 E_1' E_2 \exp\{-i[(\omega_2 + \Delta\omega)t - (k_2 + \Delta k)z\}] + c.c) . \quad (5.11)$$

119

Referring to (5.4), f_i (i = 1, 2, 3) are velocity- and frequency-dependent amplitude factors. Obviously, the first two terms on the right-hand side in (5.11) correspond to the first two terms in (5.10) and describe to the hole burning effects of the three coherent travelling waves. These terms are oscillating at ω_2, i.e., they influence the intensity of the probe beam but not the frequency deviation. In contrast, the last term in (5.11) shows that coherent terms in (5.10) due to the existence of the pumping sidebands result in frequency and phase changes of the original single-frequency probe beam. The originally unmodulated probe beam thus carries sidebands after passing through the cell as if there had been a transfer of the frequency and phase modulation from the saturating beam. The so-called modulation transfer spectroscopy [5.27] possesses the following spectral features if a double balanced mixer is used for heterodyne beats:

1) **Line center**: The signal center is on resonance at the transition frequency ω_{ab}. Due to the pure odd symmetry of the signals, however, both detectable quadratures (in-phase and out-of-phase) have zero signal at the center of the resonance when detected at the modulation frequency.

2) **Line shape**: By means of a phase discriminator the line shape can be adjusted either to reveal absorption in phase (Lorentzian) or out of phase (dispersion type).

3) **Linewidth**: For the absorption line shape the half width is $2\gamma_{ab}$. If the modulation frequency $\Delta\omega$ is smaller than the homogeneous width of an interesting transition component, the linewidth is characterized by twice the modulation frequency $2\Delta\omega$. The latter provides a further improvement of the spectral resolution over intensity modulated saturation spectroscopy.

4) **Signal intensity features**: A noticeable feature of the dispersion linewidth is that it is independent of the pumping power. This allows one to enhance the signal to noise ratio by means of increasing the intensity of the saturating beam without the loss of resolution caused by power broadening. Moreover, the opposite phases of the sidebands now contribute to the amplitude of the singal in sum. Due to the absence of the Doppler-broadening background the Doppler-free to Doppler-braodening ratio of the signal is extremely high.

Frequency modulation-saturation spectroscopy is a powerful technique which offers the advantage of zero-background phase-sensitive hetrodyne detection, and it has been used in various types of spectroscopic investigations as well as in laser-frequency locking. Radio frequency (larger than 2 MHz) optical heterodyne techniques work with the dye-laser output well enough to closely approach the fundamental photoelectron shot noise limit. With servo control of amplitude modulation in FM spectroscopy, shot-noise limited measurements have been achieved as, for example, in the determination of water-vapour pressure broadening [5.25]. It is interesting to notice additional spectral structure in FM spectroscopy when both the saturating and probe beams are frequency modulated. As an example, Fig.5.6 shows the in-phase recorded molecular hyperfine spectrum of HF. It is interesting to see the strong resonance at $\pm\omega/2$ which arises due to frequency-modulation of the probe beam and, in addition, near the center of the scan, there is a feature partly arising from the modulation transfer signals due to

120

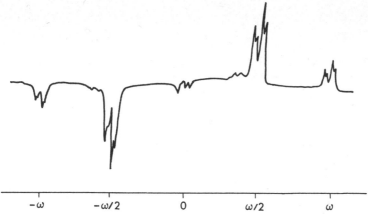

Fig.5.6. Full spectrum demodulated at $\omega/2\pi$ = 3.4 MHz from the HF spectra. The strong resonance at $\omega/2$ arises from the heterodyne of carrier and first sidebands [5.24]

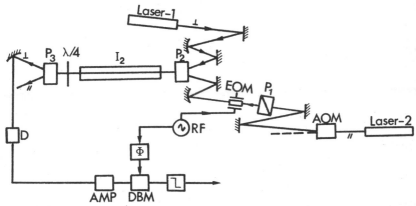

Fig.5.7. Experimental setup for four-wave mixing in I_2 for phase modulated optical heterodyne spectroscopy. AOM: Acousto-optic modulator; EOM: electro-optic modulator; RF: radio-frequency generator; Φ: phase shifter; P_1: linear polarizor; P_2: polarization beam combinator; P_3: polarization beam splitter; DBM: doubly-balanced mixer; D: photodetector; AMP: narrow band amplifier. [5.28]

frequency modulation of the saturating beam. The latter provides the transition components on resonances. Some weak cross-over resonances can be found near the main signals near $\pm\omega/2$. These three-level resonances are useful for the transition assignments [5.24].

FM saturation spectroscopy has been used to study multi-level systems. Figure 5.7 illustrates the set-up for the study of four-wave mixing in molecular iodine by phase modulated optical heterodyne spectroscopy [5.28].

5.1.3 Beam–Overlapping Modulation Labelling

One more Doppler-free spectroscopic technique is that of beam overlap modulation, i.e. common interaction volume modulation. The total non-

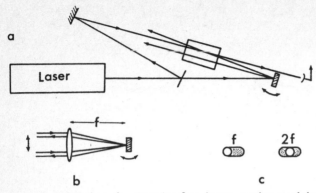

Fig.5.8. Illustrations for Doppler-free beam overlap modulation spectroscopy; a) the setup; b) beam deflecting in front of the vibration mirror; c) signal is modulated at first (left) or second (right) harmonic of the mirror vibration frequency according to the position of the stationary beam within the scan [5.29]

linear absorption signal is proportional to the product of $\Delta\alpha SL$, where $\Delta\alpha$ is given by (5.9), and SL is the overlap volume of the saturating beam with the probe. All of the above-mentioned Doppler-free techniques modulate the $\Delta\alpha$ factor. The new idea here is to alter the direction of the saturating beam without intensity, frequency or polarization modulation [5.29]. The experimental arrangement is exhibited in Fig.5.8, where the saturating beam, instead of being chopped, is displaced transversely by reflection from a vibrating mirror. The saturating and probe beams then alternately overlap and act jointly on varying regions of the sample molecules. The signal is monitored either by means of fluorescence or absorption detection. The spectral features can be described as following:

1) **Line center**: The signal center is on resonance with the transition frequency ω_{ab}.

2) **Line shape**: Typical Lorentzian profiles without Doppler-broadened background were observed. An especially valuable feature is that the main Doppler-broadened background, corresponding to the absorption of a photon from one beam, is eliminated without the need for the intermodulation technique since both beams as well as their scattered light are present at all times, whereas their common interaction with molecules occurs only during their periodic overlapping. The technique, in fact, is the small-angle modulation between the directions of saturating and probe beam. It then is necessary to show some remaining Doppler width.

3) **Modulation frequencies**: This is distinct from other saturation labeling in that the modulation frequency of the beam-overlap modulation spectroscopy depends on the position of the fixed (probe) beam in the cross-section of the sweeping (saturating) beam. If the beams just overlap at one edge of the region swept out during the mirror vibration, as shown at the left of Fig.5.8c, a signal will be received at the direction-scanning frequency f. If the beams overlap in the middle of the scan, as indicated at the right of Fig.5.8c, the moving beam will cross the fixed beam twice during the vibration, and the signal will apear at frequency 2f.

4) **Signal intensity**: The technique allows the full laser power to be used in both beams, which not only raises the signal-to-noise ratio but also allows one to study strong absorption transitions without the problem of temperature modulation of the sample molecules. Moreover, the slight Doppler-broadened signal, caused by variation of the window reflections or by stray light due to the solid angle periodically changing, can be easily eliminated by the geometric adjustments and proper phase setting of the available lock-in amplifier.

The configuration in Fig.5.8 could also be used for overlap-modulation polarization spectroscopy to provide more information in the identification of molecular transitions.

A polarized saturating beam can influence the orientation of the sample molecules so as to leave polarization labelling for the probe beam. The degree of the induced anisotropies are measures of various kinds of molecular transitions induced by distinctly polarized laser fields. Hence the polarization-labelling technique is especially useful for the identification of the molecular spectrum (as discussed in Chap.7). The special features of Doppler-free polarization spectroscopy and its further modifications will be treated in the following section.

5.2 Doppler-Free Polarization Spectroscopy and Its Development

Wieman and *Hänsch* demonstrated the method of Doppler-free spectroscopy using a polarized saturating beam, where a counter-propagating probe beam is sent through a pair of crossed polarizers, one in front of and one behind the sample region [5.30]. When a narrow-band saturating beam polarized in a selected direction preferentially excites molecules with the particular orientation at some unnormal angle with the polarization axes of the crossed polarizers, the probe beam with common interacting molecules of the saturating beam will provide a bright spectral line at the frequency on resonance even though the pair of crossed polarizers with high extinction factor. This arrangement assures that the probe beam is dark behind the analyzer whenever the two beams interact without a common package of molecules, i.e. it presents Doppler-free high resolution with high sensitivity.

The simplest treatment of the spectral features of Doppler-free polarization spectroscopy starts with the difference between the saturating effects on the two orthogonal components of the probe beam [5.30,31]. The saturating beam can be either linearly or circularly polarized. The probe beam is always linearly polarized, but can be divided into two perpendicular, linearly polarized components or right- and left-hand circularly polarized components with respect to the polarization properties of the saturating beam. With a right-hand circularly polarized pumping beam, for example, the probe beam can be expressed in the form

$$E(z=0) = \tfrac{1}{2} E_0 (\hat{x} + i\hat{y}) \exp(-i\omega t) + \tfrac{1}{2} E_0 (\hat{x} - i\hat{y}) \exp(-i\omega t) . \tag{5.12}$$

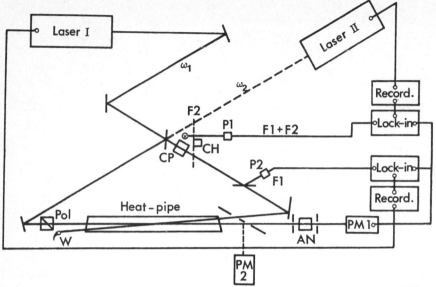

Fig.5.9. Experimental setup for Doppler-free OODR polarization spectroscopy [5.38]

Due to the presence of an effectively right-hand circularly polarized saturating beam, the absorption of the two components in (5.12) is unequal. It
produces a difference between the absorption coefficients $\Delta\alpha = \alpha_+ - \alpha_- < 0$
and a differential refractive index $\Delta n = n_+ - n_- \neq 0$. At the end of the
sample cell the transmitted probe beam will be given by

$$E(z=\ell) = \tfrac{1}{2}E_0\{(\hat{x} + i\hat{y})\exp[-i(\omega t - k_+\ell) - \alpha_+\ell/2] + (\hat{x} - i\hat{y})\exp[-i(\omega t - k_-\ell) - \alpha_-\ell/2]\}$$
$$= \tfrac{1}{2}E_0\exp[-i(\omega t - n_0\omega\ell/c) - \alpha_0\ell/2][(\hat{x} + i\hat{y})\exp(i\Delta n\omega\ell/c - \Delta\alpha\ell/2) + (\hat{x} - i\hat{y})]$$
$$\text{(5.13)}$$

where ℓ is the length of the cell. For small $\Delta\alpha$ and Δn the above formula
can be expressed, to a first approximation, by a Taylor series. Therefore,
neglecting all of the terms beyond second order in the evolution, the part of
the probe beam transmitted through a crossed analyzer is

$$E = E_x\theta + E_y$$
$$= \tfrac{1}{2}E_0\exp[-i(\omega t - n_0\omega\ell/c) - \alpha_0\ell/2](2\theta - \Delta n\omega\ell/c - i\Delta\alpha\ell/2) . \quad \text{(5.14)}$$

Its irradiation on the detector is in the form

$$I = |E|^2 = E_0^2\exp(-\alpha_0\ell)(\theta^2 - \theta\Delta n\omega\ell/c + \Delta n^2\omega^2\ell^2/4c^2 + \Delta\alpha^2\ell^2/16) , \quad \text{(5.15)}$$

where α_0 is the unsaturated absorption coefficient at the line center $\omega = \omega_{ab}$, θ is the small deviation of the crossed analyzer from an angle com-
pletely orthogonal to the polarizer. Let $\Delta\alpha_0$ signify the induced dichroism
at the line center. The induced dichroism $\Delta\alpha$ and the induced birefringence
Δn can be written

$$\Delta\alpha = \Delta\alpha_0/(1 + x^2) ,$$
$$\Delta n = xc\Delta\alpha/2\omega ;$$
(5.16)

where $x = (\omega - \omega_{ab})/\gamma_{ab}$ is the relative detuning. Substituting (5.16) into (5.14), and including the leakage ξ of the probe beam due to imperfect extinction of the crossed analyzer, and the intrinsic birefringence b of the cell's windows, we have the following formula for the irradiation intensity on the detector behind the analyzer:

$$I = I_0 \exp(-\alpha_0 \ell) \left[\xi + \theta^2 + b^2 - \frac{\theta\Delta\alpha_0 \ell x}{2(1 + x^2)} + \frac{(\Delta\alpha_0 \ell/4)^2}{1 + x^2} + \frac{b\Delta\alpha_0 \ell}{2(1 + x^2)} \right]. \quad (5.17)$$

For the case of a saturating beam polarized linearly at 45° from the polarizing direction of the probe beam the irradiation intensity is given by

$$I = I_0 \exp(-\alpha_0 \ell) \left[\xi + \theta^2 + b^2 - \frac{\theta\Delta\alpha_0 \ell}{2(1 + x^2)} - \frac{b\Delta_0 \ell x}{2(1 + x^2)} + \frac{(\Delta\alpha_0 \ell/4)^2}{1 + x^2} \right]. \quad (5.18)$$

Equations (5.17, 18) show that even without any kind of modulation the detector, initially under dark irradiation (scaled by $\xi + \theta^2 + b^2$) of the probe beam, senses the induced anisotropy of the sample molecules caused by the saturating beam whenever both beams interact with common molecules. The spectral features presented by (5.17 and 18) are as follows:

1) **Line shape**: The factors $1/(1+x^2)$ and $x/(1+x^2)$ in (5.17, 18) mean that the line shape in Doppler-free polarization spectroscopy is composed of Lorentzian and dispersion profiles. The dominant profile can be chosen by means of the proper setting of the experimental conditions. For instance, by increasing the uncomplete crossed angle θ, one may obtain a dominant dispersion shape in the case of a circularly polarized saturating beam or a dominant Lorentzian shape in the case of a linearly polarized saturating beam. On the other hand, by squeezing the sample cell one may increase the material birefringence b of the cell windows to reverse the dominant line shapes in each case.

2) **Line center**: The signal center depends on the resultant line shape. Only if a line shape is pure Lorentzian or dispersion is the signal center, corresponding to the maximum and zero value, respectively, on resonance, i.e. $\omega = \omega_{ab}$. In most cases, however, these two types of line profiles will be combined, causing the line to be asymmetric and leading to a shift of the signal's maximum or zero position from the resonance frequency. The extent of such a shift determined by b, θ and the polarization conditions of the beams, would not be more than half of the homogeneous line width.

3) **Linewidth**: Its homogeneous linewidth the same as that in Doppler-free saturation spectroscopy;

4) **Signal phase**: The signal phase in Doppler-free polarization spectroscopy is determiend by the sign of the induced dichroism $\Delta\alpha$, which is related to the sign of the polarization factor, the polarization conditions of the two interacting beams and the variations of the rotational quantum

Table 5.2. Large J values of the polarization factors ζ^C for circularly polarized pumping beam

	$J_2 = J_1+1$	$J_2 = J_1$	$J_2 = J_1-1$
$J = J_1+1$	$-3/2$	$3/2J_1$	$3/2$
$J = J_1$	$3/2J_1$	$-3/2J_1{}^2$	$-3/2J_1$
$J = J_1-1$	$3/2$	$-3/2J_1$	$-3/2$

Table 5.3. Large J values of the polarization factors ζ^L for linearly polarized pumping beam

	$J_2 = J_1+1$	$J_2 = J_1$	$J_2 = J_1-1$
$J = J_1+1$	$3/10$	$-3/5$	$3/10$
$J = J_1$	$-3/5$	$6/5$	$-3/5$
$J = J_1-1$	$3/10$	$-3/5$	$3/10$

numbers during a transition, see (7.24 and 27) for details. For the sake of convenience of the phase discussion here the large J-dependences of polarization factors are listed in Tables 5.2 and 3, where different signs correspond to opposite phases of detectable signals. Of course, the transitions between two vibration-rotatioanl levels with equal frequency saturating and probe beams refer to the diagonal terms in these tables as they can only occur between different sublevels with the same frequency. The same sign of all diagonal terms in either of the tables implies that the phase examination by phase sensitive techniques will be useless for the determination of their P,Q,R branches. On the other hand, the phase detection technique may be helpful for identification of unequal-frequency transitions. The probe transitions in the case of Optical-Optical Double Resonance (OODR) refer to the terms in one row in Tables 5.2 or 3, depending on the properties of the selected saturating beam. Moreover, if we alter the polarization of the saturating beam from circular to linear, the phase condition will now show different phases between Q branch and R,P doublets in agreement with Table 5.3. As an example, Fig.5.9 displays a typical experimental set-up for recording Doppler-free polarization signals via OODR [5.38]. P,R- branch signals then show opposite phases. With the complementary information gained with the greatly intensified signals obtained by using a linearly instead of circularly polarized saturating beam for the determination of Q-branch transitions, one was able to distinguish P, Q and R branches without ambiguity.

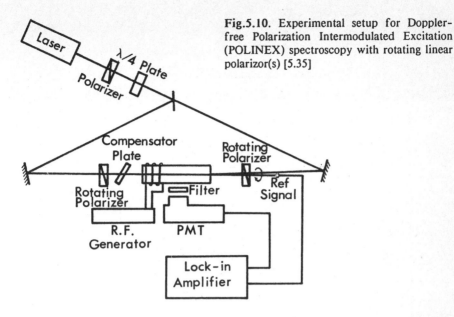

Fig.5.10. Experimental setup for Doppler-free Polarization Intermodulated Excitation (POLINEX) spectroscopy with rotating linear polarizor(s) [5.35]

It is noticeable that, although there is no need to perform any modulation in polarization spectroscopy to obtain Doppler-free signals, modulation or intermodulation enables one to use phase-locked techniques to gain useful information for spectral data analyses, especially for molecular spectroscopy, and to provide much increased sensitivity as well. A new idea recently realized, POLarization INtermodulation EXcitation (POLINEX for short) is to introduce polarization modulation of the saturating beam, or of both the saturating and probe beams, in the original arrangement of Doppler-free polarization spectroscopy [5.34]. The polarization modulation can be realized by using an electro-optical modulator to form a selectable retarder for the circular-polarization modulation [5.34] or simply by using one or two rotating linear polarizers, as illustrated by Fig.5.10 [5.35]. The latter had been introduced to study high-lying molecular states of Na_2 via equal-frequency two-photon transitions to show its usefulness in terms of the increased Doppler-free signal to background ratio and for identifying two-photon branches, as described in the followed section.

If the frequency was modulated, the Doppler-free polarization signal could be detected by optical heterodyne beats. The reported results indicated further improvement of the Doppler background elimination and, remarkably, showed opposite phases for one-photon and two-photon transitions [5.40].

These particular advantages of polarization spectroscopy, in addition to its dark background and high sensitivity, had let to its use in the study of homonuclear molecules such as Na_2, Cs_2, Li_2, etc.. The development of Doppler-free polarization spectroscopy and some examples of its applications are listed in Table 5.4.

Table 5.4 Doppler-free polarization spectroscopy, its development and some examples of applications in the study of molecular spectra and structures

Class.	Name	Modulation style	Features	Results Molecules	Results Characteristics	Ref.
I	DF polarization spectroscopy	anisotropy induced by fixed polarized pump beam	DC signals	Na_2, N_2O	distinguish $|\Delta J|=0,1$; hyperfine structure; collision terms; perturbation	5.30 5.31-33
II	polarization intermodulation excitation spectroscopy (POLENEX)	polarizations modulated at different frequencies for pump and probe beams	signal monitored at sum frequency	Na_2	distinguish $|\Delta J|=0,2$; background eliminated	5.35,36
III	optical double resonance polarization spectroscopy	intensity modulation for polarized pump and probe beams	combination of intermodulation and polarization spectroscopy	Cs_2,Li_2	distinguish $\Delta J=0,1$; Dunham const.	5.37,38
IV	polarization heterodyne spectroscopy	frequency modulation for polarized pump beam	combination of optical heterodyne and polarization spectroscopy	Na_2	observation of dipole allowed and forbidden two-photon transitions	5.39,40

5.3 Doppler-Free Two-Photon Spectroscopy and Its Development

The equation for the two-photon transition probability, (4.57), had been thoroughly discussed in Chap.4 mainly from the point of view of the structure of a molecular two-photon spectrum. In this section we shall discuss more spectral features for an individual molecular two-photon absorption line. The experimental scheme for recording Doppler-free two-photon line shapes is shown in Fig.5.11 [5.49]. Here a small amount of the laser power from the first beam splitter (BS1) was used to perform all the ancillary measurements. The output mode of a single-frequency scanning dye laser (DL) was monitored by a confocal spectral analyzer (SA). The wavelength of the laser beam was displayed by a wavemeter (WM) with an accuracy of 10^{-6} [5.55]. An iodine spectrum was recorded simultaneously, giving an accuracy of 10^{-4} cm^{-1} to the two-photon transition wave numbers. A frequency marker (FM) was used to give an accurate frequency scale to the measured intervals including the two-photon or one-photon homogeneous or Doppler widths and the offset of the intermediate enhancing level. The main laser beam was separated by the second beam splitter (BS2) into two equal-intensity beams. They passed through a stainless-steel cross oven in opposite directions. A chopper wheel with three rings of holes was placed either at A, B or B+C positions. Two photomultipliers, each connected to a lock-in amplifier, were placed at the side arms of the oven. With such a set-up and (4.57), we can discuss experimentally and theoretically the spectral features of a molecular Doppler-free two-photon absorption line as follows:

1) **Line center**: According to (4.57), the signal of a Doppler-free two-photon transition is composed of two parts. The Doppler-free part, corresponding to the first term on the right side in (4.57), has the maximum (signal center) at $\omega = \omega_{ac}/2$. The Doppler-broadened part, corresponding to the second and third terms in (4.57), with the maximum at $\omega = \omega_{ac}/2$ exhibits a saturation dip at $\omega = \omega_{ab}$ (for the case of one efficient enhancing level), which helps one to select the sign of the offset for the intermediate

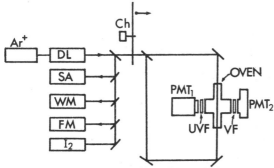

Fig.5.11. Set-up for Doppler-free two-photon spectroscopy. Ar$^+$: Argon ion laser; DL: single mode scanning dye laser; SA: optical spectral analyser; WM: wavemeter; FM: frequency marker; Ch: chopper; UVF: ultraviolet filter; VF: visible filter; PMT: photomultiplier tube. [5.49]

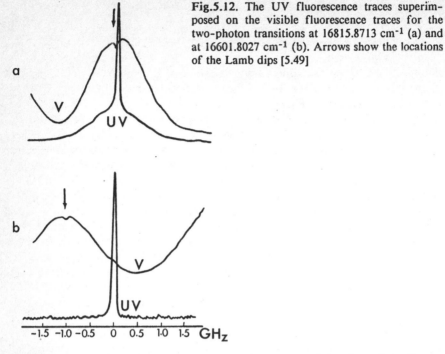

Fig.5.12. The UV fluorescence traces superimposed on the visible fluorescence traces for the two-photon transitions at 16815.8713 cm^{-1} (a) and at 16601.8027 cm^{-1} (b). Arrows show the locations of the Lamb dips [5.49]

level from the two-photon resonace. The frequency separation of the Lamp dip in the one-photon absorption profile away from the Doppler-free peak of a two-photon absorption line is a measure of the offset of the intermediate enhancing level, i.e., $\Delta = \omega_{ab} - \omega_{ac}/2$. This value can be determined by putting an ultraviolet or a red filter in front of each amplifier in Fig.5.11 to monitor ultraviolet and red fluorescence following the two-photon and one-photon transition, respectively. Examples of the ultraviolet fluorescence traces superimposed upon the visible fluorescence traces (recorded simultaneously with the chopper set at position B) are shown in Fig.5.12. In contrast, with two efficient enhancing levels, we would see extrema at ω_{ab1} and ω_{ab2} for the Doppler-broadened part of a two-photon transition line as shown in Fig.4.24.

2) **Line shape**: The saturation dip at the DF and DB signal will certainly cause some asymmetry of the two-photon line shape. In an extremely near-resonant two-photon absorption case, the line shape is essentially symmetric. It seems to be more asymmetric with increasing detuning of the intermediate level. However, due to the increase in signal ratio DF/DB a seriously asymmetric DB background will often be too weak to be noticed in the case with one efficient enhancing level.

On the other hand, due to the significant quantum interference destruction of the enhancing contributions out of a pair of near-resonant coupling levels the Doppler-free peak might greatly be reduced in certain distributions of the enhancing levels. Thus, the extraordinary asymmetric line shapes could be observed in four-level systems for spin-forbidden molecular two-photon transitions rather than in three-level systems for dipole-

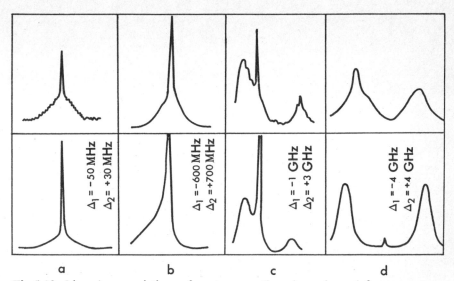

Fig.5.13. Line shape variations of near-resonantly enhanced equal-frequency two-photon transitions; upper: observed signals in Na_2, lower: calculated signals for a four-level system. Δ_1 and Δ_2 are the offsets of the two intermediate mixing levels from the middle way of a two-photon resonance. From a to d Δ_1 and Δ_2 are increasing but with different magnitudes and opposite signs [5.154]

allowed molecular two-photon transitions. The variations of the line-shape symmetry in Fig.4.24 allow one to make a comparision with the computed traces according to (4.95). The agreements between the observed (upper) and calculated (lower) curves are quite good, as can be seen in Fig.5.13.

3) **DF/DB ratio**: the DF/DB ratio of a molecular two-photon absorption line depends on the two things. Firstly, it depends on the number of enhancing levels and their offsets, which can lead to very small or very large DF/DB ratios, as illustrated in Figs.4.14,24. Secondly, the DF/DB ratio depends on the experimental arrangement. In fact, as the DB background is formed by absorbing two photons from one beam, its heigt will affected by the modulation conditions. As shown in Fig.5.14, when both of the two beams are chopped at the same frequency the background of a certain two-photon transition is doubled in comparison with that for one beam chopped but without change in the Doppler-free peak. However, if we let the two beams be chopped at different frequencies and observe the signal at the sum frequency, the background is immediately eliminated. The background can also be eliminated by using polarization modulation techniques (POLINEX, see Fig.5.10) without any intensity modulation. As an example, Fig.5.15b shows the recorded trace without background obtained by polarization modulation, while Fig.5.15a shows the profile of the same two-photon transition observed with intensity modulation.

4) **Signal phase**: The benefits of POLINEX as applied to molecular two-photon spectroscopy are not only the elimination of the DB background, but also the facilitation of transition identifications. No phase information for the identification of two-photon branches can be provided

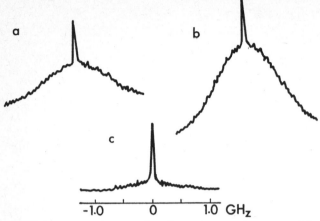

Fig.5.14. A spectral comparison between two-photon transition signals using different modulations; (a) one beam chopped at ω, signal monitored at ω; (b) two beams chopped at ω, signal monitored at ω; (c) two beams chopped at ω_1 and ω_2 individually, signal monitored at $\omega_1 + \omega_2$. The Doppler background can then be eliminated by means of intermodulation techniques [5.35]

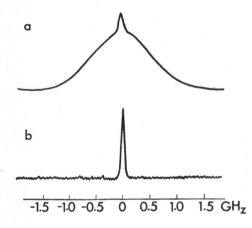

Fig.5.15. A comparison of the line shapes between aintensity modulation (a) and polarization modulation (b) [5.35]

by intensity modulation of polarized beams since all of the terms in part $\zeta^{\parallel}$, $\zeta^{\perp}$, ζ^{++} and ζ^{+-} of Table 7.5 have the same sign. On the other hand there are different signs in part ζ^C and ζ^L, offering phase information for classifying the two-photon branches. These are marked in Table 5.5 with the subscripts corresponding to the efficient enhancing transitions of the same form in Table 7.5. The comparison of data in Table 5.5 with the ρ^C and ρ^L parts in Table 7.5 enables one to recognize the various two-photon branches by altering the polarization modulation from circular to linear as follows:

a) Q_R and Q_p branches exhibit phase inversion accompanied by slight weakening,

b) S and O branches exhibit no phase changes but slight weakening;

c) P_R, R_Q branches exhibit phase inversion accompanied by considerable strengthening;

132

Table 5.5   Branch symbols and the polarization factors for equal-frequency two-photon transitions

	Branch symbols			ξ^C			ξ^L		
	$J_c=J_b+1$	$J_c=J_b$	$J_c=J_b-1$	$J_c=J_b+1$	$J_c=J_b$	$J_c=J_b-1$	$J_c=J_b+1$	$J_c=J_b$	$J_c=J_b-1$
$J_a = J_b+1$	Q_R	P_R	O_R	$-3/2$	$3/(2J_b)$	$3/2$	$3/10$	$-3/5$	$3/10$
$J_a = J_b$	R_Q	Q_Q	P_Q	$3/(2J_b)$	$-3/(2J_b^2)$	$-3/(2J_b)$	$-3/5$	$6/5$	$-3/5$
$J_a = J_b-1$	S_P	R_P	Q_P	$3/2$	$-3/(2J_b)$	$-3/2$	$3/10$	$-3/5$	$3/10$

d) R_P, P_Q branches exhibit no phase changing but are considerably strengthened;

e) measurable only in the linear polarization modulated case is the Q_Q line, which shows the same phase as Q_R and Q_P in either of the polarization modulated cases.

These analyses allow one to distinguish the two-photon branches with a certain enhancing transition (represented here by the same subscript for the branch symbol). As the ground and the first excited electronic states are usually known with good accuracy, one can then identify the branches of every two-photon lines even if the constants of the upper states are unknown.

5) **Sensitivity**: As we have seen, the offset value of an enhancing level for a molecular two-photon absorption transition can be as small as comparable with the homogeneous line width. Under such a condition the enhancing factor of a two-photon transition is close to unity (as for a one-photon transition). Because all molecules contribute to the Doppler-free peak and not just a package with a particular velocity as in Doppler-free saturation or polarization spectroscopy, it may offer very high sensitivity. Since there are numerous vibration-rotational levels in each of three electronic states involved, dozens of near-resonantly enhanced two-photon spectral lines with good signal-to-noise ratio can be found if one of the electronic states is at the mid-point between the other two [5.50].

Combinations of various high-resolution and high-sensitivity spectroscopic techniques with Doppler-free two-photon spectroscopy have further extended the range of applications. As well as the polarization modulation technique combined with two-photon spectroscopy as described above, frequency modulation has been demonstrated to show better signal-to-noise ratio and zero background [5.28]. Moreover, since the two-photon absorption of a moving molecule in a standing-wave field occurs as that of a particle at rest, separated field spectroscopy in the optical region has been successfully combined with Doppler-free two-photon spectroscopy to eliminate time-of-flight broadening [5.56]. By using two separate standing-wave fields (Sect.5.4), each one of which will provide a two-photon absorption line without Doppler-broadening, one can obtain super-high resolution with the linewidth determined by the separation between the two beams rather than by the beam diameter. Such a feature is particularly interesting for two-photon spectroscopy, which often has to use focussed beams and so cause considerable transit-time broadening. The improvement is significant for the investigation of the fine or hyperfine spectral structure of polyatomic or heavy molecules.

So far Doppler-free two-photon spectroscopy and its modifications have widely been used in the study of molecular spectra and structure, involving complementary states which are forbidden for one-photon transitions, in molecules as simple as homonuclear diatomic molecules or as complicated as large organic molecules. Some of the examples are listed in Table 5.6.

Table 5.6 Doppler-free two-photon spectroscopy, its development and some examples of its applications in the study of molecular spectra and structures

Class.	Name	Modulation*	Detection	Molecules	Results — signal features **	Ref.
I	Doppler-free two-photon spectroscopy (two colinear beams)	intensity modulation at ω_1	UV fluorescence at ω_1	Na_2	DF + DB	5.43-46
				Na_2	lineshape (DF/DB) dependence; band structure of enhanced transitions;	5.47-50
				NH_3	transition strength	5.15
				C_6H_6	decay time; intramolecular dynamics	5.51,52
				CH_3F	pressure broadening & shifts	5.53
				NO_2, SO_2 Cs_2, Li_2	Dunham const.	5.54
II	Complex two-photon spectroscopy (two colinear beams)	intensity intermodulation at ω_1 & ω_2	UV fluorescence at $\omega_1+\omega_2$	Na_2	DB eliminated	5.35
		polarization intermodulation (POLINEX for TP) at ω_1 & ω_2	UV fluorescence at $\omega_1+\omega_2$	Na_2	DB eliminated; J discriminated	5.35-36
		intensity intermodulation at ω_1 & ω_2 for distinctly polarized beams	probe beam monitored at $\omega_1+\omega_2$ behind a polarizor	Li_2	DB eliminated; J discriminated	5.38

Table 5.6 (continue)

Class.	Name	Modulation*	Detection	Molecules	Results signal features **	Ref.
II		amplitude modulation (AO) at ω for pump beam	probe beam monitored at ω behined a polarizor	Na_2	DB eliminated; shot-noise limit	5.28
		frequency & phase modulation (EO) at wm for pump beam	four-wave mixing signals at beat frequency wm	Na_2	DB eliminated; DF narrowing; shot-noise limit expected	5.40
III	separated field two-photon spectroscopy	two standing waves spatially separated	Ramsey fringes Fourier transfered	see Table 5.7	DB eliminated; DF narrowed by elimination of transit broadening	
		time separated multiple coherent light pulses	Ramsey fringes Fourier transfered			

* AO- acousto-optical modulation; EO- electro-optical modulation
** DF and DB indicate Doppler-free peak and Doppler-broadened background in the spectral profile of a molecular two-photon absorption transition, respectively; (DF/DB)- their maximum ratio

136

5.4 Superhigh-Resolution Spectroscopy with Separated Fields

The Ramsey-fringe technique, which has been used in the microwave region for decades, has recently been introduced by *Chebotayev* and co-workers to the optical region [5.56]. It has been developed in the form of both spatially-separated and time-separated methods. Detailed theoretical and experimental achievements have been discussed in the recent review [5.5] with ample references to the original literature. The principle of the Ramsey-fringe method in the long-wavelength region is as follows: One considers a bunch of collimated molecules moving at the same velocity v. When two spatially separated fields have the same phase, the phase difference of the molecules in the second zone from that in the first zone is due only to the flight-time spent in passing through the free space between the two zones of distance L, where the molecules oscillate at their transition frequency ω_0 rather than the field frequency ω. After the molecules have been labelled by the field phase in the first zone the phase difference $(\omega_0-\omega)T$ is effective for their further absorption in the second field, to an extent, proportional to $\cos[(\omega_0-\omega)T]$. This causes fringes in the absorbed energy during the scanning of the field frequency ω with a full width at half maximum, $\delta\nu$, corresponding to a phase change of π. Therefore, one has

$$\delta\nu = (2T)^{-1} = v_x/2L , \qquad (5.19)$$

where v_x is the molecules' velocity component perpendicular to the direction of propagation of the interacting fields. For a practical molecular beam there is a nonzero velocity component v_z parallel to the propagating direction of the fields, which will result in the residual Doppler width $\delta\nu_D$ given by

$$\delta\nu_D = kv_z = \delta\theta v_x/\lambda , \qquad (5.20)$$

where $\delta\theta$ is the divergence of the molecular beam. Obviously, a practical molecular beam may have a significant $\delta\nu_D$ in the short-wavelength region, while it is negligible in the microwave region. For a reasonable divergence of a molecular beam with $\delta\theta$ of 10^{-3} srad the residual Doppler width is of the order of MHz, which is about three orders larger than the fringe width $\delta\nu$ in (5.19) and so represents the limit of spectral resolution obtainable by using the method of separated fields.

Fortunately, many spectroscopic techniques have been developed for the optical region. Among these methods of nonlinear laser spectroscopy in which the influence of the Doppler effect is eliminated, allow the realization of the method of separated fields in the optical region [5.56]. As we have already described, the Doppler shift is cancelled in the molecular absorption of two photons propagating in opposite directions. With two standing-wave zones separated, for example, the resultant two-photon transition of molecules with various thermal velocities has been shown to have a super-narrow linewidth in the optical region, even when the experiments were performed in a sample cell with full thermal distribution of molecular velocities under low pressure (i.e., collision broadening minimized).

In contrast, however, the Doppler-free saturation spectroscopy for one-photon transitions is based on selecting one package of molecules interacting with two counter-propagating beams. It cannot circumvent phase diffusion for all of the molecules. The method of separated fields for the two-level system therefore requires at least three parallel and equally spaced fields, where the middle field is used to provide the phase jump for cancelling the phase diffusion in the third zone of the fields [5.5].

This means that the realization of super-high spectral resolution by the method of separated fields can take place only through essentially different optical arrangements according to the energy-level schemes of the relevant molecular transitions. A two-level transition can, in turn, only be observed by two separated fields if the signal is monitored by emission rather than absorption. Although the macroscopic polarization is the average of the dipole moments of molecules over axial velocities in the second field, it has a finite value in the separation regions and oscillates at the field frequency rather than at the transition frequency. The macroscopic polarization is therefore a coherent radiation source allowing the detection of emission signals, which does not interfere with the interaction of the molecules with the original fields.

Experiments for observing the signals in separated fields can be carried out in cells [5.63,65] or in molecular beams [5.57,61] either in spatially [5.57,63] or time [5.65] separated fields to obtain narrow transition resonances. The oscillating part of the signals obtained by the method of separated fields shows the following spectral features:

1) **Signal classifications**: there are mainly two kinds of signals which can be obtained by the method of separated fields, i.e., absorption and coherent radiation. A two-level transition can be observed either by absorption with three separated fields or, in principle, by coherent radiation with two separated fields. But the emitted coherent radiation is too weak to be detected. So an additional third field with an offset frequency is required for the super-sensitive detection of the weak signal by optical heterodyne techniques.

2) **Line center determination**: when a supersonic molecular beam is used the oscillating part of the signals obtained in separated fields is formed by a great number of Ramsey fringes, which is a serious obstacle to the location of the center frequency of a transition. However, the observable number of fringes depends on the velocity distributions of the molecules. In fact, for a molecular sample contained in a cell, higher-order fringes might be washed out by destructive inference caused by molecular thermal motion variations, leaving the few lowest orders with considerable intensities near the field frequency on resonance with a transition.

For solving the difficulty of locating the transition center, the method of time-separated fields takes advantage of the possibility of a computer-controlled series of laser pulses with an optimum duration and delay between them to average the oscillating part of the signals and to minimize the sideband oscillations, revealing hyperfine or superfine spectral components [5.66].

3) **Linewidth**: the limiting resonance widths are determined by the inverse time-of-flight between the interaction regions of the molecules with the laser fields, rather than by the transit time across one of the field zones. With the field spacing in the order of centimeters the fringe width will be in the order of kHz, which might be orders improvement compared to the transit broadening by using one interaction field zone. Of course, for realization of super-high resolution other experimental conditions should be ensured, such as a stabilized field frequency, avoidance of power and pressure broadening, and other special requirements. For example, in the case of spacially separated fields the third field introduced for probing the coherent radiation signal must keep a fixed phase relation with the exciting fields. Thus, the influence of any instability of the optical elements forming the parallel fields is appreciable. On the other hand, the instabilities of the mirrors should be unimportant for the case of time-separated fields, where the linewidth of the output of a pulsed laser might become the limiting factor in the spectral resolution.

4) **Sensitivity**: the comparison between the method of separated fields and the Doppler-free saturation spectroscopy was estimated recently [5.66]. Since the molecules for the whole Doppler distribution contribute to the signal obtained in separated fields, under conditions of optimum excitation the signal might be orders larger than that obtained by saturation spectroscopy. High contrast (~30%) absorption of as little as 60 pWatt has been detected by using a supersonic beam of SF_6 passing through four parallel separated purely travelling-wave fields [5.61].

5) **Line shape**: the shape of the Ramsey fringes can be symmetric if one ensures the intrinsic equal-time property of the separated beams as realized, for example, by cat's-eye retroreflectors [5.57]. Recently an optical imaging system for a sliced laser beam has been used to provide Ramsey fringes in CH_3F with a contrast as high as 75% [5.62]. This idea allows flexibility in the experimental conditions, such as in the number of the sliced beams and their separations. The multicrossing experiments have shown further spectral line narrowing, and the presence of gaps between the fringe peaks resembles the diffraction pattern from a grating. This result is potentially useful for high-resolution spectroscopy and for improved frequency standards as well.

The method of separated fields, a later development in super-high resolution spectroscopy, has thus proven to have attractive applications in the study of molecular spectroscopy. Some of the examples for polyatomic molecules are listed in Table 5.7. Although the types of molecules dealt with are few, the narrow optical resonances promise a wide range of applications.

Table 5.7 Method of separated fields and some examples of their applications

Class.	Name	Types of signal	Interacting fields	Detection and transformation	Molecules	Resonance-width [KHz] & contrast	Ref.
I	spectroscopy of spatially separated fields	absorption	three standing waves or two pairs of counter-propagating waves	fluorescence detected by photomultiplier	CH_4 SF_6	30 23	5.56 5.57-59 5.60,61
			two, three or five-separated fields	cryogenic bolometric	CH_3F	77 KHz with 78% contrast for two sliced interacting beams; line narrowing & contrast increasing for 5 sliced beams	5.62
		coherent radiation formed at a separated space	two separated interacting fields	heterodyne beat with an additional field	CH_4	3	5.63
II	spectroscopy of time separated fields	coherent rariation in the interaction region	two pulses or ordered multipulses	fluorescence and Fourier transformation	SF_6	10	5.65-67

5.5 Applications of Nonlinear High-Resolution Laser Spectroscopy to Studies of Molecular Spectra

It is known that molecules composed of more than two atoms exhibit subtle spectral effects, corresponding to their internal and external motions under various environmental conditions. All of the spectral phenomena are subjects of spectroscopic studies with suitable spectral resolutions. Table 5.8 lists various molecular spectral effects along with the required resolution. Whereas about half of the effects might be revealed by conventional molecular spectroscopy with Doppler-limited resolution, the others would be unresolvable due to their spectral splittings being comparable to or even smaller than the broadened linewidth of a spectral component. Fortunately, Doppler-free spectroscopy with single-mode tunable lasers is making great inroads. While some effects, which has previously been limited to study in the microwave region for molecular ground states, have now been extended to molecular excited or high-lying electronic states, others, such as various pertubation effects, could not have been explored at all previously.

After the various Doppler-free spectroscopic techniques were demonstrated for atoms one after another, they were applied in quick succession to diatomic molecules, small polyatomic molecules and recently to large polyatomic molecules. The molecular spectral effects involved are summarized in Table 5.9. Some of them will be discussed in the following sections.

5.5.1 Accurate Measurements of the Properties of Perturbed Vibration-Rotational Bands in Diatomic Molecules

There are perturbing and perturbed states in the energy range of molecular electronic excited states. Figure 2.1 illustrated that the higher the excited electronic states the more serious their overlap. The consequence of perturbations is an irregular alteration of the spectral structures of the relevant vibration-rotational bands. Doppler-free spectroscopy allows one to examine them in detail from various angles, such as:

Table 5.8. Molecular spectroscopy effects

Sources	Resolution $\Delta\nu$ [cm^{-1}]	Methods
Transition of valence electron	$10^3 - 10^4$	
Transition of nuclear vibration	$10^2 - 10^3$	
Transition of frame rotation	$10^{-1} - 10^0$	Conventional
Isotopic shift of vibration	$10^{-1} - 10^3$	spectroscopy
Isotopic shift of rotation	$10^{-5} - 10^1$	————
Nuclear electronic quadrupole interaction	$10^{-5} - 10^{-3}$	
Nuclar magnetic dipole interaction	$10^{-7} - 10^{-5}$	
Collisional broadening (per Torr)	$10^{-5} - 10^{-4}$	Doppler-free
Isomeric shift due to nuclear excitation	$10^{-7} - 10^{-5}$	spectroscopy
Nuclear recoil due to photon momentum	$10^{-7} - 10^{-5}$	

Table 5.9. Molecular spectroscopic effects and some examples of their study using various Doppler-free techniques

| | Spectroscopic effects | Investigation examples | |
		Molecules	References
I	Precise measurements of absolute frequencies of molecular transitions in the infrared and visible regions	CH_4	5.68
		$^{188,189,190,192}OsO_4$	5.69
II	Energy-level constants and wave functions:	Na_2	5.11,33,70
		Li_2	5.38,54
	Perturbation shifts and deperturbed constants	Cs_2	5.37,54,71
		N_2	5.16
		NO_2, SO_2	5.54
	Wavefunction mixing	Na_2	5.72
III	Hyperfine structures:	$^{187,198,190}OsO_4$	5.69,73
	Ground electronic states	SF_6, NH_3, PF_6	5.69
		CH_4	5.74
		$^{12}CH_3Cl$, methyl-halides	5.13,75
	Low-lying forbidden electronic states	$^{127,129}I_2$	5.9
		Na_2	5.72,76,77
	High-lying forbidden electronic states	Na_2	5.78–80
IV	Isotope effects:		
	Isotope shift	see part I	
	Isotope hyperfine structures	see part III	
V	Recoil effect and its elimination	CH_4	5.5,74,81
VI	Spectroscopic effects of external fields:	NO_2	5.82
		I_2	5.83
	Linear and quadratic Stark effect	CH_4	5.84–86
		PH_3	5.87,88
		HNO_3	5.89
		CHD_2F	5.90

Zeeman effect and anomalous	I_2	5.91
Zeeman effect	CH_4	5.92,93
	NH_2	5.94,95

VII	Intramolecular dynamics and coupling mechanisms:	C_6H_6	5.52,96
	Decay dynamics	$H_2C_2N_4$	5.42
	symmetry breaking	SF_6	5.97,98
		NH_3, PF_5	5.99,69

VIII	Molecular motion effects and their eliminations:		
	Second-order Doppler broadening and shift	CH_4	5.100,101
	Doppler phase angle; time-of-flight broadening	SF_6,CH_4, etc.	see Table 5.7

IX	Relaxation:	Na_2	5.102
	Ground states	SF_6	5.66
	Excited states	$H_2C_2N_4$	5.42

X	Molecular collision:	Na_2	5.20,79
		I_2	5.104
	Linear & nonlinear dependence of collisional broadening	H_2O	5.25
		NH_3	5.105
		CO_2	5.106,107
		SF_6	5.108,109
		CH_4	5.110
		CH_3F	5.111
	Collisional lineshift:	CH_4	5.110,112
	Involving molecular electronic ground state	CH_3F	5.111
	Involving excited states,	Na_2	5.20
	Involving Rydberg states	K_2	5.113

1) **Perturbation shifts and their J- and v-dependences**: successfull examples are the studies of homonuclear diatomic molecules. For instance, the ground states $X^1\Sigma_g^+$ of several kinds of alkali dimers have been investigated by conventional spectroscopy. The first excited singlet state, $A^1\Sigma_u^+$ of several kinds of alkali dimers has been investigated by conventional spectroscopy; it had been known for a long time that this state would be perturbed by $b^3\Pi_u$.

Half a century ago the observation of a magnetic rotation spectrum for the $A^1\Sigma_u^+$-$X^1\Sigma_g^+$ transitions of Na_2 and K_2 indicated that there are perturbations in the $A^1\Sigma_u^+$ states [5.114, 115]. No direct observations of optical transitions between the perturbing state $^3\Pi_u$ and the ground state $^1\Sigma_g^+$ have been made since this is a "forbidden" singlet-triplet transition. *Kusch* et. al. had directly observed the $A^1\Sigma_u^+$ state to be perturbed via an A-X absorption spectrum of Na_2 photographically recorded on a 9.3 m concave grating spectrometer. The observations indicated that the perturbations are caused by the $b^3\Pi_0$, $b^3\Pi_1$ and $b^3\Pi_2$ states [5.116]. The extreme experimental conditions and the tedious data analyses, however, suggested that perturbation studies with conventional molecular spectroscopic techniques were limited to a few experts.

During the last few years several new laser spectroscopic techniques have been applied to the study of the $A^1\Sigma_u^+ \sim b^3\Pi_u$ perturbations. *Kaminsky* et. al. had for the first time studied the $A^1\Sigma_u^+ \sim b^3\Pi_u$ perturbations in Na_2 by Doppler-free population modulation spectroscopy (Sect.7.2). In addition to the unusual broadening of the observed lines, their least-squares quadratic fit to the signals had shown that the perturbation shifts at $A^1\Sigma_u^+$ v' = 22, J' = 14 and J' = 12 are 0.6 cm^{-1} and 0.3 cm^{-1}, respectively [5.11]. *Teets* et. al. observed a single $b^3\Pi_u$-$X^1\Sigma_g^+$ transition using polarization labelling spectroscopy [5.117]. The $A^1\Sigma_u^+ \sim b^3\Pi_u$ perturbation of Na_2 in several high vibrational levels had been studied by Doppler-free polarization spectroscopy by *Li* et. al. [5.33]. They determined the most highly mutually perturbed rotational levels in the $A^1\Sigma_u^+$ and $b^3\Pi_u$ states for $24 \le v_A \le 34$ and $26 \le v_b \le 34$. The spin-orbit interaction matrix elements as well as the deperturbed molecular constants were given for the mutually interacting v_a=26 ~ v_b=28 and v_a=34 ~ v_b=34 vibrational levels. It is noticeable that these results were obtained by using the simplest sample cell (heat-pipe oven) and the data could be unambiguously analysed. These results, together with the data obtained by fluorescence excitation spectroscopy from a supersonic molecular beam [5.73, 118, 119], the rotationally resolved fluorescence spectra followed a perturbation facilitated optical-optical double resonance [5.120], and the Fourier-transfrom spectral data, have made the forbidden states $b^3\Pi_{\Omega u}$ (Ω = 0, 1, 2) and their perturbation effect to the $A^1\Sigma_u^+$ state in Na_2 quite well understood.

Hemmerling et. al. investigated the perturbations between Rydberg states by means of Doppler-free optical double-resonance polarization spectroscopy for alkali dimers of Cs_2 [5.37, 54, 71] and Li_2 [5.38, 54]. Figure 5.16a shows the Fortrat curve of a vibration-rotational band according to the analysis of a recorded simplified spectrum (Fig.5.16b) of Cs_2. From this, it is possible to plot the rotational perturbations as measured by the deviation of the observed transition frequencies versus the rotational quantum numbers of the upper states from the smooth Fortrat curve. For each measured vibrational level two rotational perturbations at different values of J were observed, these being due to interactions of a vibrational with two perturbing vibrational levels in succession. The phenomena confirmed that the perturbations take place between two high-lying singlet states rather than being singlet-triplet perturbations. Maximum perturbations

144

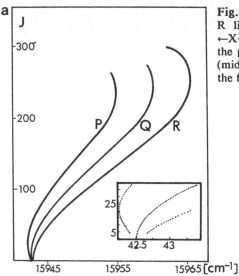

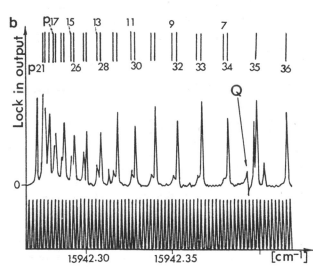

Fig.5.16 (a) Fortrat diagram of P, Q, and R lines of the $0\leftarrow0$ band in the $^1\Pi_u\leftarrow X^1\Sigma_g^+$ transition of Cs_2; (b) section of the polarization spectrum of the P-branch (middle) with the assignment (upper) and the frequency markers (lower) [5.71]

shifts beyond 1 cm^{-1} were observed. Accurate results covering a large range of J values from zero to 60 were ensured by the Doppler-free resolution, together with precise measurements of the transition wavelengths obtained with a travelling Michelson-type wavemeter with an accuracy better than 10^{-3} cm^{-1} [5.121, 122].

Many important molecular species (e.g., H_2, H_2O, N_2, CO) have their electronically excited states in the vacuum ultraviolet below 185 nm. In order to apply laser spectroscopy to study the spectra of such species and to reach the Rydberg states of most molecules, it would be necessary to develop tunable lasers for this region. Alternatively, one may use a non-optical approach for the first excitation. For example, *Suzuki* et. al. studied perturbation shifts in the Rydberg states of N_2 utilizing optogalvanic tech-

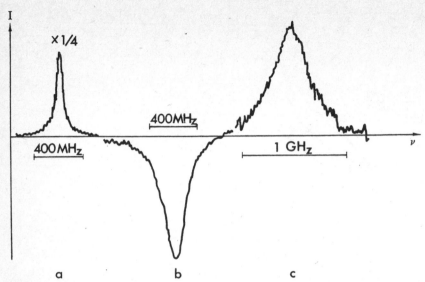

Fig.5.17. Observed lineshapes for signals via modulated population spectroscopy; (a) and (b) original signals; (c) perturbation broadened signal [5.11]

niques with a RF discharge. Doppler-free methods were applied to resolve overlapped lines, while most of the lines were studied at a Doppler-limited resolution of 0.05 cm^{-1}. Pertubation shifts versus J(J+1) values were plotted for two bands, the maximum shift being larger than 1 cm^{-1} [5.72].

2) **Perturbation variation of line widths**: while the perturbation shifts could, despite many difficulties, be measured before the invention of Doppler-free laser spectroscopy, one could not hope to explore the variation of the linewidth of individual transitions by conventional spectroscopy. Figure 5.17c shows the perturbation broadened linewidth of a molecular line in Na$_2$ compared with the unperturbed lines in (a) [5.11]. The perturbation broadening can be considered by the hyperfine splitting of the singlet state A, sharing from that of the triplet b, due to the perturbation mixing between their nominal wave functions [5.73, 78].

3) **Perturbation variation of lifetimes**: After intense optical pumping of an individual transition the time-resolved laser induced fluorescence enabled one to measure the lifetime of an upper level to allowed lower levels. Lifetime measurements for unperturbed and perturbed levels in the $A^1\Sigma_u^+$ state of Li$_2$ [5.123], and in $A^1\Sigma_u^+$ and high-lying g-parity states of Na$_2$ [5.123, 124] have been made. These results agreed with calculations of the Einstein transition probability for spontaneous emission A_{nm}, which is related to the matrix element of the transition as follows [5.125]:

$$A_{nm} = \frac{64\pi^4 \nu_{nm}^3}{3h} |R^{nm}|^2 \,, \tag{5.21}$$

where the matrix element of the electric dipole moment R^{nm} depends on the eigenfunctions ψ_n and ψ_m of the two states and would be reduced for the case when the wave function of one of the two states is partially mixing with a nearby perturbing state to show a kind of hybrid. According to the relation of

$$\tau = 1/A_{nm} \text{ (one possible transition from level n)} ,$$

or
$$\tag{5.22}$$

$$\tau = 1 / \sum_m A_{nm} \text{ (several possible transitions from level n)} ,$$

a perturbed vibration-rotational upper level of a transition would therefore show an increased lifetime compared with neighbouring levels in the same electronic state. The linewidth broadening of a transition with a perturbed level is deduced by some unresolved spectral structure rather than an acceleration of the relaxation. A perturbing level should possess quite large hyperfine splittings if the states have nonzero values of both the electronic orbital quantum number A and the total spin quantum number of the outer electrons S, such as the perturbing state $^3\Pi_u$ in Li_2 and Na_2, etc..

4) **Mixing coefficients for mutually perturbing levels**: According to the relative intensities of the series of forbidden transitions observed with the aid of wave-function mixing between the mutually perturbing levels, *Atkinson* et. al. were able to determine the mixing coefficients as a function of rotational quantum numbers in Na_2 [5.72]. The nonlinear least-square fits showed that the spin-spin interaction, the spin-rotation interaction and the orbit-rotation interaction were negligible. The most important perturbation between the first excited singlet-triplet states, $^1\Sigma_u^+$ and $^3\Pi_{\Omega u}$, is due to the spin-orbit interactions summed over both valence electrons. The results showed that the mixing coefficients between $\Delta\Omega = 0$ states, i.e., $^1\Sigma_u^+$ and $^3\Pi_{0u}$, the so-called homogeneous perturbation, might be an order or more larger than that of the heterogeneous perturbation between $^1\Sigma_u^+$ and $^3\Pi_{2u}$ with $\Delta\Omega = 2$.

5.5.2 Accurate Studies of Molecular Hyperfine Spectra

Physicists are used to measure hyperfine splittings of the molecular pure rotational levels by means of microwave spectroscopy [5.126]. The measurements were limited to the few lowest vibrational levels in the ground electronic state and were useless for levels in excited electronic states. The earliest example of a study of molecular hyperfine structures for electronic vibration-rotational transitions resolved by means of Doppler-free saturation spectroscopy was reported by *Levenson* and *Schawlow* [5.9]. Their results for large even J" and odd J" values in $^{127}I_2$ were shown in Fig.5.3a and b, respectively. Even though the tuning range of their single-mode krypton-ion laser was quite limited, it still was enough to reveal hyperfine structures with a width of about 6 MHz for 15 or 21 components which had up to that time always been obscured by the Doppler broadening of 600

MHz in the visible region even with the best spectrographs. Such a resolved structure corresponds to the difference between hyperfine energies of sub-levels of the two electronic states which have the same quantum numbers of nuclear spin I and its projection M. As the hyperfine constants for the ground state could be found, the measurements then allowed one to obtain the constants for the upper electronic states. As discussed in Chapt.2, such a spectrum provides information about the interaction between the molecular axial-field gradient and the quadrupole moments of the two iodine nuclei; it is a window for "seeing" the distribution of the electrons in the molecule. Their determinations have successfully involved the transitions with rotational quantum numbers between 10 and 117 and vibrational quantum numbers in the upper electronic state between 17 and about 62, close to the dissociation limit. Consequently, one was able to reach some very important conclusions for the molecular spectroscopy involved, such as the fact that the spin-rotation coupling coefficient (magnetic hyperfine constant) varied dramatically with vibrational energy in the electronic excited states.

Recently some attractive studies of hyperfine structures in various forbidden states of sodium dimers have been made. One of the interesting states is the low-lying triplet state $b^3\Pi_{\Omega u}$ ($\Omega = 0, 1, 2$), which is spin-forbidden for transitions from the ground state $X^1\Sigma_g^+$. *Atkinson* et. al. had reported their measurements of the hyperfine constant with dependences on the qunatum numbers Ω and J in the state $b^3\Pi_{\Omega u}$ by means of the fluorescence detection of a collimated molecular beam excited by a single-mode dye laser [5.72, 76]. *Wang* et. al. have studied the hyperfine structures for high vibrational levels in the same triplet state by using Doppler-free polarization spectroscopy [5.77]. Further more, the hyperfine structures of some vibrational-rotational levels in high-lying parity- and spin-forbidden states have been observed too. While *Li* et. al. have determined the magnetic hyperfine splittings of the $(2)^3\Pi_g$ state and the admixed $2^1\Pi_g$ state in Na_2 by means of perturbation facilitated optical-optical double resonance [5.80], Xia et. al. have achieved the direct observation of the hyperfine splittings of the $(3)^3\Pi_{\Omega g}$ state by means of Doppler-free equal-frequency two-photon spectroscopy [5.78, 79]. The advantages of the latter include: 1) thorough elimination of first-order Doppler-broadening and hence better spectral resolution than unequal-frequency OODR, 2) no interference from the hyperfine splittings of the intermediate level in the determination of the hyperfine constant for the splitting of an upper level; 3) the observable rotational quantum number J is not limited to small values.

Figure 5.18 exhibits the recorded traces of Doppler-free equal-frequency two-photon transitions in Na_2, where the traces a to c correspond to the cases $\Omega = 0, 1, 2$ for angular momentum projected on the molecular internuclear axis. According to the Pauli exclusion principle, the total wave function of a homonuclear diatomic molecule with half-integer nuclear spin (here 3/2) has to be asymmetric under exchange of nuclei. The total nuclear spin of I = 3 and 1, corresponding to symmetric spin wave functions, can be associated with asymmetric rotational wavefunctions, which in g-parity states correspond to levels with odd rotational quantum numbers, whereas I = 2 and 0 with asymmetric spin wave functions can be associated with even

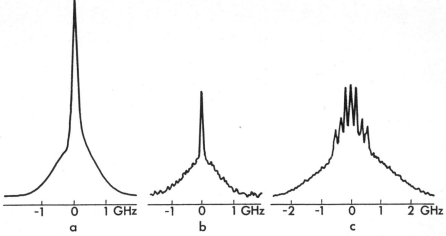

-1	0	1 GHz	-1	0	1 GHz	-2	-1	0	1	2 GHz
a			b				c			

Fig.5.18. Hyperfine structures for spin-forbidden two-photon transitions from singlet to triplet electronic states with large J values in Na_2. (a),(b) and (c) were recorded around 646 nm corrresponding to different fine electronic states. See text for details. [5.79]

J levels. The total number of hyperfine components is determined by $\Sigma(2I_i + 1) = 10$ and 6 components in odd and even J levels, respectively. The hyperfine splittings of the ground state in Na_2 are negligible. Thus the traces in Fig.5.18 are taken to reflect directly the hyperfine splittings in the high-lying triplet state. The formula for magnetic hyperfine separations is

$$\Delta\nu = C[F(F + 1) - I(I + 1) - J(J + 1)] , \qquad (5.23)$$

where $F = J+M_I$, $M_I = I, I-1, ..., (-I)$. We see that seven of the ten components $(I = 3)$ for an odd rotational level are shifted from the degenerate frequency position by an amount $6JC$, $(4J-6)C$, $(2J-10)C$, $-12C$, $(-2J-10)C$, $(-4J-10)C$, $(-6J-6)C$, whereas the three components with $I = 1$ for the same rotatioanl level are at $2JC$, $-2C$, $(-2J-2)C$. This means that i) the difference between the first and the last interval is $10C$. If the blueside intervals are smaller than the red-side intervals (Fig.5.18c), we know that the magnetic hyperfine constant C is negative, $C<0$. ii) For small J values the 10 components will be separate; in the case of large J but small C ($10C$ much less than the linewidth), the three $I = 1$ components will fall on top of other components to give the three double-strength components among a total of seven as shown in Fig.5.18c. Computer fitting gives $J \sim 61$ and $C \simeq -1.6$ MHz. The determination of J was aided by the computation of an equal-frequency two-photon transition enhanced by a pair of near-resonant singlet-triplet intermediate mixing levels. The Lorentzian profile fitting for the trace gives a half width of 25 MHz under the optimum mixing temperature of 460° C in the sodium oven with foccussed CW dye laser beams. Such a linewidth, unfortunately, prevented us from resolving the neighbouring lines, as shown in Fig.5.18a and b, corresponding to $\Omega = 0$ and 1, respectively. However, the widths of the three traces identified with similar

149

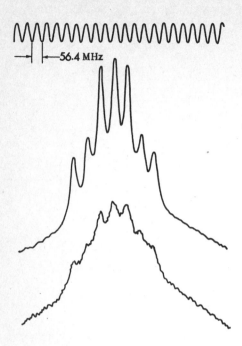

—||—56.4 MHz

Fig.5.19. Hyperfine magnetic dipole structure of a spin-forbidden two-photon absorption line of Na_2 in a weak magnetic field (about 50 Gauss, lower trace) in comparison to the original trace without external field (upper trace) [5.79]

odd J values reflect roughly the differences between the hyperfine constants among the upper fine electronic states with different Ω values. The result is interesting in that in our direct measurements we have observed that the largest hyperfine constant has a negative sign, which is in contradiction to the indirect determination by means of OODR [5.80]. Moreover, we have demonstrated that the structure of the upper level is very sensitive to the external magnetic field. The lower trace in Fig.5.19 shows the spectral effects of a weak transverse magnetic field of about 50 Gauss in comparison with the zero-field trace (upper). The Zeeman splittings in weak magnetic fields confirm, in turn, that the clustered Doppler-free peaks are indeed the hyperfine splittings of the upper level for a spin-forbidden two-photon transition. In addition to obtaining the C value, the line shape of the multipeak Doppler-free signal compared with its Doppler-broadened pedestal means that at least one out of the mutual intermediate mixing levels is at near-resonance, which helps to judge the enhancing level and to further estimate the molecular energy-level constants involved.

The hyperfine spectra for polyatomic molecules have been studied by Doppler-free spectroscopic technique with extremely high resolution. To the authors' knowledge, however, their studies have been limited to vibration-rotational transitions. The hyperfine structures of some methyl-halides have precisely been studied to provide electric quadrupole and magnetic dipole hyperfine constants with accurate measurements of the absolute frequencies in the region of 3.39 μm. Figure 5.20a and b show the hyperfine quadrupole structures of two similar symmetric top molecules CH_3I and CH_3Cl, respectively [5.13]. The observed hyperfine splittings were more than two orders smaller than their nuclear electric quadrupole interaction

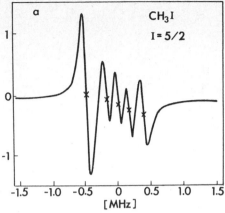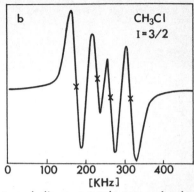

Fig.5.20. Hyperfine quadrupole structures in two similar symmetric top molecules CH_3I (a) and CH_3Cl (b) [5.13]

terms. This means that the vibrational excited states have essentially the same nuclear electric quadrupole level structures, which can therefore only be revealed by extremely-high-resolution spectroscopic techniques. For example, for nuclear Cl with $I = 3/2$ there are four components, corresponding to the $(2I+1)$ orientations of I with respect of J. The marked centers from left to right in the derivative trace in Fig.5.20b correspond to $M = F-J$ of $-1/2$, $1/2$, $-3/2$ and $3/2$. The intervals between neighbouring components with a common linewidth of about 16.7 kHz were 50.8, 36.2 and 53.5 kHz, which provided the value of the nuclear quadrupole hyperfine constant in the ground vibrational level (-74.74 MHz) and the difference between those of the levels involved in the transition (only about -0.03 MHz).

While the nuclear electric quadrupole splittings are dominant for the above-mentioned heavier methyl-halides, the hyperfine splittings of the spherical top molecule CH_4 with all nuclear spins smaller than $1/2$, had been observed with an order of magnitude weaker nuclear mangetic hyperfine structures, as shown in Fig.5.21 [5.74].

Obviously, the number of hyperfine components in Fig.5.20 allows one to determine the nuclear spins for similar molecules. For the same reason it is useful in the discrimination between isotopic molecules. As an exmaple, Fig.5.22 shows no hyperfine splitting for $^{192}OsO_4$ due to nuclear spin $I = 0$ (left trace) and hyperfine structures consisting of the paired resonance doublets for $^{189}OsO_4$ due to the nuclear quadrupole interactions with $I = 3/2$ [5.75]. Recently, *Borde* et. al. have reported the investigation of hyperfine structures as a test of intramolecular dynamics for various isotopic forms of the spherically symmetric moleculs OsO_4 and SF_6. From the observed superfine and hyperfine spectra with a halfwidth as low as 1 kHz and uncertainty of 0.1 kHz they have tested such sources of the magnetic dipole hyperfine structures as the spin-rotation and spin-vibration interactions, separated from the purely quadrupolar spectral effects, for hundreds of lines and have fully demonstrated the spontaneous symmetry breaking of

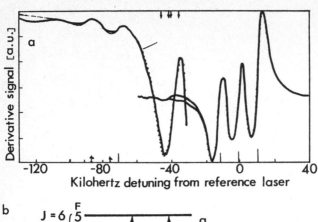

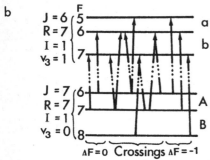

Fig.5.21. Magnetic hyperfine structure of the spherical top molecule CH_4 [5.74]

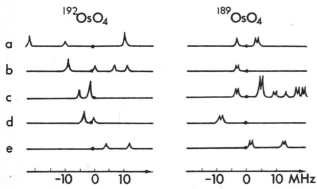

Fig.5.22. Comparison between hyperfine spectra of isotopic molecules $^{192}OsO_4$ (left) and $^{189}OsO_4$ (right) at different CO_2 laser lines (a) p(24); (b) p(22); (c) p(20); (d) p(18); (e) p(16). [5.73]

a point group into other subgroups and the evolution of hyperfine mixing towards the collapse into a single superhyperfine structure [5.69,97].

5.5.3 Quantitative Study of Molecular Collision Mechanisms

Molecular collision is a complicated physics problem. On the one hand, the collisional perturbation is quantitatively different for various molecular electronic or vibration-rotational states and for various kinds of collision partners. It is even qualitatively different for different pressure ranges and for different interaction paths between light and molecule. On the other hand, most of the spectral effects of various molecular collision mechanisms are very minute, and are unresolvable, unseparable and hence unobservable in conventional molecular spectroscopy. Collisional effects were studied extensively in the years before the advent of lasers but for this reason it was limited to the high-pressure range and to the observation of the linear dependence of broadening on pressure with the minimum broadened linewidth greater than the Doppler width, as described in a good review paper [5.127].

The development of Doppler-free spectroscopy has provided such high spectral resolution, accuracy in determining the center frequency and spectral profile of molecular transitions, and high sensitivity, that one can now observe the slight collisionally-induced linewidth broadening, line center shift and line profile function modification within the original mask of molecular Doppler-broadening. The remarkable achievements might be summarized as follows:

1) Revelation of the nonlinear broadening dependence on pressure in the low-pressure range rather than a simple extrapolation of the linear dependence in the high-pressure range. The linewidth broadening rate at low pressure (millitorrs) may be several times larger than that in the high-pressure range (Torrs or higher).

2) Revelation of the line shift of the center frequency accompanying line broadening in the low-pressure range. Some examples of the experimentally observed data for I_2, NH_3, CO_2, SF_6 and CH_4 are listed in the upper part of Table 5.10.

3) Transition-dependence of the collision effects. One example is that the different molecular transitions in Na_2 show distinct pressure broadening rates and center-frequency shifts under the same experimental conditions [5.20]. Such a phenomenon confirms the excited-state dependence of the collisional cross sections, which was previously unobtainable.

4) Exciting-path dependence of the line shape modifications due to molecular collisions. For instance, while for one-photon absorption transitions the collisional effect would cause a broad Gaussian background and a reduced Doppler-free peak [5.20], for two-photon transitions the line shape modifications, depending on the offset of the most effective intermediate enhancing level (i.e., on the relative contribution ratio between the coherent two-photon excitation and the two-step excitation), would be an enlarged and asymmetric Doppler-broadening pedestal. A good example of the latter is shown in Fig.5.23, where a represents the hyperfine structure of a two-photon vibration-rotational transiition in Na_2 at a sodium vapour pressure of 0.5 Torr. The asymmetry of the whole profile in Fig.5.23a is characterized by the height difference Δh_1 of the two side components. This is due to the superposition of the 7 (corresponding to the original 10) Doppler-free components with successively changed positive offsets form the intermediate level on resonance with the two-photon transitions, as well as to the inherent difference in the intensity factors between the two hyperfine

Table 5.10 Spectroscopic effects of molecular collision

Mol.	Upper level in the electronic states of	Pressure range [Torr]	Noticeable collisional effects	Coeff. of collisional broadening [MHz/Torr]	Observed collisional shifts	Ref.
Na_2	excited; Rydberg	10^{-1} - 10	transition-depend.; foreign gas-depend.	13 - 32	0.1-4.2 MHz/Torr	5.20 5.79
I_2	excited	0 - 20×10^{-3} (50 - 160)$\times 10^{-3}$	Nonlinear broadening	^17 ^9		5.104
CO_2	ground	10^{-2} - 2×10^{-1}	Nonlinear broadening	13.0 ± 2 4.8 ± 0.9		5.106 5.107
SF_6	ground	$<5 \times 10^{-3}$ $>30 \times 10^{-3}$	Nonlinear broadening		^ 24 KHz/mTorr ^ 3.7 KHz/mTorr	5.108 5.109
CH_4	ground	10^{-3} 10^{-2}	Nonlinear broadening	^ 30 ^ 10	400 Hz/mTorr	5.103 5.110
CH_3F	ground	10^{-3} - 10^{-1}	foreign gas-depend.	CH_3F-CH_3F: 41.3 ± 1.0 He-CH_3F: 5.0 ± 0.5	2.1 ± 0.1	5.111
$H_2C_2N_4$	excited	0 - 2	linear	290 ± 30		5.42
K_2	Rydberg $n \wedge 30$ $n \wedge 60$	3 - 30	linear		1.58 0.47	5.113

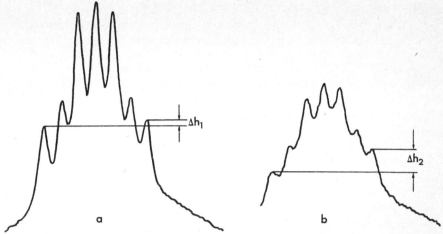

Fig.5.23. Collisional effect on near-resonantly enhanced two-photon transition components at a total pressure of (a) 0.5 Torr; (b) 3.5 Torr. The relative intensity changes are given by the difference between h_2 and h_1

components. An increased height difference Δh_2 in Fig.5.23b of almost a factor 3 obviously represents the offset-dependent strengthened asymmetry, which agrees with the theoretical description of the spectral influence of velocity-changing collisions in three-level systems [5.128].

5) Partner-sort dependence of pressure broadening and shift. Detailed experimental studies of the pressure effects for four electronic vibration-rotational transitions of the sodium dimer showed that they were all shifted to the blue when helium (He) was used a a buffer gas, but to the red when krypton (Kr) was employed. Atomic sodium (Na) as a buffer gas, however, shifted some of the transitions to the blue and others to the red [5.20]. The studies on vibration-rotational transitions in CH_3F had revealed a great disparity between the self-broadening of CH_3F-CH_3F and foreign-broadening of $He-CH_3F$ [5.111], as listed in Table 5.10.

The experimental setup for such accurate measurements usually requires the laser frequency and intensity to be stable enough for repetitive single-frequency scanning through an interesting molecular transition. To avoid frequency drifting of the laser output, the laser cavity can be locked to an external confocal cavity with its temperature being precisely controlled. The better the laser frequency stability, the lower the pressure range which can be studied. For the extremely low pressure range of microtorr, for example, frequency stability in the order of 1 kHz is required.

The experimental success has tested the theoretical explorations about the physcial mechanisms of molecular collisions. They can be summarized as follows:

(i) Collision-induced relaxation acceleration, which is obviously proportional to the collision rate (i.e., to pressure) and which exhibits line broadening in the form of a Lorentzian function.

(ii) Collision-induced phase-changes of a molecular transition, which therefore can be treated by introducing a phenomenological complex relaxation parameter to the off-diagonal elements of the density matrix, adding an additional line broadening and a line center shift to the resonance denominator. For example, the resonance denominator Δ_{ac} in the previously discussed matrix equation derivation in Chapt.4 now become

$$\Delta_{ac} = (\omega_{ac} - 2\omega + Im\{\gamma^{ph}\}) + i(\gamma_{ac} + Re\{\gamma^{ph}\}) , \qquad (5.24)$$

where $Im\{\gamma^{ph}\}$ and $Re\{\gamma^{ph}\}$ are the imaginary and real parts of the phenomenological relaxation parameter γ^{ph}, respectively [5.128].

(iii) Collision-induced velocity changes of the molecular thermal motion, which alters the velocity distribution of the molecular population in the initial level of an electronic vibration-rotational transition, i.e., in contrast with 2), it affects the diagonal elements of the density matrix. This causes the Doppler-free one-photon or two-step excitation (based on the velocity-selected first-step saturated excitation) to show a broad Gaussian background under its narrow resonance. For a Doppler-free near-resonant two-photon transition the enhancing level, with an offset Δ less than the Doppler width form two-photon resonance, will have an actual velocity-selected population, accompanying a significant fraction of the two-step excitation process in the total two-photon absorption rate. It therefore will exhibit an increasingly asymmetric broad background with increasing sample pressure as shown in Fig.5.20 [5.35].

(iv) Collision-induced repopulation or depopulation among nearby molecular levels. This causes satellite lines to appear around the transition of interest even though Doppler-free spectroscopy using single-frequency lasers is utilized to ensure single-transition excitation [5.31]. The collisional processes can only change the molecular rotational- and magnetic-quantum numbers, J and M respectively, by small amounts. Because the collisions do not affect the nuclear spins, only even changes of ΔJ are allowed. When J is large, small changes in M, corresponding to small changes in molecular orientation, mean that collisions will cause transitions to nearby rotational states without randomizing the orientation of the molecules. This maintains the validity of polarization labelling in the study of molecular collision processes [5.70, 129].

The above-listed four kinds of collisional mechanisms play varying roles in the total effect of collisions on a molecular spectral line. In the low pressure range the line broadening is due to quenching collisions (mechanism 1 above), to phase-perturbing collisions (mechanism 2) and to the fact that molecules leave the range of absorption-selected velocities (mechanism 3). In the high-pressure range the broadening is due only to the first two mechanisms [5.5, 107]. Although the fourth mechanism is effective only in the high pressure region (e.g., 100 Torrs), it increases the number of observed lines rather than broadening them. The line broadening effect of each of the mechanisms responds to increasing pressure, but the slope of line broadening at low pressure is larger than that at high pressure, i.e., there is a nonlinear dependence of linewidth versus pressure. For example, the slope for CO_2 transitions in the infrared wavelength region is 13

MHz/Torr at low pressure, which is 2.7 times that at high pressure [5.107]. More examples are listed in Table 5.10.

Doppler-free spectroscopy enabled the determination of some important molecular collisional parameters, such as the collisional cross section, the characteristic scattering angle, etc. The observation of a nonlinear dependence of pressure broadening allows one to determine the elastic and nonelastic collisional cross section by comparison of the slope at low pressure, which corresponds to the total cross section of elastic and nonelastic collisions, to the slope at high pressure, which relates to nonelasitc collisions [5.5, 107, 128]. Different sorts of molecules may have an order of magnitude difference in their broadening coefficients (compare the large polyatomic molecules, for example, s-tetrazine and the sodium dimer in Table 5.10). The measurements have been performed for the electronic ground states of many polyatomic molecules such as CO_2 [5.107], CH_4 [5.103], CH_3F [5.127] and even of homonuclear diatomic molecules such as Na_2 [5.129], as well as for the electron-excited states of some diatomic or polyatomic molecules [5.20, 42]. By means of Doppler-free multiquantum excitations molecular collision broadening in C_6H_6 [5.96] and shifts in K_2 [5.113] for high-lying electronic states have been measured, where state-dependent collisional shifts have been observed to be significant with altering blue and red shifts and with highly disparate magnitudes among transitions terminating on Rydberg states with quantum number n from 30 to 60. Depending on the collisional interaction potentials, the results for K_2 are in contrast with those in other molecules such as H_2 [5.130]. Recently, studies of the self-broadening and foreign-broadening coefficient dependences on temperature for various diatomic molecules, such as N_2, D_2, H_2 in the temperature range 100-375 K [5.131, 132], N_2 at temperatures up to 1500 K [5.133], and CO at 300-3500 K [5.134] have been reported. These measurements certainly have importance both for the theories of intermolecular interactions and for applications to the laser-stabilization technology.

5.5.4 Study of Spectroscopic Effects due to External Fields

In an external magnetic or electric field the molecular rotational level J will split into $(2S+1)$ or $(J+1)$ components, respectively. These spectroscopic effects are well known and are called the Zeeman and Stark effects, respectively. The investigation of them in molecular spectra by conventional optical spectroscopy is experimentally very difficult since even without an external field the rotational lines in a vibrational band are often too close to be resolved, and since even the individual lines split into a large number of spectral components (especially for large J). In the early years before the invention of the laser these effects used to be studied for rotational transitions by microwave measurements and were limited to polyatomic molecules and small J levels. The studies on these effects since the development of Doppler-free spectroscopy can be summarized as follows:

1) Extensive studies have been made in the infrared wavelength region for molecular vibration-rotational transitions, and some studies in the visi-

ble wavelength region for molecular electronic vibration-rotational transitions by one-photon or even two-photon absorption.

2) A variety of molecules have been studied, including various kinds of stable molecules, such as the spherical top molecule CH_4 [5.81,92,93], the symmetric top molecule PH_3 [5.87,88], the near prolate asymmetric top molecule CHD_2F [5.90], the heavier asymmetric top molecule HNO_3 [5.89], the planar molecule H_2CS [5.135], the planar molecule NO_2 [5.82], the small linear molecule DCN [5.136] and homonuclear diatomic molecules I_2 [5.83, 91] and Na_2 [5.79], as well as unstable radicals such as NH_2, etc. [5.94,95]

3) Higher levels of precision and accuracy have been achieved for the determination of the state-dependent electric dipole moments and magentic dipole moments or Lande factors for the molecular ground state and excited states. This permits one to explore such questions as whether there is any regularity in the dipole-moment functions of similar molecules (e.g., isotopic molecules [5.136,137]), and the influence of state mixing (e.g., Coriolis interaction between different vibrational modes) on the effective dipole moments of rotational levels, by looking for tiny deviations of measured values away from those expected [5.88].

4) The spectroscopic effects of very weak external fields on individual hyperfine components can be observed, such as CH_4 in magnetic field of 5 G [5.93] and Na_2 in a field of 50 G (Fig.5.19) [5.79].

5) While in the early years the observation of the Zeeman and Stark effects were limited to small J values (e.g., J<10), effects for J as large as 61 in Na_2 [5.79] and 79 in I_2 [5.91] have recently been observed;

6) Various observation techniques have been applied, such as the two-photon resonance fluorescence [5.79] or synchronus detection [5.93] with a CW single-frequency scanning laser, and quantum beats of fluorescence with a relatively broad-band short-pulse laser [5.91]. The latter permitted the determination of the rotational Landé factors of the excited level rather than the difference of the Landé factors between the ground and excited levels.

7) Some new phenomena have been found, such as the hyperfine-component dependence on the influence of an external magnetic field [5.81], and the anomalous Zeeman effect on the asymmetry of a nonlinear resonance of the hyperfine components in methane due to the influence of recoil [5.93].

8) Applications of spectroscopic effects caused by external fields have widely been extended. The significance of precise studies of these phenomena lies not only in obtaining accurate molecular parameters in various states, but also in confirming the molecular lines assignments and in identifying new molecules. In terms of practical uses, the Stark or Zeeman tuning or modulation of a molecular transition to resonantly interact with the fixed frequency of a gas laser, the study of the molecular relaxation processes via transient phenomena, and others, should be noted. Some successful examples can be found in a recent publication [5.138].

5.5.5 General Features of the Applications of Doppler-Free Spectroscopy to the Study of Molecular Spectra and Molecular Structure

Many of the applications listed in Table 5.9 have not been elaborated on here. The previous section delineated the greatly widened field of molecular spectroscopy by high-resolution laser spectroscopy, based on the nonlinear interaction of light with molecules under the presupposition of negligible reemission radiation. In its short history to date the technique has shown the following general features:

1) A tendency to use combinations of the fundamental Doppler-free spectroscopic techniques as they were developed, leading incidentally to rather complex names for various recent successes (see references listed in Tables 5.1, 4, 6);

2) A tendency to combine individual Doppler-free spectroscopic techniques with various supersensitive methods, including direct and indirect detection schemes. The direct detection schemes aim to pick up the spectroscopic signals by directly measuring the intensity or frequency changing of a transmitted probe beam by using a photodetector. The so-called indirect detection scheme, on the other hand, aims at examining one of the induced phenomena caused by resonant absorption, such as light emission (laser-induced fluorescence) from the final state of one- or two-photon absorption [5.17, 46], the induced-current effect (optogalvanic detection involving the use of electrodes inside the sample cell [5.139] or electrodeless radio-frequency detection outside the sample cell [5.140]), the induced ionizing effect (multiphoton ionization spectroscopy via unequal- or equal-frequency multiphoton absorptions) [5.141-143], the induced mechanical pressure effect (photoacoustic effect) [5.144-146], and so on. The appropriate technique depends on the excitation wavelength, the time distribution of the laser output, the interesting upper level of the excitation transition, etc. These methods allow one to further enhance the detection sensitivity, for example by using an intracavity sample cell for resonant or nonresonant fluorescence detection [5.147] or direct absorption determination [5.148].

3) A tendency to combine nonlinear high-resolution spectroscopy with other techniques, such as with supersonic molecular-beam techniques to prepare the sample either outside [5.62, 149-151] or inside a resonant cavity [5.152], with Fourier spectrometry for the analysis of hundreds of spectral lines to determine excited-state constants including the Coriolis interactions [5.41], with mass spectrometry for identifying the fragments produced by multiphoton ionization or isomers [5.153], with computer fitting of experimental spectral data for accurate determination of various molecular constants and parameters [5.42], etc..

4) Almost all of the Doppler-free spectroscopic ideas were first demonstrated by experiments in atomic systems and applied in quick succession to studies of diatomic, small polyatomic and large polyatomic molecules. Many examples of diatomic or small molecules have already been discussed. A tyical Doppler-free saturation spectrum of the ^qQ branch of the O-O band in the $\tilde{A}(^1B_{3u}) \leftarrow \tilde{X}(^1A_g)$ system of the large polyatomic

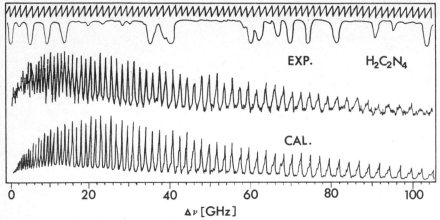

Fig.5.24. Doppler-free saturation spectrum recorded for revealing the origin of the Q-branch structure of a large polyatomic molecule (in this case S-tetrazine) [5.42]

molecule s-tetrazine ($H_2C_2N_4$) is shown in Fig.5.24 (middle trace). It shows the possibility to make Fortrat analysis of the Q branch starting from the rotationless origin of the band to obtain precise values of rotational constants for both ground and excited electronic states, and to measure the linewidth dependence on quantum states and thus analyze complicated physical or chemical processes in such complex molecules [5.42].

6. Molecular Nonlinear Coupling Spectral Effects

Once the laser-induced melecular polarization is appreciable, the molecular re-emission field is no longer negligible. In turn, it couples to the incident laser field to generate new stimulated or coherent radiation from the molecule or molecule-atom ensembles, or to cause different coherent transient spectral effects closely linked to various molecular parameters and relaxations. In this chapter we unify the theoretical treatments for different kinds of mechanisms dealing with various nonlinear coupling interactions of molecules with one or multiple laser fields. Numerous experimental results and applications are introduced as well.

6.1 Background

The nonlinear coupled interactions between molecules and a laser field can generate stimulated and coherent radiation from a molecular or molecular-atomic assembly.

By means of optical pumping of molecules both the population effect and the coherent effect can result in reemission of radiation which may, in turn, interact with the medium to generate stimulated radiation based on the transition between molecular electronic states, vibration-rotational levels or pure rotational levels as well as laser radiation through photodissociation. The reemitted field can, in turn, interact with the incident laser field nonlinearly (e.g., four-wave mixing) to generate new coherent radiation. The generation of tunable ultraviolet coherent radiation in a molecular-atomic system is an example of this. In addition to optical pumping, the energy transfer processes by collisions between atoms and molecules can generate rich stimulated emission lines.

It is known that the polarization of a medium, $\mathbf{P}$, can be written as a sum of linear and nonlinear parts, namely, adopting the convention introduced by *Bloembergen* [6.1],

$$\mathbf{P} = \mathbf{P}_{\text{linear}} + \mathbf{P}_{\text{nonlinear}} = \mathbf{P}^{(1)} + \mathbf{P}^{(2)} + \mathbf{P}^{(3)} + \ldots$$
$$= \chi^{(1)} \mathbf{E} + \chi^{(2)} \mathbf{E}\,\mathbf{E} + \chi^{(3)} \mathbf{E}\,\mathbf{E}\,\mathbf{E} + \ldots \ . \tag{6.1}$$

The nth-order nonlinear polarization $\mathbf{P}^{(n)}$ can be expressed as

$$\mathbf{P}^{(n)} = \text{Tr}\{\mu\rho^{(n)}\} = \sum_{ij} \mu_{ij}\rho_{ji}^{(n)} \ , \tag{6.2}$$

where μ_{ij} is the matrix element of the electric dipole moment of the transition from the levels i $\rightarrow$ j. By means of successive perturbation solutions for the density matrix equation with the electric dipole approximation, one can obtain solutions for the elements of each order, $\rho^{(n)}$. Substituting them into (6.2), one obtains expressions for each order of polarization. One can then obtain the expression for the field of the emitted radiation $E(\omega, r)$ by solving the following coupled-wave equation:

$$\nabla \times \nabla \times E(\omega, \mathbf{r}) - \omega^2 \epsilon(\omega) E(\omega, \mathbf{r})/c^2 = 4\pi\omega^2 P_{nonlinear}(\omega, \mathbf{r})/c^2 . \tag{6.3}$$

The monochromatic, linearly polarized plane wave propagating along the z axis, $\mathbf{E}(\omega, \mathbf{r})$, can be described by

$$\mathbf{E}(\omega, \mathbf{r}) = \hat{e} E(\omega, z) \exp(ikz) , \tag{6.4}$$

where $\hat{e}$ is the unit vector in the polarization direction, and $E(\omega, z)$ is the scalar amplitude function. Neglecting the second-order derivative, (6.3) can be simplified to

$$\partial E(\omega, z)/\partial z = i4\pi^2 \hat{e} P_{nonlinear}(\omega, z) \exp(-ikz)/\lambda |\epsilon(\omega)|^{1/2} . \tag{6.5}$$

For the various processes mentioned above involving either molecular excitation or atomic excitation, the one-photon pumping case can usually be represented by a three-level system, whereas in the two-photon pumping case we are dealing with a four-level system. For the excitation of a three-level system, the Optically Pumped Stimulated Emission (OPSE) and Stimulated Electronic Raman Scattering (SERS) generated correspond to the third-order nonlinear polarization. For the excitation of a four-level system the Stimulated Hyper-Raman Scattering (SHRS) generated corresponds to the fifth-order nonlinear polarization, whereas the four-wave mixing process can be looked at as being formed by the third-order polarization. All of these are to be individually considered the steady-state solution of the processes. In the case of short-pulse excitation for a two-level system with a relaxation time longer than the pulse duration, nonlinar coupled phenomena are to be seen as transient coherent phenomena, which have been applied to measure molecular relaxation parameters and other molecular constants.

6.2 Nonlinear Coupled Interaction in Three-Level Systems

Consider a three-level system, as shown in Fig.6.1. Energy level a can be either a vibration-rotational level in a low-lying (e.g., ground) molecular electronic state or a low-energy atomic level. After laser pumping at a suitable frequency ω_p, one of the vibration-rotational energy levels in a higher molecular electronic state (or one of the higher atomic energy levels) b can be sufficiently populated to result in so-called optically pumped stimulated

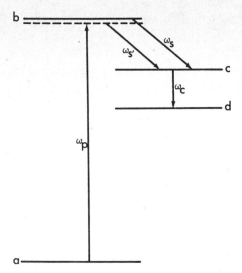

Fig.6.1. Three-level system

emission with frequency ω_s from level b to c. Detuning of the pump frequency from the resonant transition frequency between energy level a and b will result in stimulated electronic Raman scattering with frequency $\omega_{s'}$. The frequency ω_c in Fig.6.1 indicates the cascade-stimulated emission following OPSE and SERS.

If one neglects the initial thermal populations on levels b and c, the generation of OPSE and SERS as just described are third-order nonlinear coupled effects. According to (6.2), the polarization component $P_{bc}^{(3)}$, with frequency ω_s or $\omega_{s'}$, is given by

$$P_{bci}^{(3)} = 2\rho_{bci}^{(3)}\mu_{cb} , \quad i = s \text{ or } s' , \tag{6.6}$$

where i can be s and s', corresponding to OPSE and SERS processes, respectively.

To obtain the small-signal gain, it is easy to find the solution of the density matrix by successive perturbations. Hence the nonlinear interaction processes of the medium with the pump field E_p and initial stimulated radiation field E_s^0 or the Raman scattering field $E_{s'}^0$ can be written as

$$\rho_{aa}^{(0)} \xrightarrow{E_p} \rho_{ab}^{(1)} \xrightarrow{E_p} \rho_{bb}^{(2)} \xrightarrow{E_i^0} \rho_{bci}^{(3)'} , \tag{6.7}$$

$$\rho_{aa}^{(0)} \xrightarrow{E_p} \rho_{ab}^{(1)} \xrightarrow{E_i^0} \rho_{aci}^{(2)} \xrightarrow{E_p} \rho_{bci}^{(3)''} , \tag{6.8}$$

where (6.7) describes the population effect of the optical pumping, and (6.8) the coherent sum-frequency excitation effect. The total third-order matrix element $\rho_{bc}^{(3)}$ is the summation of every contribution, i.e.,

$$\rho_{bc}^{(3)} = \rho_{bc}^{(3)'} + p_{bc}^{(3)''}$$

$$= -\frac{|V_p|^2 V_i^0}{\Delta_{ab}\Delta_{ab}^*\Delta_{bci}}\rho_{aa}^{(0)} + \frac{|V_p|^2 V_i^0}{\Delta_{ab}\Delta_{aci}\Delta_{bci}}\rho_{aa}^{(0)} , \qquad (6.9)$$

where

$$\begin{aligned}
\Delta_{ab} &= (\omega_{ab} - \omega_p) + i\Gamma_{ab} , \\
\Delta_{aci} &= (\omega_{ac} - \omega_p + \omega_i) + i\Gamma_{ac} , \\
\Delta_{bci} &= (\omega_{bc} - \omega_i) + i\Gamma_{bc} , \\
V_p &= \frac{\mu_{ab} E_p}{2\hbar} , \\
V_i^0 &= \frac{\mu_{bc} E_i^0}{2\hbar} .
\end{aligned} \qquad (6.10)$$

Assume T_j and T_k to be the lifetime of level j and k, respectively. The linewidth of the transition between j and k is then $\Gamma_{jk} = \frac{1}{2}(1/T_j + 1/T_k)$. For the case where a refers to the ground state, we have $\Gamma_{ia}T_i = \frac{1}{2}$. Equation (6.9) can then be simplified to

$$\rho_{bc}^{(3)} = \frac{|V_p|^2 V_i^0}{\Delta_{ab}\Delta_{ab}^*\Delta_{aci}}\rho_{aa}^{(0)} . \qquad (6.11)$$

Substituting (6.11) into (6.6), we obtain the expression for the third-order polarization at frequency ω_i

$$P^{(3)}(\omega_i) = 2\frac{\mu_{cb}}{\Delta_{ab}\Delta_{ab}^*\Delta_{aci}}|V_p|^2 V_i^0\rho_{aa}^{(0)} . \qquad (6.12)$$

The radiated field $E_i(\omega_i, z)$ with frequency ω_i can be derived by replacing P with $P^{(3)}$ for the coupled-wave equation (6.5) as follows:

$$\frac{\partial E_i(\omega_i, z)}{\partial z} = i\frac{4\pi^2\omega_i}{cn_i}P^{(3)}(\omega_i, z)\exp(-ik_i z) . \qquad (6.13)$$

Substituting (6.12) into (6.13), we obtain

$$E_i = E_i^0\exp(g_i L) , \qquad (6.14)$$

$$I_i = I_i^0\exp(g_i' L) , \qquad (6.15)$$

where L is the length of the medium, and g_i is the gain coefficient for the radiated field with frequency ω_i. The magnitude of g_i is given by

$$g_i = \frac{2\pi\omega_i}{c(\epsilon_i)^{1/2}} \text{Im}\{P^{(3)}(\omega_i)/E(\omega_i)\}$$

$$= \frac{4\pi\omega_i}{c^2(\epsilon_i\epsilon_p)^{1/2}} \frac{\Gamma_{ac} I_p}{\hbar^3 \Delta_{ab}^2 \Delta_{ac}^2}(\mu_{ab}\,\mu_{bc}\,\mu_{cb}\,\mu_{ba})\,\rho_{aa}^{(0)}. \tag{6.16}$$

For the OPSE process, $\omega_{bc} = \omega_s$. Therefore the gain coefficient is

$$g_s = \frac{4\pi\omega_s}{c^2(\epsilon_s\epsilon_p)^{1/2}} \frac{\Gamma_{ac} I_p}{\hbar^3[(\omega_{ab}-\omega_p)^2 + \Gamma_{ac}^2]} \frac{\mu_{ab}\,\mu_{bc}\,\mu_{cb}\,\mu_{ba}}{(\omega_{ab}-\omega_p)^2 + \Gamma_{ab}^2}\rho_{aa}^{(0)}. \tag{6.17}$$

Since for SERS $\omega_{ac} = \omega_p - \omega_s$, the gain coefficient for this case becomes

$$g_{s'} = \frac{4\pi\omega_{s'}}{c^2(\epsilon_s'\epsilon_p)^{1/2}} \frac{I_p}{\hbar^3 \Gamma_{ac}} \frac{\mu_{ab}\,\mu_{bc}\,\mu_{cb}\,\mu_{ba}}{(\omega_{ab}-\omega_p)^2 + \Gamma_{ab}^2}\rho_{aa}^{(0)}. \tag{6.18}$$

The interaction channels in (6.7 and 8) show that these processes would populate the final level c. From the latter a cascade stimulated radiation to a lower level d may occur, resulting in the second generation field E_c. This can, in turn, take part in another nonlinear interaction process, such as four-wave mixing, to produce a coherent radiation field with frequency $\omega_{vi} = \omega_p - \omega_i - \omega_c$:

$$\rho_{aa}^{(0)} \xrightarrow{E_p} \rho_{ab}^{(1)} \xrightarrow{E_i^0} \rho_{aci}^{(2)} \xrightarrow{E_c} \rho_{adi}^{(3)}. \tag{6.19}$$

The relevant third-order polarization can be expressed as

$$P^{(3)}(\omega_{vi}) = \frac{2\mu_{da}}{\Delta_{ab}\Delta_{aci}\Delta_{adi}} V_p V_i^{0*} V_c^* \rho_{aa}^{(0)}. \tag{6.20}$$

6.3 Nonlinear Coupled Interaction in Four-Level Systems

For a four-level system, as depicted in Fig.6.2, the energy levels a, b, c and d can again be molecular vibration-rotational levels or atomic levels. ω_{p1} and ω_{p2} are the output frequencies from two lasers which serve as the unequal-frequency two-step or two-photon pump sources. ω_s is the OPSE frequency, $\omega_{s'}$ and $\omega_{s''}$ are Stimulated Hyper-Raman Scattering (SHRS) frequencies. Neglecting the initial population on levels b, c, d and e, OPSE and SHRS should be generated by fifth-order nonlinear effects. These can be expressed as the nonlinear coherent interaction processes in the medium among pumping fields E_{p1}, E_{p2} and the initial stimulated emission field E_s^0 or the initial SHRS fields $E_{s'}^0$ and $E_{s''}^0$ as follows:

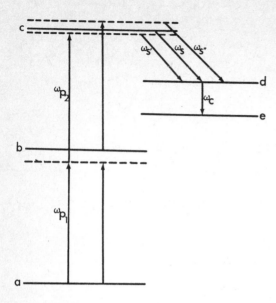

Fig.6.2. Four-level system

$$\rho_{aa}^{(0)} \xrightarrow{E_{p1}} \rho_{ab}^{(1)} \xrightarrow{E_{p1}} \rho_{bb}^{(2)} \xrightarrow{E_{p2}} \rho_{bc}^{(3)} \xrightarrow{E_{p2}} \rho_{cc}^{(4)} \xrightarrow{E_i} \rho_{cdi}^{(5)'} , \tag{6.21}$$

$$\rho_{aa}^{(0)} \xrightarrow{E_{p1}} \rho_{ab}^{(1)} \xrightarrow{E_{p1}} \rho_{bb}^{(2)} \xrightarrow{E_{p2}} \rho_{bc}^{(3)} \xrightarrow{E_i} \rho_{bdi}^{(4)} \xrightarrow{E_{p2}} \rho_{cdi}^{(5)''} , \tag{6.22}$$

$$\rho_{aa}^{(0)} \xrightarrow{E_{p1}} \rho_{ab}^{(1)} \xrightarrow{E_{p2}} \rho_{ac}^{(2)} \xrightarrow{E_{p1}} \rho_{bc}^{(3)} \xrightarrow{E_{p2}} \rho_{cc}^{(4)} \xrightarrow{E_i} \rho_{cdi}^{(5)'''} , \tag{6.23}$$

$$\rho_{aa}^{(0)} \xrightarrow{E_{p1}} \rho_{ab}^{(1)} \xrightarrow{E_{p2}} \rho_{ac}^{(2)} \xrightarrow{E_i} \rho_{adi}^{(3)} \xrightarrow{E_{p1}} \rho_{bdi}^{(4)} \xrightarrow{E_{p2}} \rho_{cdi}^{(5)''''} , \tag{6.24}$$

$$\rho_{aa}^{(0)} \xrightarrow{E_{p1}} \rho_{ab}^{(1)} \xrightarrow{E_{p2}} \rho_{ac}^{(2)} \xrightarrow{E_{p1}} \rho_{bc}^{(3)} \xrightarrow{E_i} \rho_{bdi}^{(4)} \xrightarrow{E_{p2}} \rho_{cdi}^{(5)'''''} , \tag{6.25}$$

where i can be either s, s′ or s″. Equation (6.21) corresponds to a step-by-step process, (6.22) represents a two-photon process based on the intermediate level b being populated, (6.23) the stimulated emission process following the population of the upper level c of two-photon excitation, and (6.24 and 25) the multiphoton interaction processes.

By solving the density matrix equation of motion an expression can be found for the total value of $\rho_{cd}^{(5)}$:

$$\rho_{cd}^{(5)} = \rho_{cd}^{(5)'} + \rho_{cd}^{(5)''} + \rho_{cd}^{(5)'''} + \rho_{cd}^{(5)''''} + \rho_{cd}^{(5)'''''}$$

$$= \left(\frac{1}{\Delta_{ab} \Delta_{ab}^* \Delta_{bc} \Delta_{bc}^* \Delta_{cd}} - \frac{1}{\Delta_{ab} \Delta_{ab}^* \Delta_{bc} \Delta_{bd} \Delta_{cd}} \right.$$

$$- \frac{1}{\Delta_{ab}\Delta_{ac}\Delta_{bc}\Delta_{bc}^*\Delta_{cd}} - \frac{1}{\Delta_{ab}\Delta_{ac}\Delta_{ad}\Delta_{bd}\Delta_{cd}}$$

$$+ \left.\frac{1}{\Delta_{ab}\Delta_{ac}\Delta_{bc}\Delta_{bd}\Delta_{cd}}\right) |V_{p1}|^2 |V_{p2}|^2 V_i^0 \rho_{aa}^{(0)} , \tag{6.26}$$

where

$$
\begin{aligned}
\Delta_{ab} &= (\omega_{ab} - \omega_{p1}) + i\Gamma_{ab} , \\
\Delta_{bc} &= (\omega_{bc} - \omega_{p2}) + i\Gamma_{bc} , \\
\Delta_{ac} &= (\omega_{ac} - \omega_{p1} - \omega_{p2}) + i\Gamma_{ac} , \\
\Delta_{adi} &= (\omega_{ad} - \omega_{p1} - \omega_{p2} + \omega_i) + i\Gamma_{ad} , \\
\Delta_{cdi} &= (\omega_{cd} - \omega_i) + i\Gamma_{cd} , \\
\Delta_{bdi} &= (\omega_{bd} - \omega_{p2} + \omega_i) + i\Gamma_{bd} , \\
V_{p1} &= \mu_{ab} E_{p1}/2\hbar , \\
V_{p2} &= \mu_{bc} E_{p2}/2\hbar , \\
V_i^0 &= \mu_{cd} E_i^0/2\hbar .
\end{aligned}
\tag{6.27}
$$

For (6.6 and 26) the fifth-order polarization with respect to frequency ω_i is in the form

$$P^{(5)}(\omega_i) = 2\mu_{dc}\rho_{cd}^{(5)} = \frac{2\mu_{dc}}{\Delta_{ab}\Delta_{ab}^*\Delta_{ac}\Delta_{ac}^*\Delta_{adi}}|V_{p1}|^2|V_{p2}|^2 V_i^0\rho_{aa}^{(0)} . \tag{6.28}$$

This acts as a reemitting source for the generation of a field at freqeuncy ω_i via the coupled wave equation (6.5):

$$\frac{\partial E_i(\omega_i, z)}{\partial z} = i \frac{4\pi^2 \omega_i}{c(\epsilon_i)^{1/2}} P^{(5)}(\omega_i, z) \exp(-ik_i z) . \tag{6.29}$$

Therefore the intensity of the emitted signal is

$$I_i(\omega_i, z) = I_i^0 \exp(g_i' z) , \tag{6.30}$$

where the gain coefficient g_i' is

$$g_i' = \frac{8\pi\omega_i}{c^3(\epsilon_i\epsilon_{p1}\epsilon_{p2})^{1/2}} \frac{3\Gamma_{ad}\rho_{aa}^{(0)}}{\hbar^5 |\Delta_{ab}|^2 |\Delta_{ac}|^2 |\Delta_{adi}|^2}(\mu_{ab}\mu_{bc}\mu_{cd}\mu_{dc}\mu_{cb}\mu_{ba})I_{p1} I_{p2} . \tag{6.31}$$

As mentioned above, the interaction processes (6.21 to 25) will populate the level d, which can create the conditions for the generation of the cascade-stimulated emission field E_c from level d to e. The second field corresponds to the seventh-order nonlinear polarization in the form

· 167

$$P^{(7)}(\omega_c, z) = 2\mu_{ed}\rho_{de}^{(7)} \tag{6.32}$$

$$= \frac{2|\mu_{ab}|^2|\mu_{bc}|^2|\mu_{cd}|^2|\mu_{de}|^2}{\hbar^7 \Delta_{ab}\Delta_{ab}^*\Delta_{ac}\Delta_{ac}^*\Delta_{adi}\Delta_{adi}^*\Delta_{ae}}\rho_{aa}^{(0)}|E_{p1}|^2|E_{p2}|^2|E_i^0|^2 E_c^0 .$$

As the reemitted field the OPSE or SHRS may also be nonlinearly coupled to the pump fields E_{p1} and E_{p2}, i.e., four-wave mixing, to generate coherent radiation with frequency ω_{uvi} through the following interaction

$$\rho_{aa}^{(0)} \xrightarrow{E_{p1}} \rho_{ab}^{(1)} \xrightarrow{E_{p2}} \rho_{ac}^{(2)} \xrightarrow{E_i} \rho_{adi}^{(3)} . \tag{6.33}$$

The frequency ω_{uvi} of the coherent radiation is given by

$$\omega_{uvi} = \omega_{p1} + \omega_{p2} - \omega_i . \tag{6.34}$$

The third-order polarization $P^{(3)}(\omega_{uvi})$ for generating such coherent radiation can be written

$$P^{(3)}(\omega_{uvi}) = 2\mu_{da}\rho_{ad}^{(3)} = \frac{2\mu_{da}}{\Delta_{ab}\Delta_{ac}\Delta_{ad}}V_{p1}V_{p2}V_i\rho_{aa}^{(0)} . \tag{6.35}$$

The field with frequency ω_{uvi} satisfies the following coupled-wave equation

$$\frac{\partial E_i(\omega_{uvi}, z)}{\partial z} = \frac{i4\pi^2\omega_i}{c(\epsilon_{uvi})^{1/2}}P^{(3)}(\omega_{uvi}, z)\exp(-ik_{uvi}z) . \tag{6.36}$$

This means that the field will be of the form

$$E_i(\omega_{uvi}, L) = \frac{2i\omega_{uvi}}{2\pi c(\epsilon_{uvi})^{1/2}}\frac{\mu_{ab}\mu_{bc}\mu_{cd}\mu_{da}}{\hbar^3\Delta_{ab}\Delta_{ac}\Delta_{adi}}\rho_{aa}^{(0)}E_{p1}E_{p2}E_i^0\exp(g_iL)/g_i . \tag{6.37}$$

Furthermore the cascade-stimuleted radiation can also take part in a nonlinear interaction with pump fields E_{p1} and E_{p2} to generate coherent radiation as described by

$$\rho_{aa}^{(0)} \xrightarrow{E_{p1}} \rho_{ab}^{(1)} \xrightarrow{E_{p2}} \rho_{ad}^{(2)} \xrightarrow{E_c} \rho_{ae}^{(3)} . \tag{6.38}$$

The frequency of the coherent radiation is

$$\omega_{uvi} = \omega_{p1} + \omega_{p2} - \omega_c . \tag{6.39}$$

It is thus clear that through nonlinear coupled interactions new forms of optically pumped stimulated radiation and coherent radiation can be generated in three- or four-level systems in the appropriate media. In the following sections we will see the results obtained under various experimental conditions, such as the hybrid four-level system through collisional energy transfer from atoms to molecules (Sect.6.4), the hybrid four-level system through collisional energy transfer from molecules to atoms (Sect. 6.5), the four-level system involving molecular dissociation (Sect.6.6), and the typical molecular three- or four-level system (Sect.6.8,9). They arise from the aprpopriate pumping of atoms and molecules and have been a rich source of laser or coherent radiation over the wavelength regions of vacuum ultraviolet, ultraviolet, visible, infrared and far infrared with both tunable and fixed wavelengths.

6.4 Stimulated Diffuse Band Radiation via Various Excitation Processes

The generation of stimulated diffuse-band radiation is an interesting topic, as it might be utilized to realize a kind of tunable laser covering a limited wavelength region, especially in the short-wavelength region. Diffuse-band spontaneous radiation in molecular sodium has been known for some time. Violet diffuse band fluorescence can be observed either by one-photon excitation ($X^1\Sigma_g^+ \rightarrow C^1\Pi_u$) or two-photon excitation ($X^1\Sigma_g^+ \rightarrow A^1\Sigma_u^+ \rightarrow$ high-lying g state) in Na_2, or by one-photon excitation ($3S\rightarrow3P$) or two-photon excitation ($3S\rightarrow4D$) in Na atoms [6.2-7]. Their mechanisms have been discussed in the references quoted above. Recently, violet diffuse-band stimulated radiation in Na_2, generated by the pump laser tuned to near the two-photon resonance of the 4D state in the Na atom, was reported [6.8], and the gain of the violet-band emission in sodium vapor pumped by a krypton-ion laser was measured [6.9]. Stimulated ultraviolet and violet diffuse-band radiation was produced by excitation with a laser tuned to resonance with $3S\rightarrow3P$ or half of $3S\rightarrow4D$. The detected stimulated diffuse-band radiation was in the ultraviolet region from 340.0 to 380.0 nm and in the violet from 405.0 to 460.0 nm [6.10]. In these experiments, a YAG-laser-pumped dye-laser beam with an output energy of about 20mJ was focussed to the center of a heat-pipe oven. The duration and the linewidth of the dye laser were 12 ns and 0.01 nm, respectively. The oven, containing 10 g of sodium, was 70 cm long with a heated length of 22 cm and was controlled in temperature between 400° C and 650° C. The measured divergence angle of ultraviolet and violet radiation from the oven in the forward direction was 7 mrad. The radiation was introduced into a 0.5 m grating monochromator and monitored by means of a photomultiplier. Three filters with passbands between 300.0 and 500.0 nm were placed in front of the entrance slit of the monochromator in order to absorb the residual pump beam.

When the pump wavelength is tuned to 589.0 nm ($3S\rightarrow3P_{3/2}$), 589.6 nm ($3S\rightarrow 3P_{1/2}$), 578.7 nm (two-photon transition $3S\rightarrow4D$) or 578.0 nm (two-photon transition $3S\rightarrow4F$), ultraviolet diffuse band emission coaxial with the pump beam is detected between 340.0 and 380.0 nm and with an

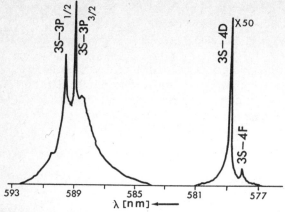

Fig.6.3. Excitation spectrum for generating the ultraviolet and violet diffuse-band stimulated radiation [6.10]

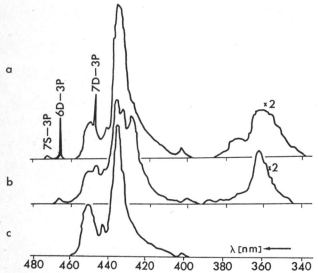

Fig.6.4. Ultraviolet and violet diffuse-band emission spectra generated by (a) one-photon excitation of 3S→3P and two-photon excitation of 3S→4D or 3S→4F in Na, (b) two-photon excitation of Na_2, and (c) one-photon excitation of Na_2 by 354.72 nm laser [6.10]

intensity of about one eighth of the violet diffuse-band emission between 405.0-460.0 nm, which is detected simultaneously. Figure 6.3 displays the excitation spectrum for generating the ultraviolet and violet band radiation. Clearly, the strongest signals correspond to the pump wavelength of 589.0 or 589.6 nm, the weaker signal corresponds to pumping at 578.7 nm, and the weakest one to pumping at 578.0 nm. The ultraviolet and violet diffuse-band spectra with laser pumping at 589.0, 589.6, 587.7 or 578.0 nm are exhibited in Fig.6.4a. Since no molecular vibrational spectral structure could be resolved by the grating spectrometer (resolution: 0.3nm), the two emission bands should not be emitted by the bound-bound transitions.

1) **One-photon excitation of atomic sodium (3S→3P)**: As pointed out in [6.1-7], the violet diffuse-band results of the transition from a high-lying triplet state (indicated by $(1)^3\Lambda_g$, here $\Lambda = \Sigma$ or Π) to the lowest triplet state $a^3\Sigma_u^+$. Stimulated ultraviolet diffuse-band radiation is assumed to result from a higher triplet state (indicated by $(2)^3\Lambda_g$) down to the ground triplet state $a^3\Sigma_u^+$.

Due to the resonant excitation of the 3P level, the high-lying triplet state $(1)^3\Lambda_g$ might be populated in two steps [6.5]. Firstly, atomic sodium in the 3P level transfers its energy to molecular sodium by collision, populating the molecular state $A^1\Sigma_u^+$ or $b^3\Pi_u$. Subsequent collision between the excited molecule and an atom in the 3P state might excite the molecule to the high-lying triplet state. It is interesting to note that the total energy for two 3P atoms is sufficient not only to populate the molecular state $(1)^3\Lambda_g$, but also $(2)^3\Lambda_g$ followed by the stimulated emission of radiation to the ground-state triplet $a^3\Sigma_u$. The above mentioned processes can be written as follows:

$$Na^*(3P) + Na_2(X^1\Sigma_g^+) \rightarrow Na(3S) + Na_2^*(A^1\Sigma_u^+, b^3\Pi_u) , \qquad (6.40)$$
$$Na^*(3P) + Na_2^*(A^1\Sigma_u^+, b^3\Pi_u) \rightarrow Na(3S) + Na_2^*[(1)^3\Lambda_g, (2)^3\Lambda_g] , \qquad (6.41)$$
$$Na_2^*[(2)^3\Lambda_g] \rightarrow Na_2(a^3\Sigma_u) + h\nu(340.0\text{-}380.0\text{nm}) , \qquad (6.42)$$
$$Na_2^*[(1)^3\Lambda_g] \rightarrow Na_2(a^3\Sigma_u) + h\nu(405.0\text{-}460.0\text{nm}) . \qquad (6.43)$$

Another possible process may be that the Na_2 in the $b^3\Pi_u$ state absorbs the 589.0 or 589.6 nm photon to populate the high-lying states directly.

There are several atomic stimulated emission lines in the violet region, corresponding to nD→3P and nS→3P, as shown in Fig.6.4a. They are strongest with on-resonance pumping. The atomic emission lines could be generated by energy pooling of high-lying molecular and atomic states, recombination of Na^++e^- or $Na_2^++e^-$, and so on [6.7]. Several atomic transitions are indicated in Fig.6.4a.

2) **Two-photon excitation (3S→4D)**: After the two-photon transition to the 4D level in Na vapour, the internal energy in the Na is enough to populate Na_2 to the $(2)^3\Lambda_g$ state by the following collisional process

$$Na^*(4D) + Na_2(X^1\Sigma_g^+) \rightarrow Na_2^*[(1)^3\Lambda_g, (2)^3\Lambda_g] + Na(3S) . \qquad (6.44)$$

In addition, ionization and subsequent recombination may be another way to populate the high-lying triplet states

$$Na^*(4D) + Na(3S) \rightarrow Na_2^+ + e^- , \qquad (6.45)$$
$$Na_2^+ + e^- \rightarrow Na_2^*[(1)^1\Lambda_g, (2)^3\Lambda_g] . \qquad (6.46)$$

3) **Molecular excitation from $X^1\Sigma_g^+$ to $C^1\Pi_u$**: By using the third harmonic (354.72nm) of a YAG laser as the pump beam, sodium dimers were excited from the ground state $X^1\Sigma_g^+$ to an energy level in $C^1\Pi_u$ at 30264 cm^{-1} [6.64]. The resulting stimulated violet diffuse-band radiation is

depicted in Fig.6.4c. The result is different from that stated above, as no atomic emission lines are observed in the wavelength region shown. Here the population in the high-lying state $(1)^3\Lambda_g$ results from the following excitation and transfer processes

$$Na_2(X^1\Sigma_g^+) + h\nu(354.72nm) \rightarrow Na_2^*(C^1\Pi_u) , \qquad (6.47)$$
$$Na_2^*(C^1\Pi_u) + Na(3S) \rightarrow Na_2^*[(1)^3\Lambda_g] + Na(3S) . \qquad (6.48)$$

We note that following the excitation of $X^1\Sigma_g^+ \rightarrow C^1\Pi_u$, only violet diffuse-band signals are detected. This is in contrast to the 3S→3P excitation, as shown in Fig.6.4a. This implies that the excited sodium dimer at about 30300 cm^{-1} does not have enough energy to efficiently populate the $(2)^3\Lambda_g$ state by the collisional transfer process.

4) **Two-photon excitation (3S→4F)**: As depicted in Fig.6.3, when the pump laser is tuned to 578.0 nm, corresponding to the 3S→4F two-photon transition, the stimulated ultraviolet and violet band signals are still detectable with intensities about two orders weaker than in the case of 3S→4D two-photon resonance. It seems that the 4F level can be populated via a 3S→4F dipole-forbidden two-photon transition. By collisional energy transfer between the molecules and atoms the molecules may populate the high-lying triplet states. The above-mentioned processes can be expressed as follows

$$Na(3S) + 2h\nu(\lambda = 578.0nm) \xrightarrow{\left[\begin{array}{c}\text{dipole-forbidden}\\\text{two-photon trans.}\end{array}\right]} Na^*(4F) , \qquad (6.49)$$

$$Na^*(4F) + Na_2(X^1\Sigma_g^+) \xrightarrow{\text{(collision)}} Na_2^*[(1)^3\Lambda_g,(2)^3\Lambda_g] . \qquad (6.50)$$

Recently the efficient population of the 4F level by a 3S→4F dipole-forbidden two-photon transition was demonstrated, and stimulated radiation corresponding to the 4D→4P transition via 3S→4F two-photon excitation and $Na^*(4F)$-$Na(3S)$ near-resonant collision transfer was observed [6.12].

5) **Two-photon excitation of molecular sodium**: When Na_2 molecules are pumped by a pulsed dye laser tuned between 640 and 650 nm with an output energy of about 30 mJ, one is able to detect stimulated-band signals, as shown in Fig.6.4b. The excitation spectra for generating the violet and ultraviolet diffuse bands are given in Figs.6.5a and b, respectively. Note that at some wavelengths signal peaks appear, but in the remaining region there are signals without apparent peaks. In addition, some of the signal peaks in Fig.6.5a correspond to peaks in Fig.6.5b (e.g., the peaks at λ_3, etc.), whereas some peaks (e.g., at λ_1, λ_2 and so on) do not. This demonstrates that there are several channels for populating the high-lying triplet states $(1)^3\Lambda_g$ or $(2)^3\Lambda_g$.

With λ_1 pumping for the first step, for example, the molecules are excited to the intermediate mixing levels, $A^1\Sigma_u^+(v'_{A1},J'_1) \sim b^3\Pi_u(v_{b1}',J_1')$, and are then excited to the $(1)^3\Lambda_g$ state by subsequent pumping. Owing to the

172

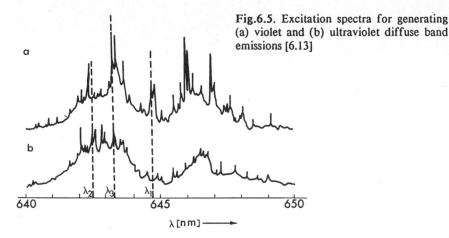

Fig.6.5. Excitation spectra for generating (a) violet and (b) ultraviolet diffuse band emissions [6.13]

lack of proper levels in both the intermediate mixing level (one-photon transition) and the upper level (two-photon transition) for the transitions λ_1, we may observe the peak signals only in the violet, and not in the ultraviolet excitation spectrum. These excitation-emission processes can be written as

$$Na_2(X^1\Sigma_g^+) + h\nu(\lambda_1) \rightarrow Na_2^*[A^1\Sigma_u^+(v_{A1}',J_1') \sim b^3\Pi_u(v_{b1}',J_1')] , \quad (6.51)$$
$$Na_2^*[A^1\Sigma_u^+(v_{A1}',J_1') \sim b^3\Pi_u(v_{b1}',J_1')] + h\nu(\lambda_1) \rightarrow Na_2^*[(1)^3\Lambda_g] , \quad (6.52)$$
$$Na_2^*[(1)^3\Lambda_g] \rightarrow Na(a^3\Sigma_u) + h\nu(405\text{-}460\,nm) . \quad (6.53)$$

Another set of processes follows when we assume that, as the first step, the pump beam at wavelength λ_2 excites the molecular sodium up to the mixing level $A^1\Sigma_u^+(v_{A2}',J_2') \sim b^3\Pi_u(v_{b2}',J_2')$ with higher vibration-rotational energy, so that the second step pumping may be effective in excitation to the $(2)^3\Lambda_g$ state. The wavelength λ_2, however, may not be so efficient for pumping the molecules to the $(1)^3\Lambda_g$ state. It should result in signal peaks appearing in the ultraviolet excitation spectrum rather than in the violet spectrum. The processes can be written as

$$Na_2(X^1\Sigma_g^+) + h\nu(\lambda_2) \rightarrow Na_2^*[A^1\Sigma_u^+(v_{A2}',J_2') \sim b^3\Pi_u(v_{b2}',J_2')] , \quad (6.54)$$
$$Na_2^*[A^1\Sigma_u^+(v_{A2}',J_2') \sim b^3\Pi_u(v_{b2}',J_2')] + h\nu(\lambda_2) \rightarrow Na_2^*[(2)^3\Lambda_g], \quad (6.55)$$
$$Na_2^*[(2)^3\Lambda_g] \rightarrow Na_2(a^3\Sigma_u) + h\nu(350\text{-}380\,nm) . \quad (6.56)$$

The wavelength λ_3 in Fig.6.5 is suitable for both channels of pumping to the $(1)^3\Lambda_g$ state via an intermediate mixing level $A^1\Sigma_u^+(v_{A3}',J_3') \sim b^3\Pi_u(v_{b3}',J_3')$ and to the $(2)^3\Lambda_g$ state via another mixing level $A^1\Sigma_u^+(v_{A4}',J_4') \sim (b^3\Pi_u(v_{b4}',J_4')$. This gives two corresponding peaks in the two excitation spectra.

Figure 6.5 also shows a weak diffuse band, which can be deduce to be caused by collisional effects. In this case the pump laser does not cause resonant excitation of the molecules to intermediate mixing levels, but to levels in the $A^1\Sigma_u^+$ state by the first-step pumping, and then to a high-lying singlet g state by the second-step pumping. Through the collision between

173

the excited molecules and other particles the excitation energy can be transfered from the singlet state to high-lying triplet states. There is also the possibility of populating the high-lying triplet states through energy transfer from $A^1\Sigma_u^+$ to $b^3\Pi_u$ and the second step excitation from the $b^3\Pi_u$ state.

6.5 Stimulated and Coherent Radiation by Hybrid Excitation in Molecule-Atom Ensembles

In this section we discuss several excitation-emission schemes in alkali molecule-atom ensembles. Using equal-frequency or unequal-frequency hybrid resonant excitation, and scanning equal-frequency hybrid excitation, a series of fixed-frequency and tunable infrared stimulated radiation lines are generated, respectively. In the last case, tunable ultraviolet coherent radiatin based on wave mixing processes is detected, too.

6.5.1 Equal-Frequency Two-Step Hybrid Resonance Pumping

When the potassium vapour contained in a heat-pipe oven is excited by a DCM dye laser pumped, for example, by a nitrogen laser, stimulated radiation at 2.71, 3.14 and 3.16 μm can be detected over the whole scanning range of the pumping laser from 625.0 to 695.0 nm (for a detailed discussion see the next section). Moreover, at pump wavelengths of 691.1 or 693.9 nm the directional signal at 3.66, 3.64, 3.16, 3.14 and 2.17 μm wavelengths rises remarkably [6.31].

Figure 6.6 depicts the excitation spectrum for generating the infrared signals. The dashed line in the figure is the output profile of the dye laser. The solid curve shows that, although the output energy of the dye laser at 691.1 and 693.9 nm is already quite low, the infrared signals detected are about four times greater than any at other wavelengths.

Obviously, the broad band in the recorded excitation spectrum in Fig.6.6 characterizes molecular excitation. It is interesting to note that the two discrete excitation lines providing the strongest signals are located at the wavelengths corresponding exactly to the atomic excitation from the first excited levels $4P_{1/2}$ and $4P_{3/2}$ to the high-lying level 6S, that the infrared laser lines at 3.66 and 3.64 μm correspond to the transitions from 6S to $5P_{3/2}$ and to $5P_{1/2}$ respectively, and that the other laser lines at 2.71, 3.14 and 3.16 μm are caused by the cascade-stimulated transitions of $5P_{3/2} \rightarrow 5S$, $5P_{3/2} \rightarrow 3D$ and $5P_{1/2} \rightarrow 3D$, respectively, as shown in Fig.6.7.

These results reveal that the energy transfer process between the excited molecular potassium K_2^* and atomic potassium K in the ground state is efficient enough to cause considerable population of the atomic 4P level. The collisional process which precedes further excitation of the atoms can be written as follows:

$$K_2^*\text{(excited state)} + K(4S) \xrightarrow{\text{(by collision)}} K_2\text{(ground state)} + K^*(4P) . \quad (6.57)$$

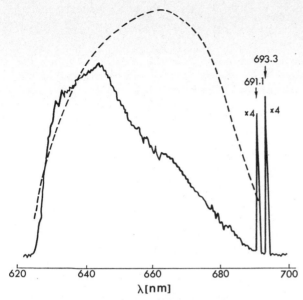

Fig.6.6. Excitation spectrum for generating the infrared laser lines in potassium vapour. The dashed line: output profile of the pump laser [6.31]

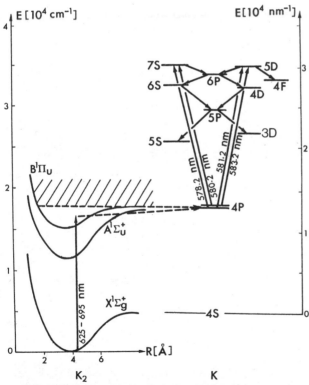

Fig.6.7. Mechanism of two-step K_2-K hybrid resonance laser [6.16]

175

Although the excitation energy of K_2^* on the left-hand side in (6.57) is more than the excitation energy required by $K^*(4P)$ on the right-hand side, the difference in energy summations of the collision partners between before and after the collision is less than the thermal motion energy $3KT/2$. Thus the collisional process is near-resonant and therefore efficient if there is a small change in the velocities concerned.

The whole equal-frequency two-step interaction process for generating the hybrid resonance laser can therefore be expressed as follows:

$$K_2(X^1\Sigma_g^+) + h\nu(691.1, 693.9\,\text{nm}) \rightarrow K_2^*(A^1\Sigma_u^+) \quad \text{or} \quad K_2^*(B^1\Pi_u),$$
$$K_2^*(A^1\Sigma_u^+) \quad \text{or} \quad K_2^*(B^1\Pi_u) + K(4S) \rightarrow K_2(X^1\Sigma_g^+) + K^*(4P_{1/2,3/2}),$$
$$K^*(4P_{1/2}) + h\nu(691.1\,\text{nm}) \rightarrow K^*(6S),$$
$$K^*(4P_{2/2}) + h\nu(693.9\,\text{nm}) \rightarrow K^*(6S),$$
$$K^*(6S) \rightarrow K^*(5P) + h\nu(3.64, 3.66\,\mu\text{m}),$$
$$K^*(5P) \rightarrow K^*(3D) + h\nu(3.14, 3.16\,\mu\text{m}),$$
$$K^*(5P) \rightarrow K^*(5S) + h\nu(2.71\,\mu\text{m}).$$

The same mechanism was found in the sodium molecule-atom system. CW laser oscillation on $4D\rightarrow4P$ $(2.34\,\mu\text{m})$, $4F\rightarrow3D$ $(1.85\,\mu\text{m})$ and $5S\rightarrow4P$ $(3.14\,\mu\text{m})$ transitions was obtained when the pump laser (a set of single-mode Rhodamine 6G dye ring laser) was tuned to the $3P\rightarrow4D$ $(\sim568\,\text{nm})$ and $3P\rightarrow5S$ $(\sim615\,\text{nm})$ resonance, respectively [6.14]. These pump wavelengths are within the allowed $X\rightarrow A$ transition band in molecular sodium. The main contribution for populating the 3P state of atomic sodium is the following collisional energy transfer process

$$Na_2^*(A^1\Sigma_u^+) + Na(3S) \rightarrow Na_2(X^1\Sigma_g^+) + Na^*(3P). \tag{6.58}$$

In the second excitation step the $Na^*(3P)$-atoms are then resonantly excited into the 4D or 5S state.

6.5.2 Unequal-Frequency Two-Step Hybrid Resonance Pumping

Clearly, the hybrid resonance laser can be set up under conditions of unequal-frequency two-step pumping, as has been demonstrated in potassium and sodium experiments, respectively.

In the potassium vapour, setting the pulsed dye laser *1* at any wavelength between 625.0 and 695.0 nm with the dye laser *2* operated at 578.2, 580.2, 581.2 or 583.2 nm, the two beams were coinciding and focussed to the center of the container, a 70 cm heat-pipe oven, without buffer gas. Stimulated radiation signals corresponding to the transitions $7S\rightarrow6P_{1/2,3/2}$ $(7.84, 7.89\,\mu\text{m})$, $5D\rightarrow6P_{1/2,3/2}$ $(8.45, 8.51\,\mu\text{m})$, $5D\rightarrow4F$ $(4.86\,\mu\text{m})$ and the cascade transitions $6P_{1/2,3/2}\rightarrow6S$ $(6.46, 6.43\,\mu\text{m})$, $6P_{1/2,3/2}\rightarrow4D$ $(6.24, 6.20\,\mu\text{m})$, $6S \rightarrow 5P_{1/2,3/2}$ $(3.64, 3.66\,\mu\text{m})$, $4D\rightarrow5P_{1/2,3/2}$ $(3.71, 3.73\,\mu\text{m})$, $5P\rightarrow5S$ $(2.71\,\mu\text{m})$ and $5P_{1/2,3/2}\rightarrow3D$ $(3.16, 3.14\,\mu\text{m})$ were generated, as shown in Fig.6.7 [6.15,16].

It should be noticed that the wavelength range of dye laser *1* (625.0-695.0nm) corresponds to the absorption band of $X^1\Sigma_g^+ \rightarrow B^1\Pi_u$ transitions

176

in molecular potassium, and the operating wavelengths of dye laser 2 to the resonant absorption of atomic potassium from the excited state $4P_{1/2,3/2}$ to the 5D or 7S state. So the populating processs for the intermediate level $4P_{1/2,3/2}$ may be considered to be the following pumping and energy transfer processes

$$K_2(X^1\Sigma_g^+) + h\nu(625.0-695.0\,nm) \rightarrow K_2^*(B^1\Pi_u),$$
$$K_2^*(B^1\Pi_u) + K(4S) \rightarrow K_2(X^1\Sigma_g^+) + K^*(4P_{1/2,3/2}).$$

Corresponding processes have been observed in sodium vapour [6.17]. In the first step, the molecule was excited from the ground state $X^1\Sigma_g^+$ to the $A^1\Sigma_u^+$ or $B^1\Pi_u$ state by a broad-band ($\sim$20.0nm) dye laser with an output energy of 3-30 μJ. Operating with 10 different dyes, its output covered the wavelength range 447.0-680.0 nm. The excited sodium molecules Na_2^* collide with the sodium atoms in the ground state and transfer their energy to them, thereby exciting them to the $3P_{1/2}$ and $3P_{3/2}$ states. A narrow-band ($\sim$0.01nm) dye laser cause the sodium atoms to take the second transitions to the 4D, 5D, 5S and 6S levels. Therefore the stimulated radiation lines from the 4D, 5D, 5S and 6S levels can be seen to follow the processes:

$$Na_2(X^1\Sigma_g^+) + h\nu(447.0-680.0\,nm) \rightarrow Na_2^*(A^1\Sigma_u^+, B^1\Pi_u),$$
$$Na_2^*(A^1\Sigma_u^+, B^1\Pi_u) + Na(3S) \rightarrow Na^*(3P_{1/2,3/2}) + Na_2(X^1\Sigma_g^+),$$
$$Na^*(3P_{1/2}) + h\nu \begin{bmatrix} 497.85,\ 514.88\,nm \\ 568.27,\ 615.42\,nm \end{bmatrix} \rightarrow Na^*(5D,\ 6S,\ 4D,\ 5S),$$
$$Na^*(3P_{3/2}) + h\nu \begin{bmatrix} 498.28,\ 515.34\,nm \\ 568.80,\ 616.07\,nm \end{bmatrix} \rightarrow Na^*(5D,\ 6S,\ 4D,\ 5S),$$
$$Na^*(5D) \rightarrow Na^*(5P) + h\nu(5.02\,\mu m),$$
$$Na^*(6S) \rightarrow Na^*(5P) + h\nu(7.52\,\mu m),$$
$$Na^*(4D) \rightarrow Na^*(4P) + h\nu(2.34\,\mu m),$$
$$Na^*(5S) \rightarrow Na^*(4P) + h\nu(3.42\,\mu m),$$
$$Na^*(4P) \rightarrow Na^*(4S) + h\nu(2.20\,\mu m).$$

6.5.3 Two–Step Hybrid Off-Resonance Pumping

The mechanism of equal-frequency two-step hybrid off-resonance pumping for generating tunable infrared stimulated radiation was demonstrated for the Na_2-Na system [6.18]. Here the pumping processes are similar to that described by (6.58) and its subsequent step. But now, despite the fact that the molecular excitation in the first step is on resonance for the X→A transition band, the pumping frequency ω_p is not in resonance with the atomic transition frequency from $3P_{1/2}$ or $3P_{3/2}$ to 5S. Thus, while the collisional energy transfer processes from the molecules to the atoms are still efficient, the second step involves a virtual upper level for both $3P_{1/2}$ or $3P_{3/2}$. This results in two stimulated emission components ω_s' and ω_s'' by SERS processes corresponding to $3P_{1/2}$-4P and $3P_{3/2}$-4P scattering, with

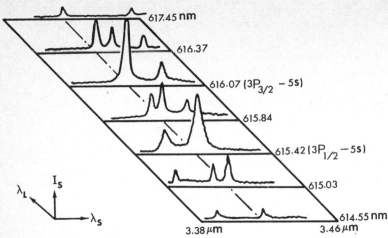

Fig.6.8. Tunable and fixed-frequency stimulated emissions based on two-step hybrid off-resonance pumping [6.18]

offset values of Δ_1 and Δ_2 for the $3P_{1/2} \rightarrow 5S$ and $3P_{3/2} \rightarrow 5S$ transitions, respectively. With reasonable offsets the optically pumped stimulated emission component ω_s corresponding to the $5S \rightarrow 4P$ transition exists, too. The frequencies of these components satisfy

$$\omega_s = \omega(5S, 4P) \ ,$$
$$\omega'_s = \omega(5S, 4P) - \Delta_1 \ ,$$
$$\omega''_s = \omega(5S, 4P) - \Delta_2 \ .$$

These equations suggest that tunable SERS can be generated by scanning the pumping laser through a certain wavelength range.

Figure 6.8 displays some of the results obtained by these mechanisms. The tunable stimulated radiation signals in the region 3.38-3.46 μm are measured by using a YAG-laser-pumped dye laser with an output of 20 mJ, scanning from 614.55 to 617.45 nm, and a PbS detector behind an infrared monochromator. Of course, one observes only two stimulated emission components ω_s and ω'_s or ω''_s for resonance excitation $3P_{1/2} \rightarrow 5S$ or $3P_{3/2} \rightarrow 5S$, respectively, since the OPSE signal is in this case degenerate with one of the SERS components. Three components ω_s, ω'_s and ω''_s are seen for the case of near-resonance excitation, while again two components are detected if one of the offsets is too large, as shown in Fig.6.8.

In Table 6.1 the measured and calculated wavelengths and relative intensity ratios for three infrared stimulated-radiation components in seven typical pumping cases with different offsets are listed.

With respect to the SERS processes only, it is reasonable to calculate the relevant of the above-mentioned mechanisms, disregarding details of how the $3P_{1/2}$ and $3P_{3/2}$ are populated. Taking $3P_{1/2}$ and $3P_{3/2}$ as the initial levels, $5S$ as the intermediate level, and $4P_{1/2}$ and $4P_{3/2}$ as the final levels, we have

178

Table 6.1 Seven typical cases for generating tunable and fixed-frequency IR stimulated radiation signals by two-step hybrid excitation in the Na_2-Na system [6.18]

Pump wavelength [μm]	offset [cm^1]	IR stimulated radiation (exp. value) [μm]	IR stimulated radiation (cal. value) [μm]	$I_s:I_{s'}:I_{s''}$ (exp. value)	$I_s:I_{s'}:I_{s''}$ (cal. value)
617.45	$\Delta_1=36.1$ $\Delta_2=53.3$	$\lambda_s=3.4155$ $\lambda_{s'}=3.4573$	$\lambda_s=3.4151$ $\lambda_{s'}=3.4577$	1:1:0.8	1.01:0.75:0.50
616.37	$\Delta_1=7.8$ $\Delta_2=25.0$	$\lambda_s=3.4155$ $\lambda_{s'}=3.4246$ $\lambda_{s''}=3.4450$	$\lambda_s=3.4151$ $\lambda_{s'}=3.4242$ $\lambda_{s''}=3.4445$	1.8:1.7:1.0	1.24:1.10:0.95
616.07	$\Delta_1=0$ $\Delta_2=17.2$	$\lambda_s=3.4155$ $\lambda_{s''}=3.4358$	$\lambda_s=3.4151$ $\lambda_{s''}=3.4353$	8.6:8.6:1.8	8.59:8.59:1.20
616.84	$\Delta_1=-6.3$ $\Delta_2=10.9$	$\lambda_s=3.4155$ $\lambda_{s'}=3.4080$ $\lambda_{s''}=3.4275$	$\lambda_s=3.4151$ $\lambda_{s'}=3.4077$ $\lambda_{s''}=3.4279$	1.9:1.6:1.3	1.70:1.20:1.02
616.42	$\Delta_1=-17.2$ $\Delta_2=0$	$\lambda_s=3.4155$ $\lambda_{s'}=3.3959$	$\lambda_s=3.4151$ $\lambda_{s''}=3.3951$	3.1:1.2:3.1	3.05:1.05:3.05
615.03	$\Delta_1=-27.6$ $\Delta_2=-10.4$	$\lambda_s=3.4155$ $\lambda_{s'}=3.3828$ $\lambda_{s''}=3.4037$	$\lambda_s=3.4151$ $\lambda_{s'}=3.3822$ $\lambda_{s''}=3.4030$	1.5:0.7:1.0	1.07:1.02:1.06
614.55	$\Delta_1=-40.4$ $\Delta_2=-23.2$	$\lambda_s=3.4155$ $\lambda_{s''}=3.3886$	$_s=3.4151$ $_{s''}=3.3882$	1.0: 0.6	0.99:0.55:0.75

$$\chi^{(3)}(-\omega_s',\omega_p-\omega_p\,\omega_s) = \chi^{(3)}_{3P_{1/2}} + \chi^{(3)}_{3P_{3/2}} \; , \tag{6.59}$$

$$\chi^{(3)}_{3P_{3/2}} = - Ne^4\rho^{(0)}_{3P_{3/2}} \, |\langle 4P_{1/2}|r|5S\rangle\langle 5S|r|3P_{3/2}\rangle|^2$$

$$+ \frac{|\langle 4P_{3/2}|r|5S\rangle\langle 5S|r|3P_{3/2}\rangle|^2}{6\hbar^3\epsilon_0\Gamma_{4P3P}\,|\Omega_{5S3P_{3/2}} - \omega_p|^2} \; , \tag{6.60}$$

$$\chi^{(3)}_{3P_{1/2}} = - Ne^4\rho^{(0)}_{3P_{1/2}} \, |\langle 4P_{1/2}|r|5S\rangle\langle 5S|r|3P_{1/2}\rangle|^2$$

$$+ \frac{|\langle 4P_{3/2}|r|5S\rangle\langle 5S|r|3P_{1/2}\rangle|^2}{6\hbar^3\epsilon_0\Gamma_{3P4P}\,|\Omega_{5S3P_{1/2}} - \omega_p|^2} \; , \tag{6.61}$$

where $\chi^{(3)}_{3P_{1/2}}$ and $\chi^{(3)}_{3P_{3/2}}$ represent the SERS susceptibilities for stimulated Raman scattering $3p_{1/2}$-4P and $3P_{3/2}$-4P, respectively. Figure 6.9 shows the calculated values for the two parts, $\chi^{(3)}_{3P_{1/2}}$ and $\chi_{3P_{3/2}}^{(3)''}$, of the SERS susceptibility $\chi^{(3)}$ as functions of ω_p, according to (6.60) and (6.61).

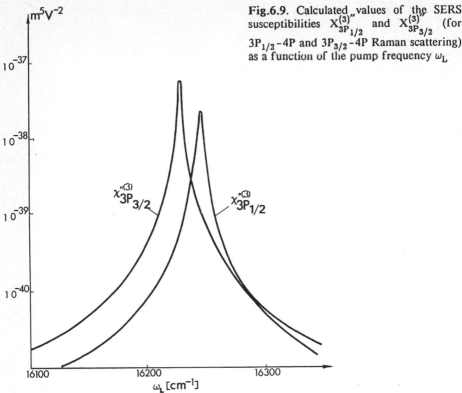

Fig.6.9. Calculated values of the SERS susceptibilities $X^{(3)''}_{3P_{1/2}}$ and $X^{(3)}_{3P_{3/2}}$ (for $3P_{1/2}-4P$ and $3P_{3/2}-4P$ Raman scattering) as a function of the pump frequency ω_L

Substituting (6.60,61) into the coupled-wave equation respectively, the SERS intensities for $\omega_{s'}$ and $\omega_{s''}$ components are found to have the form

$$I(\omega_{s'}, L) = I_{s'}(0)\exp(g_{s'},L) \ ,$$
$$I(\omega_{s''}, L) = I_{s''}(0)\exp(g_{s''},L) \ .$$

The gain coefficients $g_{s'}$ and $g_{s''}$ in the above equations are given by

$$g_{s'} = -\ \frac{3\omega_{s'}}{\epsilon_0 c^2 n_s n_p}\ X^{(3)}_{3P_{1/2}}(-\omega_{s'},\ \omega_p-\omega_p\omega_s)I_p \ ,$$

$$g_{s''} = -\ \frac{3\omega_{s''}}{\epsilon_0 c^2 n_s n_p}\ X^{(3)}_{3P_{3/2}}(-\omega_{s''},\ \omega_p-\omega_p\omega_s)I_p \ ,$$

where I_p is the pumping intensity.

By using the calculated values of the susceptibilities $X^{(3)}_{3P_{1/2}}$ and $X^{(3)}_{3P_{3/2}}$ exhibited in Fig.6.9, the relative intensity between $I(\omega_{s'},L)$ and $I(\omega_{s''},L)$ as a function of ω_L has been calculated as shown in Fig.6.10. The solid curves 1 and 2 represent the calculated values corresponding to the signal frequencies of $\omega_{s'}$ and $\omega_{s''}$, respectively, and the symbols "x" and "Δ" indicate the experimental values. The symbols "o" and the broken profile represent our

180

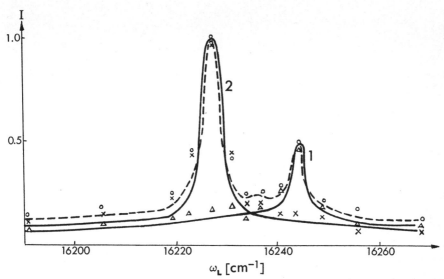

Fig.6.10. Calculated relative intensities for two SERS components 1 and 2 (solid curves) and OPSE (broken curve) as a function of pump frequency. "×" and "Δ": experimental values for SERS; "○": measured values for OPSE

measured values for OPSE and the curve calculated by means of data from reference [6.19], respectively.

6.5.4 Four-Wave Mixing Following Two-Step Hybrid Excitation

The wave-mixing processes of the pump laser beam with the OPSE, SERS, or cascade-stimulated radiation generated by the above-mentioned equal-frequency two-step hybrid excitation mechanism in the Na_2-Na system will produce tunable coherent ultraviolet radiation. This process is different from the conventional two-photon resonance four-wave mixing processes in atomic systems, as described in [6.20-22].

The two-photon resonance four-wave mixing in atomic sodium illustrated in Fig.6.11a can be represented by the following equation:

$$\omega_{uv} = 2\omega_L - \omega_s , \tag{6.62}$$

where ω_{uv} is the frequency of the ultraviolet coherent radiation generated. The following expression for the third-order susceptibility of the four-wave mixing process can be derived:

$$\chi^{(3)}(-\omega_{uv},\omega_L\omega_L-\omega_s) = \frac{N}{6\epsilon_0\hbar^3} \frac{\mu_{3S3P}\,\mu_{3P5S}\,\mu_{5S4P}\,\mu_{4P3S}}{(\omega_{4P3S} + \omega_s - 2\omega_L)(\omega_{5S3S} - 2\omega_L)(\omega_{3P3S} - \omega_L)} \tag{6.63}$$

where N is the number density of atomic sodium in the ground state 3S, μ_{ij} is the matrix element of the electric dipole moment for the transition be-

181

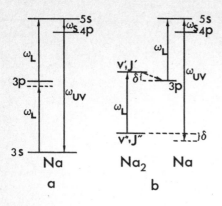

Fig.6.11. Schemes of two-photon resonance four-wave mixing in (a) atomic sodium and (b) two-step hybrid four-wave mixing in the molecule-atom system [6.11]

tween the stationary states $|i\rangle$ and $|j\rangle$, ω_{4P3S}, ω_{5S3S} and ω_{3P3S} are the transition frequencies between the states corresponding to the subscripts, ω_L is the pumping frequency, and ω_s is the frequency of OPSE generated by two-photon resonance excitation or the freqeuncy of SERS generated by two-photon off-resonance excitation.

Clearly, in the case of pure atomic sodium, if the pumping frequency is detuned from the two-photon resonance by a value of Δ, its detuning from ω_{3S5S} should be 2Δ, the ultraviolet coherent radiation due to four-wave mixing with OPSE should have a tunable range of 2Δ, while the ultraviolet coherent radiation produced by four-wave mixing with SERS should possess a fixed frequency.

The two-step hybrid resonance four-wave mixing illustrated in Fig. 6.11b is the molecule-atom hybrid resonance excitation process rather than the pure atomic approach. The first photon with frequency ω_L resonantly excites a molecule from a vibration-rotational level (v'',J'') in the ground state $X^1\Sigma_g^+$ to a vibration-rotational level (v',J') in the state $A^1\Sigma_u^+$ with an energy higher than the atomic level 3P by an amount δ. The 3P level is populated by collision between $Na_2^*(A^1\Sigma_u^+)$ and $Na(3S)$, and further resonantly excited up to the 5S level by a second photon with frequency ω_L. Of course, the third-order susceptibility expression (6.63) in the atomic system is not appropriate for the present molecule-atom system. By analyzing the correspondence of the energy levels in the molecule-atom process with that in the pure atomic process, however, the resonant terms in the denominator in (6.63) can be replaced as follows:

$$(\omega_{4P3S} + \omega_s - 2\omega_L) \rightarrow (\omega_{4Pv''J''} + \delta + \omega_s - 2\omega_L) = (\omega_{4P3P} + \omega_s - \omega_L),$$
$$(\omega_{5S3S} - 2\omega_L) \rightarrow (\omega_{5Sv''J''} + \delta - 2\omega_L) = (\omega_{5S3P} - \omega_L),$$
$$(\omega_{3P3S} - \omega_L) \rightarrow (\omega_{3Pv''J''} + \delta - \omega_L).$$

Therefore the nonlinear susceptibility for the four-wave mixing linked with two-step hybrid excitation can be expressed by

$$\chi^{(3)}(-\omega_{uv}, \omega_L\omega_L - \omega_s) =$$
$$\frac{N}{6\epsilon_0\hbar^3} \frac{\mu'_{v''J''3P}\mu_{3P5S}\mu_{5S4P}\mu'_{4Pv''J''}}{(\omega_{4P3P} + \omega_s - \omega_L)(\omega_{5S3P} - \omega_L)(\omega_{3Pv''J''} + \delta - \omega_L)}, \qquad (6.64)$$

where $\mu'_{v''J''3P}$ and $\mu'_{4Pv''J''}$ are the equivalent dipole moments, corresponding to μ_{3S3P} and μ_{4P3S} in (6.63). For example, $\mu_{v''J''3P}$ can be considered $\mu_{v''J''v'J'}k_{v'J'3P}$, which means the product of the dipole matrix element $\mu_{v''J''v'J'}$ for the molecular transition $X^1\Sigma_g^+(v'',J'') \rightarrow A^1\Sigma_u^+(v',J')$ and the energy transfer factor $k_{v'J'3P}$ for population level 3P by collision of $Na_2^*(A^1\Sigma_u^+(v',J'))-Na(3S)$. If the initial level (v'',J'') in the molecule-atom system was replaced by level 3S in the pure atomic system with $\delta = 0$, (6.64) would reduce to (6.63).

So far we see that, in contrast with the pure atomic case, in the molecule-atom nonlinear process a tuning value of Δ for the pump frequency would cause a detuning of Δ for the 5S level, so that not only can the ultraviolet coherent radiation by the four-wave mixing with OPSE be tuned over a range of 2Δ, but also the ultraviolet radiation by four-wave mixing with SERS is tunable over a range of Δ.

Because $3P_{3/2}$ and $3P_{1/2}$ are populated by collisions simultaneously, they should generate two SERS signals separated by their interval of 17 cm^{-1} and one OPSE signal corresponding to $5S\rightarrow4P$ for a given pump frequency of ω_L. Assume the detuning of the pump frequency ω_L for the transition $3P_{3/2}\rightarrow5S$ to be Δ_1 and for $3P_{1/2}\rightarrow5S$ to be Δ_2, then the four-wave mixing of ω_L with each of the two SERS or OPSE components should generate ultraviolet coherent radiation with three distinguishable frequencies, respectively:

$$\omega_{uv} = 2\omega_L - \omega(5S - 4P), \tag{6.65}$$
$$\omega'_{uv} = 2\omega_L - [\omega(5S - 4P) + \Delta_1], \tag{6.66}$$
$$\omega''_{uv} = 2\omega_L - [\omega(5S - 4P) + \Delta_2]. \tag{6.67}$$

Scanning the pump frequency ω_L causes tuning of the three ultraviolet components generated. Figure 6.12 illustrates the observed tuning, where the tuning range for the ultraviolet coherent radiation from 337.64 to 339.43 nm corresponds to the dye laser scanning from 614.55 to 617.45 nm. For example, when the pump wavelength is 616.37 nm, detuning from the transitions of $3P_{3/2}\rightarrow5S$ and $3P_{1/2}\rightarrow5S$ is small with $\Delta_1 = 7.787$ cm^{-1} and $\Delta_2 = 24.983$ cm^{-1}, respectively, and the signals at ω'_{uv} and ω''_{uv}, generated by the mixing of the pump laser with the two components of SERS, could be detected simultaneously. The signal ω_{uv} generated by the mixing of OPSE with the pump laser was also present.

Usually the above-mentioned four-wave mixing in the molecule-atom system should produce three ultraviolet coherent-radiation components. However, in practice only two of the components are detectable if either Δ_1 or Δ_2 is too large. For instance, the ultraviolet signal corresponding to Δ_2 was unobservable under our experimental conditions when the pump wavelength was 617.45 nm with $\Delta_1 = 36$ cm^{-1} and $\Delta_2 = 53$ cm^{-1}.

In the case where the pump wavelength is resonant with the transition $3P_{3/2}\rightarrow5S$ or $3P_{1/2}\rightarrow5S$, there should be only one component of SERS and one OPSE component. Therefore, the two ultraviolet coherent radiation components generated can be expressed by (6.65,67) (for $\Delta_1 = 0$) or (6.65) and (6.66) (for $\Delta_2 = 0$). For example, if the pump laser is at 616.07 nm, i.e.,

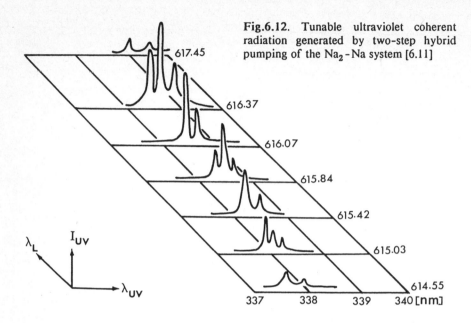

Fig.6.12. Tunable ultraviolet coherent radiation generated by two-step hybrid pumping of the Na_2-Na system [6.11]

617.45

616.37

616.07

615.84

615.42

615.03

614.55

λ_L

I_{UV}

λ_{UV}

337   338   339   340 [nm]

resonant with the transition $3P_{3/2} \to 5S$ but detuned from $3P_{1/2} \to 5S$ with Δ_2 = 17.196 cm^{-1}, the detected ultraviolet coherent signals are at 338.56 nm and 338.38 nm, generated by the mixing of ω_L with OPSE and with SERS, respectively.

The seven typical cases with different detuning conditions are summarized in Table 6.2. It shows that the experimental data for the ultraviolet coherent radiation agreed well with the calculated wavelengths with deviations less than 0.03 nm.

The mixing scheme of the pump laser beam with the infrared stimulated radiation generated by the cascade transition $4P \to 4S$ is as follows:

$$\omega_{uv} = 2\omega_L - \omega_s(4P \to 4S) . \qquad (6.68)$$

By scanning the pump wavelength from 616.07 to 618.37 nm the ultraviolet coherent radiation generated is tuned from 359.0 to 359.6 nm.

It is expected that one can generate ultraviolet coherent radiation with a wide tuning range if the second step consists of resonant or near-resonant excitation in atomic sodium with the first step wavelengths scanning over the absorption region of $X^1\Sigma_g^+ \to A^1\Sigma_u^+$ plus $X^1\Sigma_g^+ \to B^1\Pi_u$ molecular transitions by means of unequal-frequency two-step hybrid excitation.

6.5.5 Molecule-Atom Collisional Energy-Transfer Processes

To understand the collisional energy transfer processes in the above mentioned molecule-atom hybrid excitation one has to study the collisional parameters, such as collisional cross section, rate constants and so on. With the development of spectroscopic theory and technology the molecule-atom collisional dynamics have become well understood [6.23,24]. The collisional

Table 6.2 Seven typical cases for generating tunable UV coherent radiation by two-step hybrid excitation in Na_2-Na system [6.11]

Pump wavelength [nm]	Offset [cm^{-1}]	UV coherent rad. wavelength (Exp. value) [nm]	UV coherent rad. wavelength (Cal. value) [nm]
617.45	$\Delta_1=36.1$	$\lambda=339.43$	$\lambda=339.41$
	$\Delta_2=53.3$	$\lambda'=339.00$	$\lambda'=339.02$
616.37	$\Delta_1=7.8$	$\lambda=338.78$	$\lambda=338.78$
	$\Delta_2=25.0$	$\lambda'=338.68$	$\lambda'=338.67$
		$\lambda''=338.50$	$\lambda''=338.47$
616.07	$\Delta_1=0$	$\lambda=338.56$	$\lambda=338.58$
	$\Delta_2=17.2$	$\lambda''=338.35$	$\lambda''=338.38$
615.84	$\Delta_1=-6.3$	$\lambda=338.43$	$\lambda=338.44$
	$\Delta_2=10.9$	$\lambda'=338.52$	$\lambda'=338.51$
		$\lambda''=338.33$	$\lambda''=338.31$
615.42	$\Delta_1=-17.2$	$\lambda=338.20$	$\lambda=338.19$
	$\Delta_2=0$	$\lambda'=338.40$	$\lambda'=338.38$
615.03	$\Delta_1=-27.6$	$\lambda=337.92$	$\lambda=337.95$
	$\Delta_2=-10.4$	$\lambda'=338.25$	$\lambda'=338.27$
		$\lambda''=338.05$	$\lambda''=338.02$
614.55	$\Delta_1=-40.4$	$\lambda=337.64$	$\lambda=337.66$
	$\Delta_2=-23.2$	$\lambda''=337.90$	$\lambda''=337.92$

energy transfer processes for electronic transitions have been extensively studied by means of fluorescence spectroscopy or sensitive resonance fluorescence for the collisional systems of $K_2^*(B^1\Pi_u)$-$K(4S)$, $Na_2^*(A^1\Sigma_g^+$ or $B^1\Pi_u)$-$Na(3S)$, $Na_2^*(A^1\Sigma_g^+$ or $B^1\Pi_u)$-$K(4S)$, K_2-He and K_2-Ne [6.25-28], and so on. The results involve the collisional energy transfer rate and collisional cross section, etc. By the analyses of fluorescence intensities and relaxations the process of collisional energy transfer from atom to molecule causing an electronic transition had been confirmed [6.29]. The energy-transfer processes between atomic and molecular high-lying states have also been demonstrated [6.30]. In order to explain the processes we encountered in the study of hybrid pumping involving electronic transitions, we established a coupled model describing collision-induced electronic transitions, and simplified it to a two-level system. With some reasonable parameters and approximations the model has been applied to calculate the collisional cross section and rate constant [6.32].

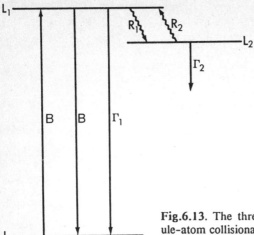

Fig.6.13. The three-level model for analysing molecule-atom collisional energy transfer

In fact, in the case of pumping by two pulsed beams, the relative time delay between the two beams will influence the intensity of the infrared stimulated radiation. Therefore we have established a set of three-level rate equations and have solved it under nonequilibrium conditions to obtain the time delay for maximum output of stimulated radiation. The three-level model for discussing the molecule-atom collisional energy transfer process is depicted in Fig.6.13. Let N_0 and N_1 represent the molecular densities on the lower level L_0 and upper level L_1, respectively, of the molecular electronic transition induced by the pumping beam with wavelength λ_1. Some of the molecules will return from level L_1 to the initial level L_0 through stimulated emission or spontaneous emission processes, while others may give up their energy to atoms to populate the first excited level L_2. Before the second step pumping for the atoms from L_2 to a higher level, we can write the rate equations as follows:

$$dN_0/dt = B_{01}\rho(\nu)(N_1 - N_0) + \Gamma_1 N_1 , \qquad (6.69)$$
$$dN_1/dt = B_{10}\rho(\nu)(N_0 - N_1) - \Gamma_1 N_1 - R_1 N_1 + R_2 N_2 , \qquad (6.70)$$
$$dN_2/dt = R_1 N_1 - R_2 N_2 - \Gamma_2 N_2 , \qquad (6.71)$$

where Γ_1 and Γ_2 are the spontaneous emission coefficients, R_1 and R_2 are the collisional relaxation coefficients from level L_1 to L_2 and its inverse process, respectively. Taking the Na_2-Na system as the sample medium, we have $R_1 = k[Na(3S)]$, $R_2 = k[Na_2(X^1\Sigma_g^+)]$, where k is the collisional rate constant and the symbols in the brackets indicate the starting levels of the respective collisions. The Einstein coefficients $B_{01} = B_{10} = B$ are given by the well-known expression

$$B = 2\pi^2 |\mu_{01}|^2/3\epsilon_0\hbar ,$$

where μ_{01} is the transition dipole moment between levels L_0 and L_1. The first-step pumping beam is considered to have uniform distribution over

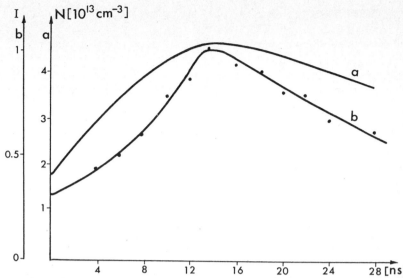

Fig.6.14. (a) Calculated curve for the atomic density on level 3P (b) measured curve for the infrared emission intensities as a function of the delay time

the laser line width of $\delta\nu$ but to vary with time with a Gaussian function. By means of an iterative computation we obtained the nonequilibrium solution. Using the following data:

$$N_0[Na_2(X^1\Sigma_g^+)] = 3.8 \cdot 10^{14} \text{ cm}^{-3} \quad (\text{at } 500° \text{ C}) ,$$
$$\Gamma_1 = 8.3 \cdot 10^7 \text{ s}^{-1} ,$$
$$\Gamma_2 = 6.25 \cdot 10^7 \text{ s}^{-1} ,$$
$$R_1 = 4.86 \cdot 10^7 \text{ s}^{-1} ,$$
$$R_2 = 3.8 \cdot 10^5 \text{ s}^{-1} ,$$
$$\mu_{01} = 1 \cdot 10^{-28} \text{ A·s·cm} ,$$

and a laser output of 10 μJ, duration 12 ns, and focussing area 0.16 mm², we computed the atomic density on level 3P as a function of delay time as shown by curve a in Fig.6.14. The curve shows a maximum of $4.7 \cdot 10^{13}$ cm⁻³ (about one thousandth of the atomic density under the experimental conditions) at 14.8 ns.

The calculated result was confirmed by experiment. The experimental data give a maximum at a time delay of 13.5±0.5 ns, as shown by curve b in Fig.6.14, which agrees with the calculated value.

Computations for different parameters, such as different metal vapour temperatures and different pulse durations, etc., reveal that curve a in Fig.6.14 will vary. In particular:

1) With a given pulse duration but increasing (decreasing) the pumping beam intensity, the optimum delay time for most efficient population of the atomic level L_2 will be slightly shortened (lengthened). For the K_2-K system, doubling the pumping intensity shortens the optimum delay time by an amount of 0.4 ns.

2) With a given pumping intensity but varying pulse duration the optimum delay time follows longer or shorter.

3) Increasing the vapour temperature so as to increase the number densities of the molecules and atoms in their thermally distributed levels will shorten the optimum delay time. For example, in the K_2-K system, increasing the temperature from 370° to 430° C will reduce it from 15.3 to 12.1 ns.

Recently, molecular-enhanced four-wave parametric generation was observed in the sodium atom [6.33]. A mechanism was proposed involving single and two-photon molecular absorption, transfer of real population to the 4P state of the atom and four-wave parametric generation enhanced by the radiative decay to the ground state.

6.6 Optically Pumped Stimulated Radiation Based on Molecular Electronic Transitions

By utilizing optical pumping techniques with homonuclear diatomic molecules (dimers) various dimer lasers have been obtained. Among them I_2 [6.34-36], Na_2 [6.37-40], Te_2 [6.41, 42], Br_2 [6.43], S_2 [6.44, 45], Li_2 [6.40], Bi_2 [6.42, 46], K_2 [6.45], etc. were the first dimers to yield optically pumped lasers based on molecular electronic transitions. Recently there have been many publications about alkali metal dimer optically pumped lasers, such as Li_2, Na_2 and K_2 lasers [6.47-71].

Figure 6.15 shows a typical energy-level diagram of a simple dimer with laser cycles between the bound electronic states A and X. Excited by an appropriate pump-laser wavelength, the molecular transition would take place from vibration-rotational level *1* in the ground state X to level *2* in the excited state A. It is easy to form a population inversion between level *2*

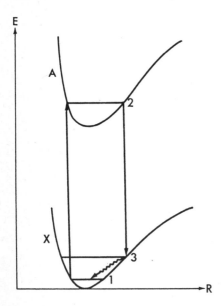

Fig.6.15. Illustration of a simple dimer laser cycle between bound electronic states A and X

and less populated high-lying vibrational levels in the ground state X (e.g., level *3* in Fig.6.15), resulting in stimulated radiation.

Subsequent collisional processes bring the molecules back to the initial level *1* to complete the laser cycle. It is to be noted that the line width of practical pump sources would usually cause several vibration-rotational levels in state A to be populated and, in turn, generate a complicated multi-line emission; even if only one level in state A were populated by optical pumping, some of the allowed fluorescence transitions would produce laser emission. For the sake of simplicity all of the stimulated emission transitions belonging to the different branches or bands are represented by a single downward arrow in Fig.6.15. On the other hand, the same vibration-rotational level in the excited state A could be populated by different pump wavelengths, given the hundreds of candidate populated levels in the ground state X.

The dimer laser is ideal for the spectroscopic and kinetic investigation of the dimers themselves, including the determination of the transition moments, relaxation rates and accurate molecular constants [6.58,59]. They are also applied to some fields where fixed-frequency laser lines are required, such as frequency standards.

In the following part of the section we will take Na_2 and Li_2 lasers as examples and discuss their various excitation and emission mechanisms. The detailed analyses of the laser properties and principles can be found in the review papers by *Wellegehausen* [6.60,61].

I) **Lasers between lower-lying states by one-photon pumping**: The first realization of a pulsed optically pumped sodium dimer laser was reported in [6.37]. Laser emission on a series of green and yellow Na_2 B-X band transitions was obtained by 473.2 nm pumping from a doubled Q-switched Nd:YAG laser, and Na_2 A-X band lasers in the near infrared were generated by pump wavelengths of 659.4, 669.1 and 679.5 nm available from a Q-switched internally doubled Nd:YAG laser.

The first CW laser oscillation with thresholds below 1 mW was observed for various B-X transitions in molecular sodium pumped by argon laser lines in the range 454-488 nm [6.39]. Output powers of up to 3 mW, and a single-pass gain of up to 0.1 cm^{-1} at a pump power of 0.5 W, were obtained even before the laser's parameters (e.g., dimer concentration, sodium vapour zone, temperature) were optimized.

Pumped by a pulsed dye laser tunable from 575.0 to 610.0 nm, numerous laser lines from a sodium dimer in a heat-pipe oven have been observed in the wavelength region between 795.0 and 810.0 nm. The pump wavelengths correspond to the electronic transitions $X^1\Sigma_g^+ \rightarrow A^1\Sigma_u^+$ in the sodium dimer, whereas the stimulated emissions correspond to $A^1\Sigma_u^+ \rightarrow X^1\Sigma_g^+$ transitions. An optical cavity was formed by a spherical mirror and a plane mirror at the incident and exit ends, respectively. The spherical mirror was coated for 99% reflectivity in the 800 nm region and for 98% transmissivity for the pump wavelengths. The output coupling mirror had 80% reflectivity at 800 nm and high transmissivity for the pump wavelengths. For measuring single-path gain the output coupling mirror could be removed or tilted.

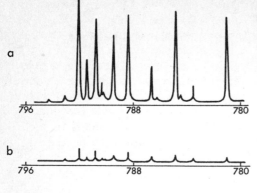

Fig.6.16. Na$_2$ laser emision spectra (a) with resonator, (b) without output mirror and (c) detuning output mirror. Pump wavelength: 585.0 nm

Figure 6.16 shows the recorded emission spectra of a sodium-dimer laser pumped by a pulsed dye laser operated at 585.0 nm. Figure 6.16 shows the output spectrum (a) with the above-mentioned resonant cavity; (b) without the output mirror; (c) with the output mirror tilted. Comparison of the cases (a) and (b) demonstrates that a resonance cavity might strengthen stimulated radiation signals by as much as one order of magnitude. The signal intensity in (c) is about one fifth of that in (b) because the transmissivity of the output mirror for Na$_2$ laser wavelengths is 20%.

As mentioned above, due to the multiplicity of molecular energy-level structures one output line out of a Na$_2$ laser might be pumped by distinct wavelengths. Figure 6.17 shows the excitation spectra in the regions of 587.0-591.5 nm and 577.0-580.0 nm for generating the Na$_2$ laser line at 797.5 nm. They show rich efficient pump wavelengths, except at the wavelengths corresponding to the atomic one-photon transition 3S→3P and two-photon transition 3S→5S, at which wavelengths the pump power is reduced by absorption.

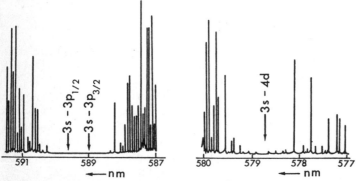

Fig.6.17. Excitation spectra (a) in the 587.0-591.5 nm region and (b) in the 577.0-580.0 nm region for generating 797.5 nm Na$_2$ laser output

A similar situation exists in the lithium dimer. With a pump dye laser operated in the region 570.0–610.0 nm numerous laser lines in the region 874–912 nm, corresponding to electronic transitions from $A^1\Sigma_u^+ \rightarrow X^1\Sigma_g^+$ in Li_2, can be obtained out of a highly heated pipe oven (about 900° C). By using a resonant cavity consisting of the sample cell with Brewster windows and appropriate mirrors, the output signal was nine times stronger than was observed in a single path.

II) **Lasers between high-lying states by two-photon pumping**: By means of two-photon pumping the molecules can reach even-parity high electronic states and then produce stimulated radiation between high-lying states. Utilizing a flashlamp-pumped dye laser operated in the 585.0–610.0 nm region, infrared laser action in the region of 4–5 μm was observed with sodium dimers [6.62]. Obviously this was not due to emission transitions starting from $A^1\Sigma_u^+$ or $B^1\Pi_u$ states. By analyzing the output spectra and the time dependence of the multiline infrared laser, especially the evident vibrational structures which seemed to belong to the $C^1\Pi_u$ state, the infrared lasers were recognized to be transitions from the high-lying $^1\Sigma_g^+$ state to the $C^1\Pi_u$ state following population inversion by two-photon pumping, where $A^1\Sigma_u^+$ is the resonant or near-resonant intermediate state. Figure 6.18 shows the relevant potential curves with the excitation and emission transitions indicated. The total multiline infrared laser output power has been estimated to be approximately 2 mW at a pump wavelength of 597.5 nm and an average pump power of 40 kW. Narrowing the pump linewidth from 0.3 to 0.01 nm increased the output power by two orders of magnitude.

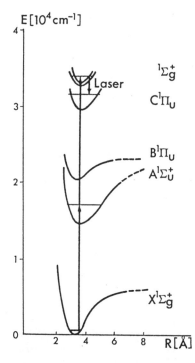

Fig.6.18. Energy state diagram of molecular sodium, indicating the two-photon pumped Na_2 laser scheme [6.62]

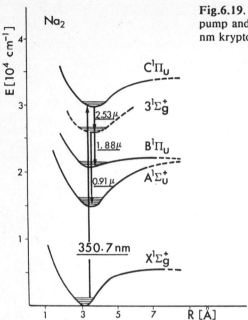

Fig.6.19. Na$_2$ molecular state scheme, indicating pump and laser transitions for excitation with 350.7 nm krypton ion laser [6.63]

III) Cascade stimulated radiation in sodium dimer: A cascade laser generated by the optical pumping of Na$_2$ molecules has been proposed and demonstrated [6.63-67]. Intense infrared stimulated radiation lines around wavelengths of 2.5 μm, 0.91 μm and 1.88 μm have been observed by excitation of the sodium dimers by ultraviolet radiation from a nitrogen laser, a krypton ion laser, a frequency-doubled dye laser or a third-harmonic Nd:YAG laser, respectively.

Figure 6.19 shows Na$_2$ molecular level scheme with relevant pump and cascade laser transitions for excitation by a 350.7 nm krypton ion laser. The cascade laser lines were attributed to the $C^1\Pi_u \rightarrow (3)^1\Sigma_g^+ \rightarrow A^1\Sigma_u^+$ (or $B^1\Pi_u$) cascade transitions after the pumping $X^1\Sigma_g^+ \rightarrow C^1\Pi_u$.

The set of stimulated radiation lines generated by pumping of the sodium dimer with a third-harmonic Nd-YAG laser are exhibited in Fig.6.20. It can been found that the spacings of the observed stimulated

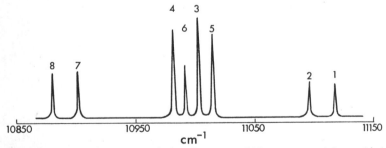

Fig.6.20. Some stimulated emission lines near 910 nm pumped by a third harmonic Nd:YAG laser [6.64]

Table 6.3 The wavelengths, observed and calculated vibrational and rotational spacings, and respective quantum numbers for the lower state $A^1\Sigma_u^+$ of the laser emissions [6.64]

Line number	Wavenumber (exp.) [cm⁻¹]	Vib. and rot. quantum number for $A^1\Sigma_u^+$ (v,J)	Rot.spac. (exp.) [cm⁻¹]	Rot.spac. (cal.) [cm⁻¹]	Vib.spac. (exp.) [cm⁻¹]	Vib.spac. (cal.) [cm⁻¹]
1	11118.56	(2,49)				
			21.72 (1-2)	21.71		
2	11096.84	(2,51)				
					113.98 (2-4)	113.72
3	11004.28	(3,49)				
			21.43 (3-4)	21.60		
4	10982.85	(3,51)				
5	11015.77	(3,50)				
			21.93 (5-6)	22.01		
6	10993.84	(3,52)				
					113.20 (6-8)	112.95
7	10902.48	(4,50)				
			21.84 (7-8)	21.89		
8	10880.64	(4,52)				

emission lines are in agreement with the calculated spacings between vibrational and rotational levels in the $A^1\Sigma_u^+$ state [6.72], as listed in Table 6.3. Therefore, the lower state of the laser emission is believed to be the state $A^1\Sigma_u^+$, and the vibrational and rotational quantum numbers can be determined.

Some improved and some new molecular parameters for the excited $(3)^1\Sigma_g^+$ and $C^1\Pi_u$ states have recently been reported [6.65,73,75]. They are helpful for more accurate identifications of the pump transitions from $X^1\Sigma_g^+$ to $C^1\Pi_u$ and the emission transitions from $C^1\Pi_u$ to $(3)^1\Sigma_g^+$.

IV) **Stimulated emissions from the double-minimum state** $(2)^1\Sigma_u^+$ **and higher lying state** $C^1\Pi_u$ **by UV one-photon pumping:** Stimulated emission lines around 2.5 μm were generated by pumping of the sodium dimer with ultraviolet radiation from a nitrogen laser [6.67,70]. The pumping and emission processes for the 2.5 μm laser lines in these experiments are $X^1\Sigma_g^+ \rightarrow C^1\Pi_u \rightarrow (3)^1\Sigma_g^+$.

Figure 6.21a,b and d display the laser emission spectra around 2.5 μm pumped by a nitrogen laser at the sample temperatures of 520, 540 and 600°C, respectively. The infrared laser emission spectrum at the vapour temperature of 520°C exhibits only three main emission lines (3rd, 5th and

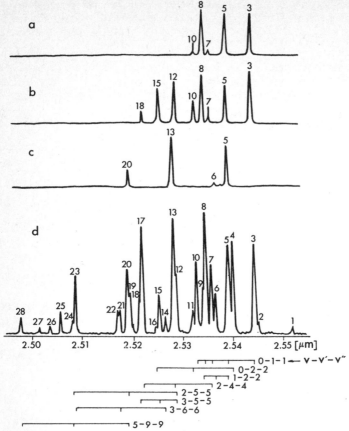

Fig.6.21. Laser emission spectra around 2.5 μm with a nitrogen laser pump under different experimental conditions: (a) unpolarized pump, sample temperature of 520°C; (b) unpolarized pump, 540°C; (c) pumped by linearly polarized N_2 laser beam, 600°C; (d) unpolarized pump, 600°C [6.70]

8th) with two weaker lines (7th and 10th), as shown in Fig.6.21a. As the temperature rose to 540°C, another three lines were detected, as indicated in Fig.6.21b marked by 12, 15 and 18, respectively.

For the separations of the PQR triplets in Fig.6.21a with the rotational constant of $(3)^1\Sigma_g^+$ [6.73], the rotational quantum number J was estimated to be 40, which agreed with the J value of the maximum thermal population according to the Boltzman distribution in the initial state $X^1\Sigma_g^+$. Compared with the calculated transition frequencies from $X^1\Sigma_g^+$ to $C^1\Pi_u$ according to the constants from [6.74, 75], respectively, the J values were in reasonable agreement with the possible selection of the lowest vibrational quantum numbers of the relative electronic states of the 337.18 nm pumping. Due to the wide bandwidth of the N_2 laser output, the candidates for the possible excitation transitions should take into account all of the calculated frequencies within 1 cm^{-1} of the center frequency of the strongest N_2 laser line at 29657.96 cm^{-1}. The example is illustrated in Fig.6.22a, where

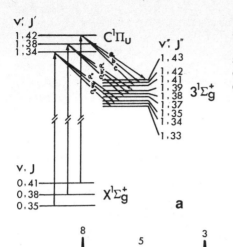

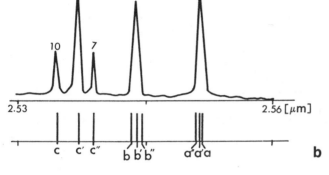

Fig.6.22. An example of the laser emission with broad-band pumping. (a) identified energy-level diagram; (b) detected signals at a temperature of 520°C. The calculated wavelength positions are marked at the bottom of the figure [6.70]

the three pumping transitions have frequencies with separations less than 0.8 cm⁻¹. The frequency separations of the laser emission lines from ν_a to ν_x could be expressed as follows:

$$\nu_x - \nu_a = [J''_a(J''_a + 1) - J''_x(J''_x + 1)]\Delta B$$

$$+ \begin{cases} +8B'' \text{, for transitions marked a' and a'' with } J''_x = J'_x + 1 \\ 0 \text{, for b, b' and b'' with } J''_x = J'_x) \\ -8B'' \text{, for c, c' and c'' with } J'' = J'-1 \text{,} \end{cases}$$

where x could be either a', a'', b, b', b'', c, c' or c''; $\Delta B = B''-B'$ is the difference between the rotational constants of the lower and upper states, having a small negative value in our case. Therefore, with a suitable J value group a of the laser lines has very close frequency positions, while group c has quite large separations, as shown in the bottom of Fig.6.22b. Another triplet in Fig.6.21b and the strongest triplet in Fig.6.21d with the series numbers of 8, 13 and 17, and so on were identified using the same scheme.

The infrared laser spectrum pumped by the linearly polarized N₂ laser beam were recorded, as shown in Fig.6.21c. The result can be easily understood by the two-step polarization labelling calculation [6.76] for the $\Sigma \rightarrow \Pi \rightarrow \Sigma$ transitions under conditions of linearly polarized pumping. The

Q-branch should dominate over the others in both the excitation and emission processes, and would show up as the only one with rather weakly polarized pumping energy. It is pointed out that the 5th, 6th, 13th and 21th lines should be Q-branch lines. Since the 6th and 21th lines did not appear at a lower temperature, they should correspond to large v and J values. The corresponding vibrational quantum numbers are labelled in the bottom of Fig.6.21d.

Recent calculations and experiments have demonstrated [6.73, 77] that there is a $(2)^1\Sigma_u^+$ electronic state with a double-potential-minimum near the $C^1\Pi_u$ state but that the energy of the top of the potential barrier is slightly lower than the minimum of the potential of the $C^1\Pi_u$ state in the sodium dimer. The high-gain stimulated emission lines around 1.03 μm were observed by the pumping of the sodium dimer with the pulsed third harmonic of a Nd:YAG laser at a wavelength of 354.706 nm (Fig.6.23). These near-infrared laser were deduced to be due to transitions from the double-minimum state $(2)^1\Sigma_u^+$ to the state $(2)^1\Sigma_g^+$. Some of the doublets have been identified as indicated in Fig.6.23 using the molecular constants of $X^1\Sigma_g^+$ [6.74], $(2)^1\Sigma_u^+$ [6.77] and $(2)^1\Sigma_g^+$ [6.73]. The identified laser lines have upper levels all belonging to the inner-potential well.

V) $A^1\Sigma_u^+ \rightarrow X^1\Sigma_g^+$ laser transitions by two-photon pumping of the 5S and 4D levels of atomic sodium.

The first observation of the laser emissions between $A^1\Sigma_u^+$ and $X^1\Sigma_g^+$ states by collisional energy transfer caused by two photon excitation of the 5S or 4D levels of Na has been reported [6.71]. A nitrogen laser pumped dye laser (pulse energy: 120μJ; pulse duration: 3ns; spectral width: ≤0.02nm) was used for the excitation.

Two spectra (Fig.6.24a and b) show the laser lines around 800.0 and 790.0 nm generated by two-photon resonance excitations of 3S→5S and 3S→4D of Na, respectively.

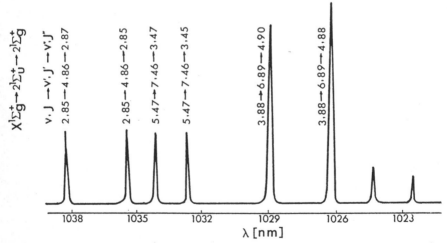

Fig.6.23. Laser emission lines around 1.03 μm, corresponding to the doublet transitions between double-minimum state $(2)^1\Sigma_u^+$ and state $(2)^1\Sigma_g^+$ [6.69]

196

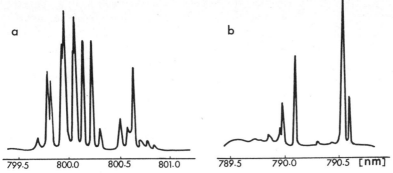

Fig.6.24. The infrared laser emission spectra excited by (a) $\lambda_1 = 602.23$ nm and (b) $\lambda_1 = 578.73$ nm [6.71]

One of the following processes for forming the high-lying excited Na_2^* dimers in the bound states could be assumed:

1) Collision with other Na atoms in the ground state:

$$Na(4D,5S) + Na(3S) + M(Ar,Na(3S), ...)$$
$$\rightarrow Na_2^*(^1\Sigma,^1\Pi, ...) + M(Ar,Na(3S), ...) .$$

2) Radiative transition in the collision complex $Na(4D,5S)-Na(3S)$.
3) Energy pooling:

$$Na_2 + Na(4D,5S) \rightarrow Na_2^*(^1\Sigma,^1\Pi, ...) + Na(3S) .$$

The state $A^1\Sigma_u^+$ can then be populated via cascade transitions

$$Na_2^*(^1\Sigma,^1\Pi, ...) \xrightarrow{\text{(cascade transitions)}} Na_2^*(A^1\Sigma_u^+) .$$

6.7 Lasers Based on Molecular Photodissociation

Since laser action in iodine utilizing molecular photodissociation of CF_3I or CH_3I was first demonstrated [6.78], more than fifty types of molecules have successfully been operated in the photodissociation laser. A series of atomic lasers based on the photodissociation of metal halides (for examples, SnI_2, SbI_3, GeI_4, PbI_2, $PbBr_2$, NaI, CsI, InI, TlI, etc.) are of considerable interest [6.79-88] particularly when laser action occurs on the resonance transitions. They possesses very low thresholds, and some of the photodissociation resonance laser transitions are potentially applicable to trace element detection.

Usually a suitable one-photon or two-photon excitation can be used to photodissociate a molecule:

$$AB + h\nu \rightarrow A^* + B$$
$$AB + 2h\nu \rightarrow A^* + B .$$

197

Most of the atomic lasers utilizing molecular photodissociation investigated up to now are generated by optical pumping in the wavelength region below 200 nm [6.89-104]. The pumping processes can be performed by one-photon or two-photon excitation utilizing, for example, an eximer laser ArF, or by two-photon or multiphoton excitation with a dye laser. The photodissociation lasers utilizing two-photon pumping in alkali dimers covering a wide wavelength region have been demonstrated, as described as follows.

With the photodissociation of molecular potassium, the optically pumped infrared lasers at 3.14 μm ($5P_{3/2}$-3D), 3.16 μm ($5P_{1/2}$-3D) and 2.71 μm (5P-5S) in atomic potassium were simultaneously generated by two-photon or two-step excitation of the potassium dimers covering a wide range of pumping wavelengths from 625.0 to 690.0 nm [6.31]. In rubidium vapour, when a nitrogen pumped DCM and Nile Blue dye laser with output energy of 10-100 μJ was scanned from 625.0 to 705.0 nm, the laser lines at 2.25 μm ($6P_{3/2}$-4D), 2.29 μm ($6P_{1/2}$-4D) and 2.73 μm (6P-5S) could be measured all over the pump region. Figure 6.25 exhibits spectra 1, 2, and 3 for generating the 2.25, 2.29, and 2.73 μm laser lines, respectively. The dashed curve represents the output profile of the DCM dye laser used in the experiment.

The scheme for generating K_2 photodissociation laser radiation by two-photon pumping over a wide wavelength region is illustrated in Fig.6.26. The infrared laser lines at wavelengths of 2.71, 3.14 and 3.16 μm can be detected over the whole pump region from 625.0 to 690.0 nm, which coincides exactly with the absorption band of molecular potassium from the $X^1\Sigma_g^+$ to the $B^1\Pi_u$ state. Thus state $B^1\Pi_u$ is the intermediate state of the molecular two-photon or two-step transition process. The energy for K_2 dissociation to K(4S)+K(5P) can be expected to be about 29240 cm^{-1}. Since the two-photon energy between 625.0 and 690.0 nm in the molecular transitions lies above the dissociation level the 5P level can be populated. Thus far we may express the mechanism of the double quantum photodissociation laser with a wide pumping wavelength region:

$$K_2(X^1\Sigma_g^+)+2h\nu(625.0-690.0\,\text{nm}) \rightarrow \text{high-lying g states} \rightarrow K(4S)+K^*(5P) \, .$$

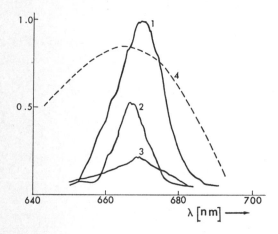

Fig.6.25. Excitation spectra 1, 2 and 3 for generating 2.25, 2.29 and 2.73 μm, respectively, in rubidium vapor. The dashed curve 4 represents the output profile of the DCM dye laser

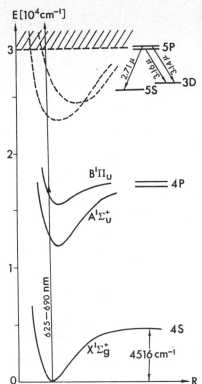

Fig.6.26. Mechanism of the photodissociation laser by two-photon pumping covering a wide wavelength region [6.104]

The two-photon process may be replaced by the two-step dissociation process as

$$K_2(X^1\Sigma_g^+) + h\nu(625.0-690.0nm) \rightarrow K_2^*(B^1\Pi_u) \ ,$$
$$K_2^*(B^1\Pi_u) + h\nu(625.0-690.0nm) \rightarrow \text{high-lying g states} \rightarrow K(4S) + K^*(5P) \ ;$$
$$K^*(5P) \rightarrow K^*(3D) + h\nu(3.14, 3.16\mu m) \ ,$$
$$K^*(5P) \rightarrow K^*(5S) + h\nu(2.71\mu m) \ .$$

Recently, ultraviolet and vacuum ultraviolet anti-Stokes Raman laser action was achieved based on molecular photodissociation and atomic anti-Stokes Raman effects. This is particularly attractive on account of the extremely short output wavelengths, tunability, high inversion efficiency and intense output energy.

The first anti-Stokes Raman laser was developed in thallium [6.105]. The photodissociation was performed by the optical pumping of TlCl molecules with an ArF laser at 193 nm:

$$\text{TlCl} + h\nu(193\,nm) \begin{cases} \rightarrow Tl^*(6p^2P_{3/2}^0) + Cl \\ \rightarrow Tl^*(6p^2P_{3/2}^0) + Cl^* \ . \end{cases}$$

199

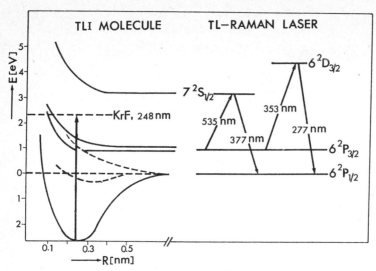

Fig.6.27. TlI photodissociation scheme (left part) and Tl anti–Stokes Raman laser cycle [6.110]

The above processes are considered mainly to populate the metastable state $6p^2 P_{3/2}^0$ rather than the ground state $6p^2 P_{1/2}^0$. A population inversion between the atomic metastable and ground states is then formed. Raman scattering from the Tl metastable state to the ground state using 532 and 355 nm pump lasers results in the nonresonant stimulated anti-Stokes Raman emissions at 376 and 278 nm, respectively.

Anti-Stokes Raman lasers have been extensively realized over the last few years. Up to now the laser lines from I, Br, Sn, Pb, Se, S and so on have been obtained [6.106-113]. Some of these are in the region between 100 and 200 nm. For instance, following the molecular photodissociation of OCSe by the first step pumping with an ArF excimer laser, laser radiation at 169 and 146 nm was generated from atomic Se with conversion efficiencies of up to 2.3% and output energy up to 350 μJ in a 2-ns pulse [6.113].

Figure 6.27 depicts the molecular photodissocation scheme of TlI and the anti-Stokes Raman laser cycle of Tl [6.110]. In the first step excitation the molecules TlI are photodissociated by a KrF excimer laser (around 248 nm), yielding the metastable atoms $Tl(6^2 P_{3/2})$. The latter results in the population inversion to $6^2 P_{1/2}$. Taking $Tl(7^2 S_{1/2})$ or $Tl(6^2 D_{3/2})$ level as the intermediate level of a Raman-type interaction with the second step excitation at 535 or 353 nm, the anti-Stokes Raman radiation is generated near 377 and 277 nm, respectively.

The typical experimental set-up is shown in Fig.6.28 [6.110]. Part of the radiation of an excimer laser is used for the photodissociation, the dye laer beam pumped by the other part of the same excimer laser being used for exciting $Tl(6^2 P_{3/2})$ atoms. The latter is opitcally delayed and overlaps with the former in the sample cell. The anti-Stokes Raman radiation generated is detected in the direction of the dye-laser beam and analyzed by means of a spectrometer.

200

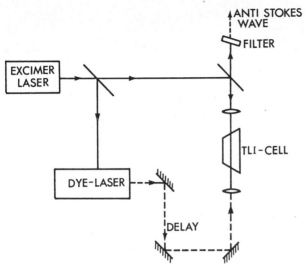

Fig.6.28. A typical experimental set-up for generating anti-Stokes Raman radiation [6.110]

Using a near-resonant three-level approximation the anti-Stokes Raman gain cross section σ_R, may be calculated as follows [6.105, 114]:

$$\sigma_R = e^4 f_1 f_2 \frac{\nu_R}{32\pi^2 \epsilon_0^2 m^2 hc^2 \nu_1 \nu_2 (\Delta\nu)^2 \gamma} \; ,$$

where f_1 and f_2 are the oscillator strengths connecting the initial and final states to the intermediate Raman level, ν_1 and ν_2 are the corresponding frequencies, ν_R is the anti-Stokes Raman frequency, $\Delta\nu$ is the detuning, and γ is the Raman linewidth. The single pass Raman gain is $\exp(\Delta N \sigma_R IL)$, where ΔN is the metastable population inversion density, I is the pump laser intensity, and L is the length of the Raman medium. Obviously, high pump intensity is required over a wide tuning range.

The tuning curve for the Tl $7S_{1/2}$-$6P_{1/2}$ anti-Stokes emission is shown in Fig.6.29 [6.107]. As the pump wavelength is tuned in the region of 534 nm, the anti-Stokes tunable radiation follows around 377 nm. The tuning range increases with pump energy.

6.8 Optically Pumped Far-Infrared Lasers Based on Pure Rotational Molecular Transitions

Until the discovery of the optically pumped far-infrared laser based on pure rotational transitions in 1970 [6.115], the laser was deficient in the long-wavelength region. Since 1970 more than four thousand optically pumped far infrared laser lines have been generated in the wavelength

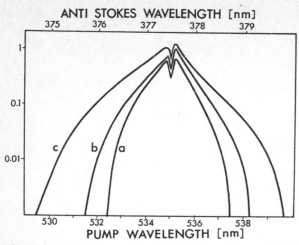

Fig.6.29. Tuning curves for the Tl $7S_{1/2}$-$6P_{1/2}$ anti-Stokes emission. Length of the sample cell: 25 cm; temperature: 730° K. Curves a-c correspond to pump energies 5, 10 and 20 mJ, respectively. Pump beam cross-section: 6 mm² [6.107]

region 0.03-3 mm, and about a hundred molecules have been used as active materials [6.116]. For example, from CH_3OD and CD_3OD molecules pumped by the CO_2 laser a total of several hundred laser lines are known with wavelengths ranging from tens of 1 μm to 1.6 mm [6.116-122]. CH_3OH is the best molecule with 453 lines. Many of them have been obtained with hundreds of kilowatts of pulsed peak power or a few hundreds of milliwatts of continuous power. Far-infrared molecular lasers have found wide applications in some fields of physics, for example, the spectroscopy of the laser molecules themselves, atomic and molecular spectroscopy, astrophysics and metrology, etc. [6.123-125]. Detailed discussions of the theory, experimental techniques and applications of far-infrared lasers can be found in review papers [6.126-135]. The following discussions are restricted to deal with optical excitation and emission schemes only.

Figure 6.30 illustrates a part of the energy-level diagram related to the optical pumping and the pure rotational transitions in CH_3F molecules [6.115]. The P(20) line (at 9.55μm) of a Q-switched CO_2 laser pumps CH_3F from the vibratioanl ground level to an excited level with rotational quantum number J = 12 unchanged. This produces a molecular population inversion on the sublevels (J',K') = (12,1) and (12,2) to (11,1) and (11,2), and is followed by the laser emissions B and b and the cascade laser emissions C and c, as indicated in Fig.6.30 with wavelengths of 496 and 541 μm, respectively. On the other hand, the densities on the lower vibrational levels, (12,1) and (12,2), are strongly decreased due to the pumping, resulting in the laser transitions (13,1)→(12,1) and (13,2)→(12,2), as indicated by A and a in Fig.6.30 with a wavelength of 452 μm. The frequency intervals between the two components for K = 1 and K = 2, i.e., between A and a, B and b, C and c are about 30-40 MHz.

The above-mentioned processes in symmetric-top molecules are also applicable to linear molecules, except for the absence all of the K≠0 rota-

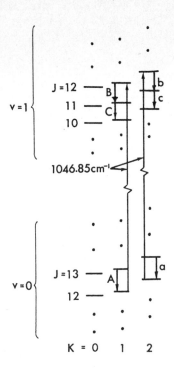

tional levels. The selection rules for asymmetric-top molecules allow several laser transitions with discrete wavelengths and unequal intensities from a common upper level, and also allow a given pump laser line to excite the molecules to several different rotational levels, which are not limited by $\Delta J = 0, \pm 1$ and $\Delta K = 0$ as in the symmetric-top case.

The tunable CO_2 laser is a good pump source for far infrared lasers since its dozens of output lines around 9.4 and 10.4 μm coincide with absorption lines for many molecules. Many other lasers, such as N_2O, CO, HF, and HCl lasers, are also used as pump sources.

Infrared/Radio-freqency (IR/RF) double-resonance pumping for increasing far-infrared laser output has been reported [6.136]. The lower level of the pumping transition is connected with other rotational levels in the vibrational ground state by RF or microwave transitions. Therefore, by means of RF resonance as shown in the left part in Fig.6.31 one can transfer molecules to the lower level of the pumping transition from other rotational levels so as to enhance the pump rate. On the other hand, an external RF field will accellerate the depopulation of the molecules on the lower FIR laser level, which again enhances the output of FIR laser. These two enhancing effects are shown in the right part of Fig.6.31.

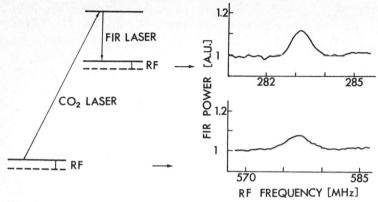

Fig.6.31. Output power increase of the CH_3OH 251 μm FIR laser with IR-RF double resonance pumping [6.136]

6.9 Optically Pumped Mid-Infrared Laser Based on Molecular Vibration-Rotational Transitions

Optically pumped molecular lasers operated in the mid infrared region (5-50μm) have been extensively developed for the requirements of photo-chemistry and spectroscopy. In such lasers excitations and emissions are based on vibration-rotational transitions. The pulsed peak power is greater than 1 MW, while a continuous-wave output power of 10 W in NH_3 has been obtained [6.137]. There are good review papers for the details [6.126, 138-140].

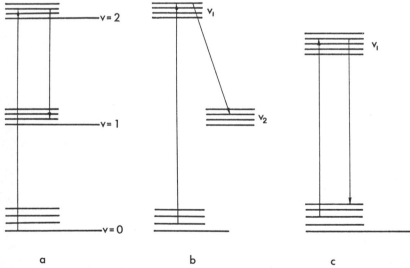

Fig.6.32. Energ-level schemes for optically pumped laser based on vibration-rotational transitions

These optically pumped laser transitions can be classified into three groups according to their pumping and emission schemes. The first group corresponds to a normal vibrational mode of the active molecule, as shown in Fig.6.32a. For the second group the laser transitions correspond to the difference between two normal vibrational modes (Fig.6.32b). In the third group the excitation and the lasing transitions take place in the same vibrational band (Fig.6.32c). Table 6.4 lists some examples for each of the three classes of lasers.

Laser action has been observed for the first time in ammonia subsequent to high-power pulsed two-photon pumping [6.143]. As depicted in Fig.6.33, by using a pair of TEA CO_2 lasers operating on the P(34) and P(18) lines in the 10.4 μm band, the $2\nu_2^-(5,4)$ state in $^{14}NH_3$ was populated by two-photon excitation. The observed optically pumped laser lines in the ν_2 mode of $^{14}NH_3$ ranged between 12.11 and 35.50 μm.

Table 6.4 Some examples of optically pumped mid infrared lasers

Pump laser	Pump wavelength [μm]	Active molecule	IR laser wavelength	Scheme (Fig.6.32)	Ref.
pulsed CO_2	~10.4	$FClO_3$	16.3-17.7	(a)	6.141
CO_2	~10.4	NOCl	~16	(a)	6.142
CO_2	~10.4	NH_3	6-35	(a)	6.143
CO_2	~9.4 or ~10.4	CF_4	~16	(a)	6.143-145
CO_2	~9.4	HCCD	17.7-21.4	(a)	6.146
CO_2	~9.4	OCS	~19	(a)	6.147
HBr	~4.3	CO_2	~16	(a)	6.148
HF	~2.7	N_2O	~4.6	(a)	6.149
HF	~2.7	HCOOH	~5.7	(a)	6.149
HF	~2.7	CS_2	~6.9	(a)	6.149
CW CO_2	~9.4	CF_4	~16	(a)	6.150
Pulse CO_2	~9.4	OCS	~8.3	(b)	
		CO_2	~10.6	(b)	
		N_2O	~10.8	(b)	6.151
		C_2H_2	~8	(b)	
		CS_2	~11.4	(b)	
HBr	~4.3	N_2O	~10.5,~10.8	(b)	6.152,153
HBr	~4.3	CO_2	~10.6	(b)	6.153,154
Pulse CO_2	~9.4	NH_3	9.3-13.8	(c)	6.155-158
HBr, CO_2	~4.3,~9.4 (or 10.4)	CO_2	~16 or 14	(c)	6.159
CW CO_2	~9.4	NH_3	~12	(c)	6.160-162

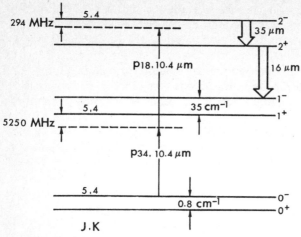

Fig.6.33. Partial energy-level diagram of the ν_2 mode of $^{14}NH_3$ illustrating the two-photon pumping utilized to populate the $2\nu_2^-$ (5,4) state and the laser transitions at 35.50 and 15.88 μm [6.143]

6.10 Applications of Coherent Transient Spectroscopy in the Measurement of Molecular Parameters

The nonlinear coupled interaction of the molecules with the laser field may occur on a time scale which is short even compared with the energy-level relaxation times of the molecules. The main characteristics of the coherent transient phenomena are well explained by semiclassical theory for a two-level system. In this case the reemitted radiation caused by the nonlinearly induced polarization of the molecules does not constitute a new light source for the direct study of molecular spectra, but interferes with the incoming laser fields to generate various superimposed transient signals, containing significant information about molecular parameters. In fact, such optical transient techniques reveal the response of molecular assembles to a sudden resonant excitation or to the sudden cessation of a resonant excitation, offering information about molecular structure and molecular spectra which may be difficult to glean by steady-state spectroscopic studies. The transient technique also allows one to achieve simultaneous high resolution in the frequency and time domains, which is of particular benefit in the study of molecular spectroscopy.

6.10.1 Determination of Dipole Moments for Molecular Transitions

Transition dipole moment is an important parameter in spectroscopy. The conventional determination is carried out by means of absorption measurements. In that case, however, the quantity obtained is always in the form of the product of the square of the dipole moment with the population difference between the transition levels, $|\mu_{ab}|^2(N_a^0-N_b^0)$. To determine μ_{ab} one has to measure $(N_a^0 - N_b^0)$ as well, which is not an easy task for the case of molecules. As numerous vibration-rotational levels are thermally populated, one is in difficulties to know exactly the value of the population difference between two given levels. Fortunately, one kind of coherent transient signal, optical nutation, provides a precise and direct measure of μ_{ab}. As is well known, the optical nutation signal has the form of an oscillation with Rabi frequency χ, which can be precisely defined by finding the zero-signal points in a recorded curve without difficulty. According to the definition of the Rabi frequency,

$$\chi = \mu_{ab} E_0 / \hbar ,$$

μ_{ab} can be determined after E_0 is measured.

Therefore the accuracy of the μ_{ab} determination depends on the accuracy of E_0, which is determined by the stability of the incident laser. Thus although first measurements were made with a pulsed laser [6.163], the Stark-switch technique with a CW laser source [6.164] improved the accuracy of the μ_{ab} measurement to a few thousandth of a Debye.

Recently high contrast Rabi oscillations in the ν_3 band of CH_3F have been observed [6.165]. Using Stark tuning to bring the molecular line into resonance with a CW CO_2 laser, and varying the intensity of the laser, five complete oscillations were clearly observed before the phenomenon was washed out by rapid-passage effects and damping mechanisms. Analysis of the measured data yielded a rotationless transition dipole moment of (0.21 ± 0.01)D.

6.10.2 Lifetime of the Upper Level of a Molecular Transition

Since the nutation signal amplitude is proportional to the initial population difference between the two levels involved, with two driving pulses the amplitude difference between the two nutation signals will reflect the relaxation process of the population distribution. Assuming the time interval between th two pulses to be τ_D, which is smaller than the life time of the upper level, the relation between the two maxima of the nutation signals is in the form:

$$S(\infty) - S(\tau_D) \propto \exp(-\tau_D / T_b) ,$$

where $S(\infty)$ indicates the signal after an earlier pulse an infinite time ago, and T_b is the life time of the upper level [6.163, 166, 167].

6.10.3 Determination of Transition Relaxation

Another kind of optical transient phenomenon, photon echo, is useful for measuring the transition relaxation. This is because the amplitude of the photon echo signal is inversely proportional to the exponential function $\exp(-4\gamma_{ab}\tau_{32})$, where τ_{32} represents the duration of free-induced decay between two driving pulses and γ_{ab} is the inverse value of the transition relaxation time. By changing τ_{32}, one can obtain a series of photon echo signals with different amplitudes I_{max}. The slope of the curve $\ln(I_{max}) \sim \tau_{32}$ is $4\gamma_{ab}$, allowing $T_b = 1/\gamma_{ab}$ to be determined.

With this technique other parameters, such as the pressure broadening coefficient and the collisional cross section between molecules, can also be obtained by varying the sample pressures.

The curve $\ln(I_{max}) \sim \tau_{32}$ does not remain linear for very large values of the time delay τ_{32} [6.168]. This phenomenon was explained by introducing two relaxation terms into the optical Bloch equations for describing the elastic phase-changing collisions. This reveals that the mechanism of the elastic collision is to change the speeds of the collision partners rather than the molecular energies. Based on the analyses one was able to measure the parameters for $^{13}CH_3F$ molecules as

$$\sigma = 430 \text{ Å}^2 ,$$
$$\Delta v = 85 \text{ cm/s} ,$$

where σ indicates the total collision cross section and Δv the characteristic speed-changing step.

6.10.4 The Fine or Hyperfine Splitting of Molecular-Spectral Lines

A suddenly imposed Stark field will cause different displacements for the individual components of a spectral line from their original degenerate levels. They then will present different amounts of detuning from the laser frequency. Therefore their free-induced decay will oscillate at frequencies separated by particular intervals. The interference signal together with the propagating laser field shows a time behaviour similar to the pattern formed by multiple beams interferencing in space. By means of its Fourier transform one can observe very fine molecular-spectral structures. For example, a Doppler-free spectrum within a spectral interval less than the Doppler width (about 66 MHz) and with a linewidth of about 0.5 MHz has been obtained for $^{13}CH_3F$ [6.169]. The transient spectroscopic technique may be used to achieve high resolution in the time domain and in the frequency domain simultaneously. For the study of molecular spectroscopy it therefore has special importance for monitoring samples which have rapid time variations. Recently it has been demonstrated that with incoherent light (i.e., zero-point photon amplifiers) the photon echo signal is still large and still displays echo modulation due to coherent beating of the hyperfine levels, so that high-resolution spectroscopy with time resolution on the order of 100 ns has been achieved using thermal radiation sources [6.170].

6.10.5 Study of Weak Intramolecular Coupling

When the Stark pulses are of large amplitude, the weak interaction mixing levels can be in isolated pairs, one of which is excited by the laser. The levels are mixed only during free precession in small or zero Stark field. In that case the interacting energy-level scheme is of the Raman type, and the signal envelope of the photon echo will show anomalous Raman beat. By measuring the decrease in signal of the echo maximum, the unresolved beat caused by weak coupling (e.g., spin-rotation interaction) may be obtained. The effect has been observed, for example, in an infrared vibration-rotational transition from $(J, K) = (4, 3)$ to $(J, K) = (5, 3)$ of the ν_3 fundamental frequency in gaseous $^{13}CH_3F$, and results from state mixing among spin-rotation states due to a small spin-rotation interaction interaction [6.171]. The Raman beat, corresponding to the spin-rotation splittings in CH_3F molecules, is of the order of only a few tens of KHz, corresponding to a coupling constant of $C \simeq 20$ KHz. One can even measure the increasing of the parameter with external magnetic field [6.172].

6.10.6 Measurement of Ultrafast Inter- or Intra-Molecular Processes by Novel Transient Techniques

While dilute gas samples typically have a relaxation time above 1 ns, that in condensed matter is often far below 1 ps. For the study of such ultrafast relaxation processes ultrahigh-time-resolution coherent transient spectroscopy has been demonstrated during the last few years, which shows the remarkable advantage that the time resolution achieved depends on the correlation time of the beams, rather than on the duration of the light as in all of the other well developed nonlinear spectrsocopic methods in both the frequency and time domains. For exmaple, dephasing times over a thousand times shorter (picosecond) than their pulse duration (10ns) with a bandwidth of 1.5-10 nm were clearly resolved [6.173]; subpicosecond time resolution was obtained by using two delayed nanosecond dye laser beams [6.174]; and picosecond and femtosecond transient spectroscopy yielding information about ultrafast relaxation processes could be performed with a broad band incoherent light [6.175]. These new techniques have provided extremely simple means to meet the needs of the studies of high pressure (buffer gas) low density samples.

7. Simplification and Identification of Molecular Spectra

From Chaps.1 and 2 we have seen that molecular energy levels and the corresponding transition spectra are very complicated. In fact, even the simplest diatomic molecule will give rise to a spectrum far more complicated than an atomic spectrum. An atomic electronic level can take evolution to form two or more molecular electronic states, corresponding to different couples between the valences' spins when the atoms are getting closer. Each of the molecular electronic states contains tens of vibrational levels, each of which in turn contains about a hundred rotational levels.

Figure 2.1 displays a section of the potential curves of the electronic energy states of the alkali dimer Na_2, compared with the corresponding levels of Na. As the spins of the two valence electrons can be parallel or antiparallel, each atomic configuration, e.g., 3s+3s gives rise to two molecular electronic states, the triplet state $a^3\Sigma_u$ (dashed curves) and the singlet state $X^1\Sigma_g^+$ (solid curves). As can be imagined, there is serious overlap between molecular levels.

The complexity of molecular spectra is further increased by the molecular thermal distribution of the large number of levels. Taking Na_2 for example, under common experimental conditions the molecules populate the singlet ground state $X^1\Sigma_g^+$; the population in the first singlet excitation state $A^1\Sigma_u^+$ is negligible (the bottom of the state curve being at 14680 cm^{-1}). In fact, however, the number of populated vibration-rotational levels is amazing. Table 7.1 lists the relative population at 500°C of the levels with the vibrational quantum number $v_x = 0$-6 and the rotational quantum number $J = 0$-100 in the ground state, evaluated from the Maxwell-Boltzmann distribution law

$$N(v, J) \propto g_J \exp[-E(v, J)/KT]$$

$$= (2J + 1)\exp\{-[G(v) + F(J)]hc/KT\} , \tag{7.1}$$

where the spectral terms $G(v)$ and $F(J)$ were calculated with the molecular constants from [7.1]. The table reveals that the conventional absorption transitions can start on hundreds of vibration-rotational levels in the ground electronic state. The resulting molecular absorption spectra would be fairly complicated and have great overlaps, so that some close lines might even fall within the Doppler profile of a transition line. Thus even small molecules produce spectra complicated enough that their analysis is complex and difficult.

If one confines oneself to the study of only molecular ground states, far infrared or infrared spectroscopy is quite effective. For the study of

210

Table 7.1. Relative thermal population on the vibration-rotational levels of the ground electronic state in Na_2 at 500° C

J	v = 0	v = 1	v = 2	v = 3	v = 4	v = 5	v = 6
0	0.86	0.64	0.48	0.36	0.27	0.20	0.15
20	31.36	23.40	17.51	13.13	9.88	7.45	5.64
40	43.76	32.72	24.53	18.44	13.90	10.51	7.97
60	37.02	27.77	20.89	15.76	11.92	9.05	6.88
80	22.59	17.03	12.87	9.76	7.42	5.66	4.32
100	10.62	8.05	6.12	4.67	3.58	2.74	2.11

molecular excited states, however, it is necessary to analyze the molecular spectra in the visible and ultravisible regions. Due to the overlap of most spectral bands, together with inaccurate knowledge of molecular constants or of the perturbation status of the two related electronic states, the identification of every spectral line is excessively tedious and difficult. Extreme efforts must be undertaken even for a part of the conventional visible or UV molecular spectra. For example, *Kusch* and *Hessel* have measured some 11 000 absorption lines in analyzing the sodium (Na_2) $X^1\Sigma_g^+ \rightarrow B^1\Pi_u$ system [7.2]. Even for molecules as simple as Na_2 only six excited singlet u–parity electronic states have been identified after more than 50 years of great effort by many spectroscopists using methods of conventional spectroscopy. To explore effective approaches for the simplification of complicated spectra and straightforward methods to fill in the wide gaps in molecular energy level constants without ambiguity is still a target of great significance for molecular spectroscopists.

The physical basis for the simplification of molecular spectra is the selective excitation of one or a few transitions. A necessary substantial condition is to find a suitable light source, while ingenious spectral techniques are often richly rewarding. *Wood*, for example, discovered that resonance fluorescence could be used to simplify molecular spectra [7.3]. However, the simplification with conventional sources is limited by (i) the low monochromatic brightness. Almost always when the spectral width of a conventional source is narrowed to excite one or few molecular transitions, the radiated energy from the source is too low to observe a simplified spectrum; (ii) atomic lamps might have enough monochromatic brightness but only suit the occasional coincidental molecular transition; (iii) the energy level information obtained from resonant fluorescence is limited to molecular ground states.

Lasers, with their incomparable monochromaticity, beam collimation and brightness, provide new possibilities for selective and nonlinear studies of complicated molecular spectra. Soon after the appearance of the wide tunable dye laser [7.4], great progress has been made in various spectroscopic approaches to the simplification, identification and analysis of molecular spectra. Not only has the method found by *Wood* been further expanded, which is to obtain the ground-state information from a given final level of a molecular transition, but also the inverse methods, i.e., to obtain information about molecular excited states from a given initial level, have been proposed and realized. Furthermore the method of utilizing a chosen intermediate level to obtain the molecular constants in the high-lying states has become a powerful technique [7.5]. *Schawlow* and his students have sought systematic ways to use lasers to simplify molecular spectra so as to identify their various states. For example, by means of two-step polarization labelling spectroscopy they have, in a few years, observed and identified more than 24 high-lying electronic states in Na_2. For larger values of the principal quantum number in the electron configuration these are molecular Rydberg states. This work has made a great contribution to the volume of data on molecular constants. None of the recently developed approaches and splendid results would have been achieved without the invention of the tunable laser, together with clever spectroscopic techniques.

7.1 Laser-Induced Fluorescence

7.1.1 Formation and Classification of Resonant Fluorescence

The so-called resonant-fluorescence technique involves the analysis of the spontaneous emission spectra from a certain upper level, chosen by a resonant absorption transition, to all of the levels allowed by the selection rules. The classification of the type of resonant fluorescence refers to the excitation processes. Figure 7.1 illustrates three possible kinds of resonant-fluorescence processes induced by laser sources, where the thick arrows represent the excitation processes and the downward arrows indicate the formation of the resonant fluorescence spectra. Figure 7.1a is the most basic of the three schemes. Following a single transition excitation, there is one v''-progression of bands for a given v', satisfying the following selection rules (e.g., for a diatomic molecule):

$$\Delta v = v' - v'' = 0, \pm 1, \pm 2, \dots , \tag{7.2}$$

$$\Delta J = J' - J'' = \begin{cases} \pm 1 \ (\Sigma \longleftrightarrow \Sigma \text{ transitions}) \\ 0, \pm 1 \ (\text{other transitions}) . \end{cases} \tag{7.3}$$

These rules mean that for a given v' and J' the v''-progression of bands is long whereas each band is short, containing only one line for each branch of a band. For a homonuclear diatomic molecule there are additional selection rules for the change in symmetry of the rotational wave function:

$$+ \longleftrightarrow - , \quad + \longleftrightarrow\!\!\!/\longrightarrow + , \quad - \longleftrightarrow\!\!\!/\longrightarrow - , \tag{7.4}$$

and the parity of electronic wave functions

$$u \longleftrightarrow g , \quad u \longleftrightarrow\!\!\!/\longrightarrow u , \quad g \longleftrightarrow\!\!\!/\longrightarrow g . \tag{7.5}$$

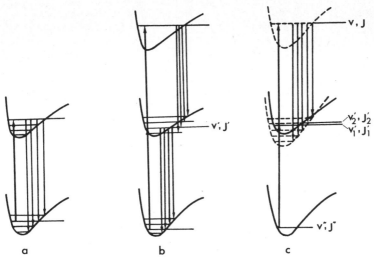

Fig.7.1. The schemes of the possible fluorescence processes. (a) Following a single transition excitation; (b) following a near-resonant molecular two-photon transition; (c) following an optical-optical double resonance

These two sets of constraints might restrict the branches which appear in a given v''-progression. A good example was given by *Demtröder* et. al. [7.6] and is depicted in Fig.7.2. They had recorded the v''-progression in Na_2, containing as many as 20 bands. With the 488.0 nm line from an argon laser as the source the pumping excitation is a Q-branch transition, and the selection rules (7.3,4) allow the ensuing Q-branch emission (Fig.7.2a); whereas with 476.5 nm as the source (P-branch excitation), P and R partners are allowed (Fig.7.2b). However, the frequency intervals between successive lines in a given branch are similar in both cases, corresponding to the molecular constants of the electronic ground state.

Figures 7.1b and c show the laser-induced resonant-fluorescence processes following a near-resonant molecular two-photon transition or Optical-Optical Double Resonance (OODR). This also gives a simplified fluorescence spectrum consisting of the bands with one line in a branch. The difference between the two-photon fluorescence spectra from the above-mentioned one-photon fluorescence spectra is that there are two band progressions rather than one. The graphical representations of part of a two-photon fluorescence sepctrum with the laser frequency at (a) 16704.169 cm^{-1} and (b) at 16601.8027 cm^{-1} in Na_2 are exhibited in Figs. 7.3a and b, respectively. Owing to the possibility of a real population in the intermediate state in the near-resonant enhanced case, the photographs consisted of both v'- and v''-progressions from a chosen upper level (v,J) and the chosen intermediate level (v',J'), respectively. Therefore, in addition to providing the information on the molecular constants in the ground state as does one-photon laser-induced fluorescence, it also provides this information on the intermediate electronic state. Moreover the two progressions have one line in common, which is right at the laser wavelength. Due to the

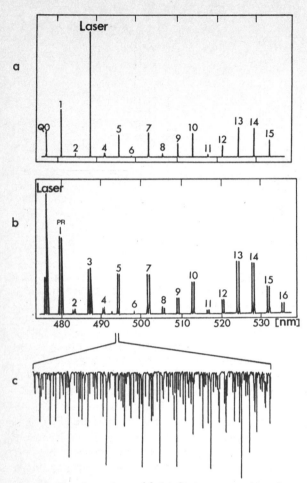

Fig.7.2. The dependence of the fluorescence spectral structure on the excitation species in Na_2. (*a*) Q-branch excitatin upper level is $v' = 6$, $J' = 43$; (*b*) P-branch excitation, upper level is $v' = 6$, $J' = 27$. (*c*) The original spectral pattern of the selected vibration-rotational band [7.6]

line overlap and the laser light scatterd from the spectrometer, the common line in the two progressions can be easily recognized. Referring to the location of the common line, we can then determine the properties of the two-photon excitation branch. In the case of Fig.7.3a the two-photon transition at 16704.169 cm^{-1} belongs to an S branch because both partners of a common branch, located near the laser-freqeuncy position, are on the longer-wavelength side, being P-branches in emission because of the one-photon absorption and the first decay step of the two-photon absorption. The procedure was repeated for the 16601.8027 cm^{-1} excitation, which was described in detail by *Woerdman* [7.7, 8]. As can be seen in Fig.7.3b the two partners near the laser line are on opposite sides, indicating a Q_2-branch transition (as defined in Chap.4). Examination of all of the lines appearing in pairs reveals that the upper state is a Σ state.

214

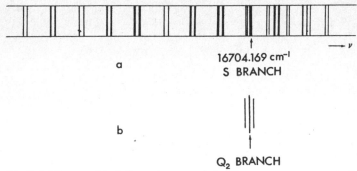

b

Q_2 BRANCH

Fig.7.3. The graphical fine structure of the fluorescence spectra in Na_2 following the near-resonant molecular two-photon excitations with the laser frequency at (a) 16704.169 cm⁻¹ as a S-branch excitation and (b) 16601.8027 cm⁻¹ as a Q_2-branch excitation [7.27]

In the wide overlap area of the singlet and the triplet states, as shown in Fig.2.1, there are dozens of levels with mixing wave fucntions. Provided a mixing level is chosen to be the intermediate level of a two-photon or OODR transition, the final level may be a quite pure triplet level, which normally cannot be reached by an allowed transition from a singlet level in the ground state. Therefore the laser-induced fluorescence spectra from the upper level might offer simplified spectra of a band progression, corresponding to the intermediate perturbating state, which was not clearly known yet as it is spin forbidden from the ground state. For example, recently *Li* et. al. have directly observed the $b^3\Pi_u$ state in Na_2 by OODR $^3\Lambda_g$ $-b^3\Pi_u$ fluorescence spectroscopy [7.9]. As mentioned in Chap.2, there is an additional quantum number Ω introduced to describe an S≠0 state. For the $^3\Pi$ state Ω takes the values 2, 1, 0. Therefore a transition dealing with S≠0 states should obey the selection rule

$$\Delta\Omega = 0, \pm 1 . \tag{7.6}$$

Thus there are three different kinds of fine structures. A complete fine structure of a simplified band consists of three pairs of R-P partners, provided the upper level is chosen to be $\Omega = 1$ as shown in Fig.7.4. The distinguishing fine structure from the laser-induced fluorescence spectra depicted in Figs.7.2,3 puts its recognition beyond doubt. The advantage of this over previous experiments utilizing fluorescence excitation spectroscopy is that it does not require that all observed levels of the $b^3\Pi_u$ state contain a significant admixture with a nearby $A^1\Sigma_u^+$ level. The v = 0 level of the $b^3\Pi_u$ state was observed for the first time, the absolute vibrational numbering of the $b^3\Pi_u$ state in molecular sodium Na_2 has been established, and more accurate molecular constants of the forbidden state have been obtained.

The laser-induced resonant fluorescence from high-lying states provides abundant information on molecular energy levels. The complete fluorescence spectra consist of all of the allowed spontaneous emissions to the

215

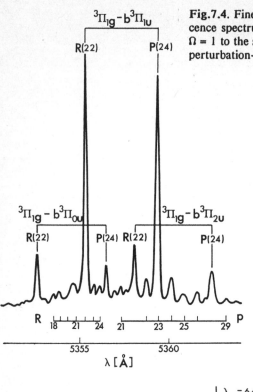

Fig.7.4. Fine structure of the resonance fluorescence spectrum from the upper state $^3\Pi_{0g}$ with $\Omega = 1$ to the states $b^3\Pi_{0u}$ ($\Omega = 0, 1, 2$) following a perturbation-facilitated OODR [7.9]

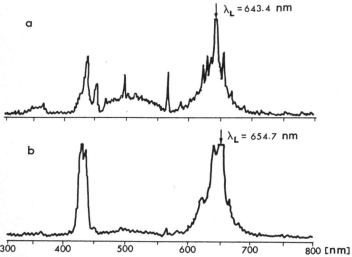

Fig.7.5. Low resolution fluorescence spectra from 300.0 to 800.0 nm in Na_2 with excitation wavelengths at (a) 643.4 nm and (b) 654.6 nm [7.10]

many lower electronic states and cover a wide wavelength region. As an example, Figs.7.5a and b exhibit two-photon fluorescence spectra recorded at low resolution from 300.0 to 800.0 nm with the excitation wavelengths at 643.4 and 654.6 nm, respectively, using a YAG pumped dye laser [7.10].

216

The different regional signals in Figs.7.5a and b imply different properties of the corresponding upper levels. According to the wavelength ranges of the various signals, and the known electronic state distributions, we can conclude that in Fig.7.5b the upper level is a quite pure triplet state while in Fig.7.5a the upper level exhibits singlet-triplet mixing. In principle, provided with enough signal-to-noise ratio, one can analyze the simplified spectra of the fluorescence regions one by one. From each region we can obtain the molecular constants of the respective electronic states, as explained in Sect.7.7.

7.1.2 Measurements and Applications

For one-photon or equal-frequency two-photon excitation one laser is enough, while in the OODR excitation approach two laser sources are necessary. The experimental arrangement in common use is to collect fluorescence signals at right angles to the exciting laser beams, so a side window is needed for an opaque sample cell. For example, in one case a cross oven was employed, which leaves one of the side windows available for monitoring the wavelength shift through signal fluctuations. The fluorescence signals from the other side window were fed to a spectrometer. Because the spectral line densities are reduced to about a hundredth of conventional levels, a medium size spectrometer is sufficient for the observation of the fine structure of a fluorescence spectrum. With the sample cell held in the horizontal direction a Dove prism is inserted between the side window and the spectrometer to rotate the fluorescence image to fit the slit of the spectrometer to improve the fluroescence collection, which is particularly important for recording weak two-photon fluorescence spectra [7.11].

The benefit of laser-induced fluorescence spectroscopy is obvious, especially for the molecules with smaller force constants (smaller vibrational constants) but larger moments of inertia (smaller rotational constants). The analysis of the simplified spectra is so easy that even an inexperienced student can do it. Moreover, the energy-level information provided by this form of spectroscopy is rich. Vibrational quantum numbers v'' as high as 84 (in I_2) and the $v' = 0$ level of a lower electronic state have been observed. Therefore accurate energy-level constants including higher order corrections can be determined. Easy analysis, rich information and inexpensive spectral apparatus form the main advantages for its wide application. Dozens of homonuclear or heteronuclear diatomic molecules and many kinds of polyatomic molecules have been studied to measure their electronic excitation energy, vibrational and rotational constants, dissociation energy, potential curves, Franck-Condon factors, perturbation positions and shifts, electronic-state fine splittings, energy transfer by collisions, and other important parameters and mechanisms.

7.2 Population Labelling

The above-discused laser-induced fluorescence is an important method for sufficiently complex molecules, but the lower levels which it can explore are usually the ones most likely to be already known.

Instead of chosen an upper level as in laser-induced fluorescence spectroscopy, a new method to label a low-lying level was demonstrated by *Kaminsky* et. al. [7.12]. They have made use of the high monochromatic brightness of lasers to pump a considerable fraction of molecules from a lower level to an upper level and have found a way to pick up the signals corresponding to the transitions form the chosen *lower* level to all of the allowed levels satisfying the selection rules. The energy-level scheme for population labelling, which is illustrated in Fig.7.6, looks like the inverse of that in Fig.7.1a. However, the investigation of the lower level is performed by using two laser beams rather than one as in the one-photon or two-photon fluorescence cases depicted in Figs.7.1a and b. When light from an intense pump laser is absorbed by the medium, it excites molecules from some particular lower level and thereby reduces the population of that level. Consequently, the absorption coefficient will be reduced for all absorption lines originating from the chosen lower level. If the pump laser is turned on and off by a mechanical chopper, a probe beam from a tunable laser will be modulated whenever it is tuned to any absorption line from the level whose population is being modulated. Therefore, by using a lock-in amplifier one can pick up the signals from the transmitted probe beam which is constant before entering the sample cell and thus obtain a simplified spectrum consisting of a v'-progression.

The scheme shown in Fig.7.6 can be visualized as two laser beams interacting with a three-level system. Let a and b represent the lower and upper level of the pumping transition, respectively, and c the upper level of the probe transition. Then Fig.7.6 can be reduced to Fig.7.7(*1*). There are other processes following the pumping which cause population modulation in some levels, as indicated in Fig.7.7 (*2*)-(*3*). Different types of population

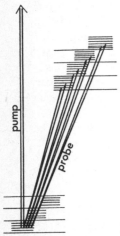

Fig.7.6. Illustration for population labelling spectroscopy

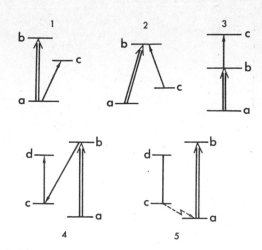

Fig.7.7. Various schemes of population-modulated spectroscopy. (*1-3*) three-level system; (*4,5*) four-level system. The dotted line in (*5*) represents a nonradiating transition [7.12]

labelling schemes can be distinguished by examining the phases and linewidths of the signals. Clearly, in the schemes of both Fig.7.7(*1*) and (*2*) the existence of the pumping beam will decrease the population difference between the lower level and the upper level of the probe transition, and increase the transmission of the probe beam. Thus the modulated component of the probe beam will be in phase with the pumping beam. Case (*2*) should not appear often if one uses two lasers with a wide wavelength gap. In the case of Fig.7.7(*3*), on the other hand, the modulated component of the probe beam is out of phase with the pumping beam. Fig.7.7(*4*) and (*5*) show the four-level system cases. By means of fluorescence from the upper level b of a pumping transition to the lower level c of the probe transition, the population in c increases and enhances the absorption from c to d. Therefore this scheme will give a probe signal with an opposite phase to the pumping beam. Due to the potential for rapid equilibrium of the population in ground states the decrease of the number of molecules in level a through optical pumping will be replenished by the nonradiative transition from a few neighbouring levels represented by c if the pressure is high enough. This causes the probe absorption from c to d to be weakened with a modulation factor in phase with the pumping beam. However, the last two cases deal with relaxation processes in which the transition linewidth of the probe absorption is increased. Within the scanning range of a cw dye laser from 560.0-617.5 nm 119 lines (an extremely simplified molecualr spectrum compared with the conventional one) have been observed [7.12]. 114 of these 119 lines have been recognized to belong to type (*1*) and thus to the same electron-excited state. The whole spectral appearance of the v′-progression should be similar to the v″-progression depicted in Fig.7.2. However, the v′-progression might include perturbation effects in line shapes and frequency positions at the upper levels of the probe transitions. As an example, Fig.7.8 displays the deviations of the observed R,P doublets and the least square quadratic fitting curve (solid curve) from the calculated frequencies for the observed progression by using the molecular constants published by *Kusch* and *Hessel* at that time [7.13]. The gradually increasing

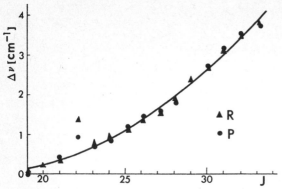

Fig.7.8. The observed line shifts from calculated values in Na_2; the chosen lower level is $X^1\Sigma_g^+$ (3.13) [7.2]

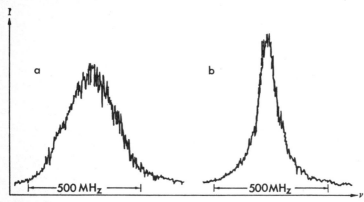

Fig.7.9. Line-shape distortion of scheme type 1 in Fig.7.7a, (a) upper level c perturbed; (b) upper level c normal [7.12]

deviations of the curve obviously means that the known molecular constants used in the calculation needed a further adjustment, especially for high v' levels. Note that the figure shows an abnormal deviation from the smooth curve at the $v' = 22$ level; the $(v', J') = (22,14)$ level is shifted even more than the $(22,12)$ level with a perturbation shift beyond 1 cm^{-1}. The line shape associated with a perturbated level is distorted, as indicated in Fig.7.9. Thus the perturbated lines are easily recognized even though they are far from the expected positions.

This method of labelling a lower level by modulating its population is quite similar to the saturation-spectroscopic technique, except that the probe beam comes from a second laser which is tuned to wavelengths other than the pumping line. Even more complicated molecular spectra can be simplified by labelling techniques. In a polyatomic molecule the number of vibrational modes linearly increases with increasing number of atoms composing the molecule. Consequently the spectrum can be so dense that the spacing between lines is less than the natural linewidth determined by the life times of the excited state. Then one could not avoid pumping more

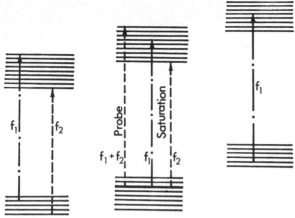

Fig.7.10. A proposed multi-modulated population labelling spectroscopy for polyatomic molecules. (*Right:* One pump beam modulated at frequency f_1, and *left:* Two pump beams modulated at different frequencies f_1 and f_2. In either case each of the pump beams may be in resonance with numerous vibration-rotationaltransitions. But few of them may have common levels which are therefore labeled at the sum frequency (f_1+f_2). *M:* Two beams modulated at different frequencies pump one level, whose population is then altered at the sum frequency. The probe beam monitored at (f_1+f_2) will provide a selectively simplified spectrum [7.14]

than one level, no matter where the pump laser is tuned. The labelled spectrum would then contain lines from the two or more pumped levels. To separated them A.L. Schawlow has suggested the use of a second laser to pump one of the absorption lines labelled by the first laser [7.14]. Any line that is labelled by both pump wavelengths must then originate on their common lower level. The two lasers could be used simultaneously with different chopping frequencies, so that the line having a lower level in common would be modulated at the sum of the chopping frequencies. The idea is illustrated in Fig.7.10, where the heavy horizontal line means the common lower level and the left dashed line in the middle scheme represents the probe signal detectable at the sum frequency (f_1+f_2).

7.3 Polarization Labelling

The principle of polarization labelling is based on the laser-induced anisotropy of a sample medium, as discussed in Chapt.5. The term describing the induced anisotropy is $\Delta\alpha$ in (5.16), corresponding to the difference in absorption between the two polarization components divided from a probe beam. A detailled discussion of the transition dependence of $\Delta\alpha$ is required before one can contemplate the identification of molecular transitions by means of polarization labelling.

Of course, an intense pump beam is needed to obtain an appropriate anisotropy. But one should note that with an unduly high intensity the pump beam might saturate all of the sublevel transitions. For optimum in-

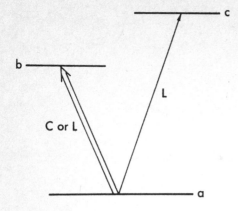

Fig.7.11. Energy level scheme for polarization labelling spectroscopy, The pump beam can be circularly (c) or linearly (L) polarized, whereas the probe beam is al-

duced anisotropy, the perturbation treatment is valid. For the polarization-labelling technique the energy-level scheme is as exhibited in Fig.7.11, which obviously refers to the case of a three-level system. The general motion equation of the molecules is written as (4.24,25). The counter-propagating radiation fields can be written as the combination of two independent travelling fields with complex amplitudes

$$E = E_1 \exp[i(\omega_1 t + k_1 z)] + E_2 \exp[i(\omega_2 t - k_2 z)] + \text{c.c.} , \tag{7.7}$$

where the complex amplitudes E_1 and E_2 correspond to linear or circular polarization with identical or different modulation conditions.

For a three-level system interacting with such a combined field it is more convenient to expand the elements of the density matrix into a double-power series $\rho_{ij}^{\alpha\beta}$. Here α and β indicate the interacting times of E_1 and E_2 fields, respectively [7.15].

For the polarization-labelling process depicted in Fig.7.11, the interaction approach can be written as (for the sake of brevity, $\rho_{bb}^{00} = \rho_{cc}^{00} = 0$ are assumed):

$$\rho_{aa}^{00} \xrightarrow{E_1} \rho_{ab}^{10} \xrightarrow{E_1} (\rho_{aa}^{00} - \rho_{aa}^{20}) \xrightarrow{E_2} \rho_{ac}^{21} . \tag{7.8}$$

This implies that the pump light is linearly absorbed, and should encounter an absorption coefficient similar to (5.8). For the usual case the pump frequency is in resonance, $\omega_1 = \omega_{ab}$. Then the formula is reduced to

$$\alpha(\omega_1 = \omega_{ab}) = \frac{4\pi^{1/2}}{\hbar\nu} N_a^0 |\mu_{ab}^{(1)}|^2 . \tag{7.9}$$

The absorption cross section of the pump beam is

$$\alpha_1 = \alpha/(N_a^0 - N_b^0) = A |\mu_{ab}^{(1)}|^2 . \tag{7.10}$$

According to quantum mechanics, the matrix elements of the dipole moment in (7.10) can be calculated generally by

$$|\mu_{ij}|^2 \infty |<n_j \Lambda_j J_j M_j| \mathcal{T}(q)|n_i \Lambda_i J_i M_i)|^2 , \tag{7.11}$$

where $\mathcal{T}(q)$ is the tensor operator for dipole transitions. The rotational eigenfunction, e.g., for transitions between singlet electronic states in the Hund case (a) approximation, can be represented by [7.15]

$$|n\Lambda JM) = (2J + 1/8\pi)^{1/2}|n\Lambda)D_{\Lambda M}^J(\omega) , \tag{7.12}$$

where $D_{\Lambda M}^J(\omega)$ is the Edmond's D Function. By means of Edmond's treatments, (7.11) can be rewritten as

$$|\mu_{ij}|^2 = A'S_{J_i J_j}^{\Lambda_i \Lambda_j} \begin{pmatrix} J_j & 1 & J_i \\ -M_j & M_j - M_i & M_i \end{pmatrix}^2 , \tag{7.13}$$

where A' is a factor independent of the angular momentum of energy level; $S_{J_i J_j}^{\Lambda_i \Lambda_j}$ is the spectral strength factor (Honl-London factor) using the 3-j symbol, which can be expressed as

$$S_{J_i J_j}^{\Lambda_i \Lambda_j} = (2J_i + 1)(2J_j + 1) \begin{pmatrix} J_j & 1 & J_i \\ -\Lambda_j & \Lambda_j - \Lambda_i & \Lambda_i \end{pmatrix}^2 . \tag{7.14}$$

The 3-j symbol $\begin{pmatrix} J_j & 1 & J_i \\ -M_j & M_j - M_i & M_i \end{pmatrix}$ is the only factor in (7.13) related to the polarization properties of the light fields, where the selection rules of the magnetic quantum numbers are

$$q \equiv M_j - M_i = \begin{cases} +1 & \text{for right circularly polarized field} \\ -1 & \text{for left circularly polarized field} \\ 0 & \text{for linearly polarized field .} \end{cases} \tag{7.15}$$

Substituting (7.11-15) into (7.10) and summing over all M values, we can obtain the M dependence of the absorption cross section for the pump beam

$$\sigma_{J_a M_a J_b M_b} \equiv \sigma_1 = C(2J_a+1)(2J_b+1) \begin{pmatrix} J_b & 1 & J_a \\ -\Lambda_b & \Lambda_b - \Lambda_a & \Lambda_a \end{pmatrix}^2 \begin{pmatrix} J_b & 1 & J_a \\ -M_b & q & M_a \end{pmatrix}^2 \tag{7.16}$$

where C is a factor which does not depend on the angular momentum. The last two squared factors do arise in the so-called 3-j symbols instead of ordinary vector-coupling coefficient for the sake of briefness during rec-

223

ursion calculation of the transition probabilies. Each of the middle quantum numbers in a 3-j symbol corresponds to the difference of its right from its left one. Their computation relations with the ordinary vector-coupling coefficients are tabulated in [7.15]. Eq.(7.16) shows that under irradiation by a polarized pump beam the allowed absorptions from the sublevels will be unequal, so that the population on the related sublevels will be nonuniform. This is the so-called induced orientation or anisotropy of molecules. The last 3-j symbol, defined as polarization factor of the absorption cross section $\sigma_{J,J+\Delta J}$, can be wirtten in more explicit forms:

For *circularly polarized light*

$$\sigma^{\pm}_{J,(J+\Delta J),M} = \begin{cases} \sigma^0_{J,J+1}(J \pm M + 1)(J \pm M + 2) & \text{for R branch} \\ \sigma^0_{J,J}(J \pm M)(J \mp M + 1) & \text{for Q branch} \\ \sigma^0_{J,J-1}(J \mp M)(J \mp M + 1) & \text{for P branch}, \end{cases} \quad (7.17)$$

and for *linearly polarized light*

$$\sigma_{J,(J+\Delta J),M} = \begin{cases} \sigma^0_{J,J+1}(J + 1)^2 - M^2 & \text{for R branch} \\ \sigma^0_{J,J}M^2 & \text{for Q branch} \\ \sigma^0_{J,J-1}(J^2 - M^2) & \text{for P branch}, \end{cases} \quad (7.18)$$

where the superscripts $\pm$ in (7.17) refer to the cases with right (rhc) - and left (lfc) - circularly polarized pump light, respectively, and M indicates the magnetic quantum number of the lower level of the transition. The M dependences of σ^+, σ^- and σ are illustrated in Fig.7.12, corresponding to rhc, lhc and y (linear) polarization, respectively. The inducing of anisotropy can be explained in more detail as follows:

1) In the case of a circularly polarized pumping beam

Assume the pumping beam is a right circularly polarized beam and is resonant with a molecular R-branch transition ($\Delta J = J' - J'' = +1$). According to (7.15), we have $\Delta M = +1$. From (7.17) and Fig.7.12a we know that the absorption increases with M, and therefore is such as to cause population nonuniformity among sublevels, as illustrated in Fig.7.13a, where the thicknesses of the sublevels (M levels) present the initial population distributions on them and the thickness of the transition arrows present the magnitudes of the absorption cross sections. In fact, the resultant nonuniformity of the populations on sublevels corresponds to the nonuniform variations of population density, ρ^{20}_{aa}, in (7.8). Following the steps in Chap.5 we obtain the solution for ρ^{20}_{aa} from (7.8)

$$\rho^{20}_{aa} = d_1 |E_1|^2, \quad \text{where} \quad (7.19)$$

$$\begin{aligned} d_1 &= (\text{Im}\{\rho^{00}_{aa}/\Delta_1 \hbar^2 \gamma_a\}) |\mu^{(1)}_{ab}|^2 \\ &= [2n^0_a \gamma_{ab}/\gamma_a \hbar^2 (\omega_1 - \omega_{ab} - k_1 v_z)^2 + \gamma_{ab}^2] |\mu_{ab}|^2. \end{aligned} \quad (7.20)$$

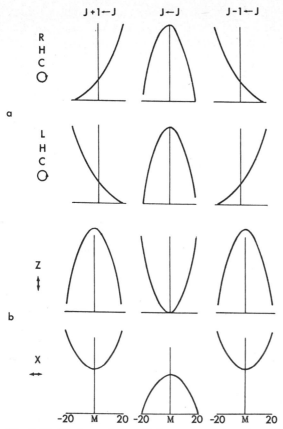

Fig.7.12. M-dependence of the cross section absorbing (a) circularly poalrized light and (b) linearly polarized light

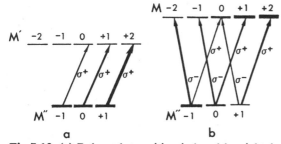

Fig.7.13. (a) R-branch transition induced by right-hand circularly polarized pumping beam; (b) R-branch probe transition following the pump transition in (a). The linearly polarized probe beam was divided into right- and left-hand polarized components. The widths of the arrows represent the magnitudes of the absorption cross sections; the widths of the horizontal lines represent the relative population of sublevels

Now let us consider the absorption of the probe beam. We divide the probe beam into two orthogonal components with respect to the pump beam, e.g., for a circularly polarized pump beam, the linearly polarized probe

225

beam should be divided into right and left circularly polarized beams with equal amplitudes. As expressed in (7.8), the probe transition starts on sublevels of the common level with populations exhibiting partial anisotropy on account of the pump beam given by $(\rho_{aa}^{00} - \rho_{aa}^{20})$ (partial anisotropy). Assume that the probe beam is also R-branch resonant, then the two upper curves in Fig.7.12a are available for probe components. The mirror symmetry of the M dependence of the relevant absorption cross sections means that only the nonuniform component ρ_{aa}^{20} among $(\rho_{aa}^{00} - \rho_{aa}^{20})$ will cause an appreciable absorption difference between the two probe components, so that the third-order solution is required. Following the steps in Chap.5, we obtain

$$\rho^{21} = V_2(\rho_{aa}^{00} - \rho_{aa}^{20})/\Delta_2 . \tag{7.21}$$

The absorption of the probe beam goes as $\mathrm{Im}\{\chi^{(1)} + \chi^{(3)} E_1{}^2\}$. Therefore, similar to the form of (5.9), we obtain the modified formulas for the absorption coefficients of the two probe components:

$$\alpha^+ = D|\mu_{ab}{}^+|^2 |\mu_{ac}{}^+|^2 ,$$
$$\alpha^- = D|\mu_{ab}{}^+|^2 |\mu_{ac}{}^-|^2 , \tag{7.22}$$

where $\Delta\alpha^+$ is the nonlinear absorption of the σ^+ component of the probe beam, and $\Delta\alpha^-$ is that of σ^- caused by a σ^+ pump beam; D is proportional to $(\rho_{aa}^{00} - \rho_{aa}^{20})$ referring to line shape and is independent of the angular momentum of the molecule. Therefore the difference in absorption coefficients of the probe components

$$\Delta\alpha = \alpha^+ - \alpha^- = D|\mu_{ab}{}^+|^2 (|\mu_{ac}{}^+|^2 - |\mu_{ac}{}^-|^2) . \tag{7.23}$$

Substituting (7.13,15) into (7.23) and summing up over all M levels we have

$$\Delta\alpha = \frac{F}{2J+1} + \sum_M S_{J_a J_b}^{\Lambda_a \Lambda_b} S_{J_a J_c}^{\Lambda_a \Lambda_c} \begin{pmatrix} J_b & 1 & J_a \\ -M_b & 1 & M_a \end{pmatrix}^2$$

$$\times \left[\begin{pmatrix} J_c & 1 & J_a \\ -M_c & 1 & J_a \end{pmatrix}^2 - \begin{pmatrix} J_c & 1 & J_a \\ -M_c & -1 & M_a \end{pmatrix}^2 \right] . \tag{7.24}$$

D in (7.23) and F in (7.24) are independent of angular momenta. The polarization factor $\varsigma_{J_a J_b J_c}^c$ is introduced to represent the result of summing over all the M levels, which is given by

$$\varsigma_{J_a J_b J_c}^c = 9(2J_a + 1) \sum_{M_a M_b M_c} \begin{pmatrix} J_b & 1 & J_a \\ -M_b & 1 & -M_a \end{pmatrix}^2$$

$$\times \left[\begin{pmatrix} J_c & 1 & J_a \\ -M_c & 1 & M_a \end{pmatrix}^2 - \begin{pmatrix} J_c & 1 & J_a \\ -M_c & -1 & M_a \end{pmatrix}^2 \right]. \tag{7.25}$$

Substituting it into (7.24), we have

$$\Delta\alpha = \frac{F}{(2J_a + 1)(2J_b + 1)} S_{J_a J_b}^{\Lambda_a \Lambda_b} S_{J_a J_c}^{\Lambda_a \Lambda_c} \varsigma_{J_a J_b J_c}^c < 0. \tag{7.26}$$

The above formula means that the probe beam experiences depolarization, depending on the polarization of the pump laser as well as on the angular momenta for the pump and probe transitions. The result, $(\alpha^+ - \alpha^-) < 0$, is illustrated in Fig.7.13b, where the thicknesses of the arrows for the two components σ^+ and σ^- of the probe beam are distributed symmetrically (Fig.7.11a), but the nonuniform thicknesses of the lower levels resulting form the pump beam cause an absorption difference between the components σ^+ and σ^-, i.e. the total absorption of σ^- will be larger than that of σ^+. Thus the transmission of the probe beam out of the region of overlap with the pump beam in a sample cell is composed of two opposite-circularly polarized components with unequal amplitudes rather than equal amplitudes as in the incident case. Hence the probe beam is elliptically polarized, and is therefore different from other probe transitions without a common level with the pumping beam. This process is illustrated in Fig.7.14. The elliptically polarized probe beam can partially pass through the crossed analyzer and reach the photodetector, while all the other transitions of the probe beams starting from any isotropic levels will be eliminated.

Other cases, such as when the pumping beam is left circularly polarized or resonant with a P-branch transition, can be analyzed in a similar fashion. Note the case of circularly polarized light resonant with a Q-branch transition. Checking the middle curves in Fig.7.12a we know that the pump beam will result in the molecules being partially aligned (same

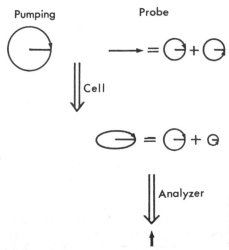

Fig.7.14. Illustration of the influence of the circularly polarized pumping beam on the polarization of a probe beam and of its discrimination

227

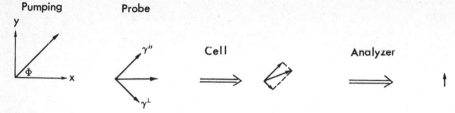

Fig.7.15. Illustration of the influence of the linear polarized pumping beam on the polarization of the probe beam and of its discrimination

absolute value of $|M|$, same population). Thus the probe resonant with neither branch could produce an appreciable absorption difference and would in no way give a simplified molecular spectrum. This implies that the saturating transition $\Delta J \neq 0$ is preferable when the saturating (pumping) beam is circularly polarized.

2) In the linearly-polarized pumping-beam case:

If a linarly polarized pumping beam is used, one should decompose the probe beam into two orthogonal components, one parallel and one perpendicular to that of the pump beam (Fig.7.15). Obviously, due to the depleting effect of the pump beam on the population of the lower levels, the absorption of $\gamma^{\parallel}$ will be less than that of $\gamma^{\perp}$. Consequently the recomposed probe beam behind the sample cell will present a polarization deviating from the original direction, and it will partially pass a crossed analyzer in front of the detector. Following the same steps as were used for a circularly polarized pump laser, the detectable signal intensity I^L of a probe beam due to the linearly polarized pump beam has an expression similar to (7.26), where ς^C is replaced by ς^L, i.e.,

$$\alpha^{\parallel} - \alpha^{\perp} = \frac{F}{(2J_a + 1)(2J_b + 1)} S^{\Lambda_a \Lambda_b}_{J_a J_b} S^{\Lambda_a \Lambda_c}_{J_a J_c} \varsigma^L_{J_a J_b J_c} . \tag{7.27}$$

Here the S represents the line strengths given by (7.14), ς^L is the polarization factor representing the depolarization experienced by the probe due to linearly polarized beam pumping, which is expressed by [7.17]:

$$\varsigma^L_{J_a J_b J_c} = 9(2J_a + 1) \sum_{M_a M_b M_c} \begin{pmatrix} J_a & 1 & J_a \\ -M_b & 0 & M_a \end{pmatrix}^2$$

$$\times \left\{ \begin{pmatrix} J_a & 1 & J_c \\ -M_a & 0 & M_c \end{pmatrix}^2 - \frac{1}{2} \left[\begin{pmatrix} J_a & 1 & J_c \\ -M_a & 1 & M_c \end{pmatrix} - \begin{pmatrix} J_a & 1 & J_c \\ -M_a & 1 & M_c \end{pmatrix} \right]^2 \right\} .$$

From Fig.7.12b we know that using a linearly polarized saturating beam resonant with either of the transition branches (J = 0 or ±1), there are smaller or larger differences between the absorption cross section $\sigma^{\parallel}$ and $\sigma^{\perp}$

228

Table 7.2. The dependence of the polarization factor ς^c on J with circularly polarized pump beam [7.16]. Subscript 1 indicates the common level

	$J_2=J_1+1$	$J_2=J_1$	$J_2=J_1-1$
$J=J_1+1$	$-3J_1/2(J_1+1)$	$3/2(J_1+1)$	$3/2$
$J=J_1$	$3/2(J_1+1)$	$-3/2J_1(J_1+1)$	$-3/2J_1$
$J=J_1-1$	$3/2$	$-3/2J_1$	$-3(J_1+1)/2J_1$

Table 7.3. The dependence of the polarization factor ς^L on J with linearly polarized pump beam [7.16]. Subscript 1 indicates the common level

	$J_2=J_1+1$	$J_2=J_1$	$J_2=J_1-1$
$J+J=J_1+1$	$\dfrac{3J_1(2J_1-1)}{10(J_1+1)(2J_1+3)}$	$\dfrac{3(2J_1-1)}{10(J_1+1)}$	$\dfrac{3}{10}$
$J=J_1$	$-\dfrac{3(2J_1-1)}{10(J_1+1)}$	$\dfrac{3(2J_1+3)(2J_1-1)}{10J_1(J_1+1)}$	$-\dfrac{3(2J_1+3)}{10J_1}$
$J=J_1-1$	$\dfrac{3}{10}$	$-\dfrac{3(2J_1+3)}{10J_1}$	$\dfrac{3(J_1+1)(2J_1+3)}{10J_1(2J_1-1)}$

of the two components $\gamma^{\parallel}$ and $\gamma^{\perp}$, respectively. Thus a simplified spectrum can be obtained without exception.

The intensity of the detected probe signal is given by (5.17). On the line center, i.e., $x = 0$ in (5.17), neglecting the background terms in that formula, we obtain the expression for the signal intensity

$$I^{C,L} \simeq I_0(\Delta\alpha\ell/4)^2$$

$$\propto \left[\frac{1}{2J_a+1}(2J_b+1)S_{J_aJ_b}^{\Lambda_a\Lambda_b}S_{J_aJ_c}^{\Lambda_a\Lambda_c}\varsigma_{J_aJ_bJ_c}^{C,L} \right]^2 . \tag{7.29}$$

Table 7.2 and 3 list the J dependence of the polarization factors ς^C and ς^L of the labelling signal for circularly and linearly polarized pumping beams according to (7.25 and 28), respectively [7.17]. These tables reveal that the strength of the polarization labelling depends in a complex way on quantum numbers (Λ, J, and M) and their changes in the pump or probe transitions. This has the following consequences: even if the energy level schemes are of the same class (e.g., Fig.7.7a) different kinds of saturating transitions will cause great disparity in transition strength between the same probe transitions; on the other hand, the same kind of saturation transition will have very different signatures for different transitions. These differences often reach orders of magnitudes in size. The great fluctuation of the signal strength in polarization labelling is the fundamental difference from

Table 7.4. Strength factor S of a molecular spectral line

	$\Lambda_1 = \Lambda+1$	$\Lambda_1 = \Lambda$	$\Lambda_1 = \Lambda-1$
$J = J_1+1$	$\dfrac{(J_1-\Lambda_1+1)(J_1-\Lambda_1+2)}{2(J_1+1)}$	$\dfrac{(J_1+\Lambda_1+1)(J_1-\Lambda_1+1)}{J_1+1}$	$\dfrac{(J_1+\Lambda_1+2)(J_1+\Lambda_1+1)}{2(J_1+1)}$
$J = J_1$	$\dfrac{(J_1-\Lambda_1+1)(J_1+\Lambda_1)(2J_1+1)}{2J_1(J_1+1)}$	$\dfrac{(2J_1+1)\Lambda_1^2}{J_1(J_1+1)}$	$\dfrac{(J_1+\Lambda_1+1)(J_1-\Lambda_1)(2J_1+1)}{2J_1(J_1+1)}$
$J = J_1-1$	$\dfrac{(J_1+\Lambda_1-1)(J_1+\Lambda_1)}{2J_1}$	$\dfrac{(J_1+\Lambda_1)(J_1-\Lambda_1)}{J_1}$	$\dfrac{(J_1-\Lambda_1-1)(J_1-\Lambda_1)}{2J_1}$

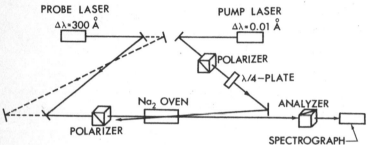

Fig.7.16. Apparatus for simplifying spectra via polarization labelling [7.18]

other simplification methods of molecular spectrum. Table 7.4 lists the dependence of strength factor S of a molecular spectral line on the electronic angular momentum Λ and the rotational angular momentum J calculated according to (7.14).

Based on the principles described above, it is possible to record the labelling spectrum with the aid of a probe laser that emits in a broad band of wavelengths or which is scanned during the exposure. An apparatus for the polarization labelling method is exhibited in Fig.7.16. A polarized beam from a repetitively pumped dye laser is used to pump molecules of a particular orientation from a chosen lower level, and to leave the lower level with the complementary orientation. A broad-band probe from a second laser is directed through two crossed polarizers, before and after the sample, and then into a photographic spectrograph. If a linearly polarized pumping beam is required, one need only remove the quarter-wave plate from the optical path of the pump beam.

As an example, Fig.7.17 depicts the spectra observed by *Teets* et. al. [7.18]. The respective quantum numbers of pumping transitions from (v'', J'') to (v', J') and the polarization properties of the pump beam were varied for the rows in the spectrum. A long vertical line in the middle of the lower diagram links all of the locations of the pumping wavelengths for each respective V'-progressions. We see that despite very close pumping wavelengths the spectra are quite different. The left-most pair of the partners in each row show the energy of the lowest vibrational levels of the electronic state containing the upper level of the probe transition. Referring

230

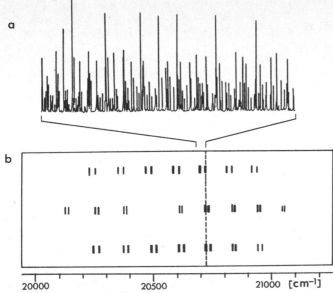

Fig.7.17. A small section of the Na$_2$ spectrum revealed by (a) conventional spectroscopy and (b) polarization labelling taken from a photograph. The dashed vertical line links the pumping wavelengths [7.18]

to the neon lines in the same diagram (omitted in Fig.7.17) the wavelengths of each of the signals of a v'-progression can be determined. As the molecular constants of the ground state are known, it is not difficult to assign the spectra, to determine the excited electronic state, and to obtain its energy level constants (Sect.7.7).

The experimental setup is flexible. The probe laser can be either pulsed or cw and have a bandwidth either as wide as the full gain width of the lasing medium or as narrow as a single-mode scanning laser. With the former source a spectrograph can be more conveniently used as the display equipment as shown in Fig.7.16. If the latter source is used, the detection system should consist of a photodetector, dc amplifier and x-y chart-recorder. When pulsed lasers are used, by altering the time delay of the probe beam relative to the pump beam we can determine the type of energy level scheme, i.e. whether it is type (*1*) or (*3*) in Fig.7.7 can be determined by means of checking the lifetime of the common level.

If a linearly polarized pump beam is used, it is better to set the polarization direction of the probe beam at 45° to that of the pump beam to give a maximum absorption difference between the two probe components. For an enhancement of the detection sensitivity the overlapping volume in the sample cell should be maximized, and the windows of the cell should be carefully selected so as to minimize their carried birefringence. Because the detection system for polarizaiton labelling works under conditions of dark background (a typical analyzer has an extinction ratio of 10^{-5}), the signals can be greatly amplified, so that its sensitivity used to be much higher than that of the population modulation method.

231

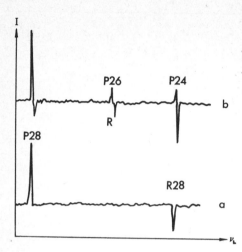

Fig.7.18. OODR polarization spectra in Na$_2$. Pumping laser: single mode Ar$^+$ laser stabilized on the X$^1\Sigma_g^+$ ($v''=0$, $J''=28$) $\rightarrow$ B$^1\Pi_u$ ($v'=6$, $J'=27$) transition with sample pressure of (a) 0.01 Torr and (b) 0.1 Torr [7.19]

Note the symmetric distribution of absorption cross-section of the R and P branches in Fig.7.12. This means that signal phases of the R and P branches should be opposite if a cw circularly polarized pumping beam is used. By means of intensity modulating the pump beam and of phase-locked techniques, *Demtröder* et. al. have differentiated between P and R branches and studied the molecular collision transfer processes at different pressures [7.19]. As shown in Fig.7.18 several satellite lines appear in Fig. 7.18b with increased pressure rather than in Fig.7.18a with low pressure.

Of course, the pump beam simultaneously causes the molecular orientation in upper sublevels to reflect that in lower levels, as indicated in Fig.7.13a by nonuniform thickness lines. Therefore the probe beam will also possess a rotation of its polarization plane, provided it is resonant with the transitions from the upper level of the pumping transition to the high-lying states (so-called two-step polarization labelling with a common intermediate level, see Sect.7.4 for details). Thus it is necessary to discriminate the transition types before obtaining molecular constants. As well as examining lifetimes (using pulsed lasers), as mentioned above, the appearance of the spectrum yields important information. There should be no clear starting band at the longer-wavelength side in the probed band progression for the type (2) scheme, which can therefore be distinguished from type (1) and (3) in Fig.7.7.

7.4 Two-Step Polarization Labelling

Recently *Carlson* et. al. [7.20] employed polarization labelling to explore highly excited electronic states in sodium dimers. The method used which involves a common intermediate level, is called two-step polarization labelling, as shown in Fig.7.19. It is especially useful because it gives information about molecular states with the same symmetry as the ground state, called "gerade" (even) states. One-step optical transitions occur between a gerade state, such as the ground state, and an "ungerade" (odd) state. Thus absorp-

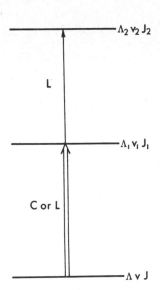

$\Lambda_2 v_2 J_2$

L

$\Lambda_1 v_1 J_1$

C or L

$\Lambda v J$

Fig.7.19. The scheme of two-step polarization labelling. Pump source: linearly (L) or circularly (C) polarized laser; probe source: linearly (L) polarized laser

tion from the ground state provides information only about ungerade states, and almost nothing was known about gerade states other than the ground state.

The experimental setup is depicted in Fig.7.16. A polarized pump laser tuned to resonance with a molecular transition preferentially pumps molecules with a chosen orientation to the first excited state. A pulsed pump laser is used, so that many molecules are oriented during and just after the short (e.g., 5ns) pump pulse. As for the case of ordinary polarization labelling, a broad dye laser probe is used. At the wavelengths of absorption transitions from either of these oriented molecular states, the light is depolarized in the sodium dimers and so can reach the spectrograph. These wavelengths then appear as bright lines on a dark-background, as in ordinary polarization labelling.

We can follow the steps from (7.8) to calculate the signal intensity for two-step polarization labelling. Comparing the energy level scheme in Fig. 7.19 to that in Fig.7.11, we can write the following interaction approach instead of (7.8)

$$\rho_{aa}^{00} \xrightarrow{E_1} \rho_{ab}^{10} \xrightarrow{E_1} \rho_{bb}^{20} \xrightarrow{E_2} \rho_{bc}^{21} \tag{7.30}$$

where the initial condition $\rho_{bb}^{00} = \rho_{cc}^{00} = 0$ is assumed. It had been mentioned above that the uniform distributions of molecules on M levels such as P_{aa}^{00} do not contribute to the labelling signals. Since ρ_{bb}^{20} is equal to ρ_{aa}^{20} but is of opposite sign, the M dependence of ρ_{bb}^{20} is opposite in sign to that of ρ_{aa}^{20}. Thus the signal-intensity expression for two-step polarization is [7.20]

$$I^{C,L} \simeq I_0(\tfrac{1}{4}\Delta\alpha\ell_1)^2 \propto \frac{1}{(2J_a + 1)(2J_b + 1)} S_{J_a J_b}^{\Lambda_a \Lambda_b} S_{J_b J_c}^{\Lambda_b \Lambda_c} \zeta_{J_a J_b J_c}^{C,L} \tag{7.31}$$

233

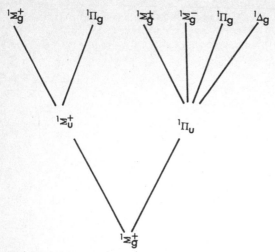

Fig.7.20. Two-step excitation approaches from ground state $X^1\Sigma_g^+$ to allowed high-lying excitation states

which is similar to (7.29); however, the polarization factor $\zeta_{J_a J_b J_c}^{C,L}$ in (7.31) has the same terms as those listed in Tables 7.2 and 3, but with opposite signs.

According to the selection rules in the electric-dipole approximation for various polarizations of pump and probe, several kinds of molecular high-lying states that might be found by two-step excitations are indicated in Fig.7.20. Not only is there a multitude of molecular high-lying states, but they overlap in a complex fashion. Hence it is necessary to identify the upper states individually. Figure 7.20 indicates that there are two candidates for the intermediate state, $^1\Sigma_u^+$ or $^1\Pi_u$, which can easily be distinguished by the distinct pumping wavelengths. For example, pump wavelengths longer than 600.0 nm correspond to $A^1\Sigma_u^+$; shorter than 500.0 nm to $B^1\Pi_u$. Using Tables 7.2,4 and with regard to the various approaches in Fig.7.20, one can obtain clues for the identification of the upper electronic states. In the limit of large J (>20 reasonable values under most experimental conditions), the line strength and polarization factors are simplified to be in the form shown in the ζ^c and ζ^L columns in Table 7.5. One can use these large J values along with the dipole selection rules to determine the relative intensities for both linearly and circularly polarized pumps. Figure 7.21 displays the relative intensity with the data given by *Carlson* et. al. [7.20] using different thicknesses of the arrows for transitions to every allowed higher excited state when the pump laser is circularly and linearly polarized, respectively. It explains why the intensity features of the polarization-labelling signals provide the criteria for identification of both the various excitation approaches and every related electronic state. For example, if $^1\Sigma_u^+$ was chosen as the intermediate state, and if one observed intense doublets with an apparatus as in Fig.7.16, but one order weaker than usual once the quarter-wave plate was taken away, the two-step excitation should be $^1\Sigma_g^+ \rightarrow {}^1\Sigma_u^+ \rightarrow {}^1\Sigma_g^+$ (Fig.7.21a); if the doublet became a triplet with only the

Table 7.5. Polarization factors $\rho^P_{J_a J_b J_c}$ for large J values (P=C,L,∥,++) [7.21]

J_a \ J_c	ζC			ζL			$\zeta \parallel$			$\zeta ++$		
	J_b+1	J_b	J_b-1	J_b+1	J_b	J_b-1	J_b+1	J_b	J_b-1	J_b+1	J_b	J_b-1
J_b+1	$-\dfrac{3}{2}$	$\dfrac{3}{2J_b}$	$\dfrac{3}{2}$	$\dfrac{3}{10}$	$-\dfrac{3}{5}$	$\dfrac{3}{10}$	$\dfrac{6}{5}$	$\dfrac{3}{5}$	$\dfrac{6}{5}$	$\dfrac{3}{10}$	$\dfrac{9}{10}$	$\dfrac{9}{5}$
J_b	$\dfrac{3}{2J_b}$	$\dfrac{-3}{2J_b{}^2}$	$\dfrac{-3}{2J_b}$	$-\dfrac{3}{5}$	$-\dfrac{6}{5}$	$-\dfrac{3}{5}$	$\dfrac{3}{5}$	$\dfrac{9}{5}$	$\dfrac{3}{5}$	$\dfrac{9}{10}$	$\dfrac{6}{5}$	$\dfrac{9}{10}$
J_b-1	$\dfrac{3}{2}$	$\dfrac{-3}{2J_b}$	$-\dfrac{3}{2}$	$\dfrac{3}{10}$	$-\dfrac{3}{5}$	$\dfrac{3}{10}$	$\dfrac{6}{5}$	$\dfrac{3}{5}$	$\dfrac{6}{5}$	$\dfrac{9}{5}$	$\dfrac{9}{10}$	$\dfrac{3}{10}$

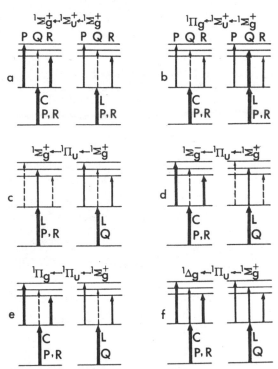

Fig.7.21. Relative intensities of two-step polarization labelling. C: Circular polarization; L: Linear polarization. The arrow widths represent the intensity of the probe signal passing through the analyser [7.20]

middle line (Q-branch) intense, the approach should be $^1\Sigma_g^+ \rightarrow {}^1\Sigma_u^+ \rightarrow {}^1\Pi_g$ (Fig.7.21b). Other approaches drawn in Fig.7.21 can be interpreted by analogy with these. The total electronic orbital angular momenta of molecular high-lying states have then been determined.

The resulting simplified spectra have made it possible to study 24 new excited states whose energies reach about 39000 cm^{-1}. The states are identified as low-lying members of several molecular Rydberg series, as mentioned in Sect. 2.9.

Note that an observed polarization labelling spectrum usually includes lines from both the upper and the lower states of the pump transition. They should be examined before calculation. It can be done experimentally because molecules stay in the upper electronic state for only a few nanoseconds before they give up their excitation energy by spontaneous radiation to lower states. If the probe is delayed relative to the pump for a time comparable to the radiation life time, the upper state will relax before the probe pulse arrives. In that case, only the probe signals from the lower pump level can occur. Thus two–step signals can be identified by comparing a spectrum obtained by a delayed probe with that obtained by a simultaneous pump and probe.

7.5 Modulated Polarization Two-Photon Spectroscopy

The mechanism of two-photon polarization labelling without a level chosen by the pump transition is different from the various above-mentioned labelling techniques. The application of polarization labelling techniques to the study of a two-photon excitation spectrum nevertheless gives important criteria for its identification, especially for the determination of distinct branches of two-photon bands.

Two-photon excitation used to be monitored by the total fluorescence from the final state c. Therefore the interaction to the fourth order $\rho_{cc}^{(4)}$ is needed rather than to the third order as in (7.3 or 8). With consideration of all the population pathways the two-photon excitation processes can be divided into four parts [7.21]:

$$\int \rho_{cc}^{(4)}(v_z)dv_z = \int [\rho_{cc}^{22}(v_z)_1 + \rho_{cc}^{22}(v_z)_2 + \rho_{cc}^{40}(v_z) + \rho_{cc}^{04}(v_z)]dv_z , \qquad (7.32)$$

where the first term $\rho_{cc}^{22}(v_z)_1$ represents the final-level population caused by the fact that the field E_1 interacts with dipole moments μ_{ab}, μ_{ba}, and E_2 with μ_{bc}, μ_{cb}. Their interaction pathways include

two photon excitation:

$$\rho_{aa}^{00} \xrightarrow{E_1} \rho_{ab}^{10} \xrightarrow{E_2} \rho_{ac}^{11} \xrightarrow{E_1} \rho_{bc}^{21} \xrightarrow{E_2} \rho_{cc}^{22} , \qquad (7.33)$$

two-step excitation:

$$\rho_{aa}^{00} \xrightarrow{E_1} \rho_{ab}^{10} \xrightarrow{E_1} \rho_{bb}^{20} \xrightarrow{E_2} \rho_{bc}^{21} \xrightarrow{E_2} \rho_{cc}^{22} . \qquad (7.34)$$

Referring to Chap.5, we know that the first term in (7.32) can be written as

$$\int \rho_{cc}^{22}(v_z)_1 dv_z = L_1 |\mu_{ab}^l|^2 |\mu_{bc}^2|^2 . \qquad (7.35)$$

where L_1 means Lorentzian lineshape factor; the superscripts l and 2 correspond to E_1 and E_2, respectively.

By analogy with this the second term $\rho_{cc}^{22}(v_z)_2$ in (7.32) represents the population in level c contributed by the interactions of E_2 with μ_{ab}, E_1 with μ_{bc}, and is in the form

$$\int \rho_{cc}^{22}(v_z)_2 dv_z = L_2 |\mu_{ab}^2|^2 |\mu_{bc}^l|^2 . \qquad (7.36)$$

The last two terms $\rho_{cc}^{40}(v_z)$ and $\rho_{cc}^{04}(v_z)$ in (7.32) correspond to the population in level c due to the absorption of two photons from E_1 and E_2, respectively. They then are

$$\int [\rho_{cc}^{40}(v_z) + \rho_{cc}^{04}(v_z)] dv_z = D_1 |\mu_{ab}^l|^2 |\mu_{bc}^l|^2 + D_2 |\mu_{ab}^2|^2 |\mu_{bc}^2|^2 , \qquad (7.37)$$

where D_1 and D_2 are the Doppler-broadened lineshape factors, $|\mu_{ij}^m|^2$ ($m = 1, 2$) has a pattern similar to (7.13).

In polarization-modulation labelling the different complex amplitudes of fields, E_1 and E_2, in (7.7) lead to distinct results.

(i) *Circular polarization modulated case*:

Suppose E_1 is the right circularly polarized field, with the complex amplitude in the form

$$E_1 = \tfrac{1}{2}\sqrt{2}(x + iy)E_{10} ; \qquad (7.38)$$

E_2 is a counter-propagating field with the polarization modulated at 2Ω without intensity modulation. So its amplitude can be written as

$$E_2 = \{x\sin(\Omega t + \pi/4) + iy[\cos(\Omega t + \pi/4)]\}E_{20} . \qquad (7.39)$$

The circular-polarization-modulated two-photon-excitation signals can then be picked up at the frequency 2Ω with a certain fixed phase difference. The signal intensity is proportional to the probability difference between para-circular and counter-circular two-photon excitations. The reason of E_1 being without any modulation and E_2 being with polarization modulation along with an isotropic medium explains that the last two terms in (7.32) would not contain the modulation factor and the signals were picked out of the first two terms in (7.32), i.e., [7.21]

$$I^C = \left[L_1 \varsigma_{J_a J_b J_c}^{C_1} + L_2 \varsigma_{J_a J_b J_c}^{C_2} \right] \frac{1}{2J_a + 1} \frac{1}{2J_b + 1} S_{J_a J_b}^{\Lambda_a \Lambda_b} S_{J_b J_c}^{\Lambda_b \Lambda_c} , \qquad (7.40)$$

where

237

$$\zeta^{C_1}_{J_a J_b J_c} = 9(2J_b + 1)$$

$$\times \sum_{M_b} \begin{pmatrix} J_b & 1 & J_a \\ M_b & 1 & M_b-1 \end{pmatrix}^2 \left[\begin{pmatrix} J_c & 1 & J_b \\ 1-M_b & 1 & M_b \end{pmatrix}^2 - \begin{pmatrix} J_c & 1 & J_b \\ 1-M_b & -1 & M_b \end{pmatrix}^2 \right],$$

(7.41)

$$\zeta^{C_2}_{J_a J_b J_c} = 9(2J_b + 1)$$

$$\times \sum_{M_b} \begin{pmatrix} J_c & 1 & J_b \\ -M_b-1 & 1 & M_b \end{pmatrix}^2 \left[\begin{pmatrix} J_b & 1 & J_a \\ -M_b & 1 & M_b-1 \end{pmatrix}^2 - \begin{pmatrix} J_b & 1 & J_a \\ -M_b & -1 & M_b-1 \end{pmatrix}^2 \right].$$

Using the 3-j symbol operators we obtain

$$\zeta^{C_1} = \zeta^{C_2} \equiv \zeta^C .$$

(ii) *Linear polarization modulated case*:
Assume E_1 is a linearly polarized field with the amplitude

$$E_1 = (x\sin\phi_0 + y\cos\phi_0)E_{10} ,$$

and the polarization direction is fixed in space; E_2, on the other hand, has the polarization direction rotating in space at the frequency Ω, i.e.,

$$E_2 = [x\sin(\Omega t + \phi_0) + y\cos(\Omega t + \phi_0)]E_{20} .$$

Again, there is no contribution from the last two terms in (7.32) for the signals picked at 2Ω. So the linear polarization modulated two-photon excitation signals are proportional to the difference between parallel and perpendicular polarized fields in the form of

$$I^L = (L_1 + L_2) \frac{1}{2J_a + 1} \frac{1}{2J_b + 1} S^{\Lambda_a \Lambda_b}_{J_a J_b} S^{\Lambda_b \Lambda_c}_{J_b J_c} \zeta^L_{J_a J_b J_c} . \tag{7.42}$$

However, if we use intensity modulation for the E_2 beam instead of the above-mentioned polarization-modulation conditions, the signal picked at the frequency Ω is

$$I^{\parallel} = (L_1 + L_2 + D_2) \frac{1}{2J_a + 1} \frac{1}{2J_b + 1} S^{\Lambda_a \Lambda_b}_{J_a J_b} S^{\Lambda_b \Lambda_c}_{J_b J_c} \zeta^{\parallel} , \tag{7.43}$$

$$I^{++} = (L_1 + L_2 + D_2) \frac{1}{2J_a + 1} \frac{1}{2J_b + 1} S^{\Lambda_a \Lambda_b}_{J_a J_b} S^{\Lambda_b \Lambda_c}_{J_b J_c} \zeta^{++} , \tag{7.44}$$

where $I^{\parallel}$ and I^{++} correspond to the polarization statuses of parallel linearly polarized and para-circularly polarized fields, respectively. Note that D_2

238

appears in (7.43 and 44), indicating that there is Doppler background in the intensity modulation case, whereas in the polarization modulation case only Doppler-free terms are encountered, provided E_1 and E_2 are for the counter-propagating condition.

The polarization factors ζ^C, ζ^L, $\zeta^{\|}$ and ζ^{++} for large J are listed in Table 7.5, where the different branches of two-photon absorption correspond to distinct variations for the modulation conditions from one to another. Therefore it is useful for the branch assignment of two-photon lines by changing the polarization conditions of the pump beam during the experiment. According to the ratio of $\rho^C_{J_a J_b J_c} / \rho^L_{J_a J_b J_c}$ for circular- to linear-polarization modulation) we have

$$+5 \qquad\qquad\qquad\qquad \text{for } |J_c - J_a| = 2 \ (0, S) \ ,$$

$$-5/2J \begin{cases} J_b + 1 \to J_b \to J_b & \text{for } J_c - J_a = -1 \ (P) \ , \\ J_b \to J_b \to J_b + 1 & \text{for } J_c - \geq_a = +1 \ (R) \ , \end{cases}$$

$$+5/2J \begin{cases} J_b \to J_b \to J_b - 1 & \text{for } J_c - J_a = -1 \ (P) \ , \\ J_b - 1 \to J_b \to J_b & \text{for } J_c - J_a = +1 \ (R) \ , \end{cases} \qquad (7.45)$$

$$-5/2J^2 \ \ J_b \to J_b \to J_b \qquad\quad \text{for } J_c - J_a = 0 \ (Q) \ ,$$

$$-5 \qquad J_b \pm 1 \to J_b \to J_b \pm 1 \quad \text{for } J_c - J_a = 0 \ (Q) \ .$$

According to the ratio of $\rho^{++}_{J_a J_b J_c} / \rho^{\|}_{J_a J_b J_c}$ for the intensity-modulated signal of a two-photon transition we have

$$3/2 \qquad\qquad\qquad\qquad \text{for } J_c - J_a = 1, 2 \ (O, P, R, S) \ ,$$

$$\left. \begin{array}{ll} 3/2 & J_b \to J_b \to J_b \\ \\ 1/4 & J_b \pm 1 \to J_b \to J_b \pm 1 \end{array} \right\} \quad \text{for } J_c - J_a = 0 \ (Q) \ . \qquad (7.46)$$

The above results can be used even for the identification of equal-frequency two-photon transitions with two laser beams from one laser source. For unequal-frequency two-photon transition only the first term in (7.32) would be effective; if we could keep the frequency ω_1 of the E_1 field nearly resonant with ω_{ab}, the lines shape is then Lorentzian and with relative intensities listed in Table 7.6. This means that polarization modulated two-photon spectroscopy will be useful for the determination of intermediate or upper electronic states of a resolved two-photon absorption transition as well as for full elimination of the wide Doppler background to perform a satisfactory Doppler-free spectral trace.

Table 7.6. Relative intensities of unequal-frequency polarization-modulated two-photon signals [7.21]

	J_a / J_c	CPMTPS			LPMTSP		
		J_b+1	J_b	J_b-1	J_b+1	J_b	J_b-1
$X^1\Sigma_g^+ \to {}^1\Sigma_u^+ \to {}^1\Sigma_g^+$	J_b+1	$-\frac{3}{8}$	0	$\frac{3}{8}$	$\frac{-3}{40}$	0	$\frac{3}{40}$
	J_b	0	0	0	0	0	0
	J_b-1	$\frac{3}{8}$	0	$-\frac{3}{8}$	$\frac{-3}{40}$	0	$\frac{3}{40}$
$X^1\Sigma_g^+ \to {}^1\Sigma_u^+ \to {}^1\Pi_g$	J_b+1	$\frac{-3}{16}$	$\frac{3}{8J}$	$\frac{3}{16}$	$\frac{3}{80}$	$\frac{-3}{20}$	$\frac{3}{80}$
	J_b	0	0	0	0	0	0
	J_b-1	$\frac{3}{16}$	$\frac{-3}{8J}$	$\frac{-3}{16}$	$\frac{3}{83}$	$\frac{-3}{20}$	$\frac{3}{80}$
$X^1\Sigma_g^+ \to {}^1\Pi_u \to {}^1\Sigma_g^+$	J_b+1	$\frac{-3}{32}$	0	$\frac{3}{32}$	$\frac{3}{160}$	0	$\frac{3}{160}$
	J_b	0	$\frac{-3}{8J^2}$	0	0	$\frac{3}{10}$	0
	J_b-1	$\frac{3}{32}$	0	$\frac{-3}{32}$	$\frac{3}{160}$	0	$\frac{3}{160}$
$X^1\Sigma_g^+ \to {}^1\Pi_u \to {}^1\Sigma_g^+$	J_b+1	0	$\frac{3}{16J}$	0	0	$\frac{3}{40}$	0
	J_b	$\frac{3}{16J}$	0	$\frac{-3}{16J}$	$\frac{-3}{40}$	0	$\frac{-3}{40}$
	J_b-1	0	$\frac{-3}{16}$	0	0	$\frac{-3}{40}$	0
$X^1\Sigma_g^+ \to {}^1\Pi_u \to {}^1\Pi_g$	J_b+1	$\frac{-3}{16}$	$\frac{3}{8J^3}$	$\frac{3}{16}$	$\frac{3}{80}$	$\frac{-3}{20J^2}$	$\frac{3}{80}$
	J_b	$\frac{3}{8J}$	$\frac{-3}{4J^4}$	$\frac{3}{8J}$	$\frac{-3}{20}$	$\frac{3}{5J^2}$	$\frac{-3}{20}$
	J_b-1	$\frac{3}{16}$	$\frac{-3}{8J^3}$	$\frac{-3}{16}$	$\frac{3}{80}$	$\frac{-3}{20J^2}$	$\frac{3}{80}$
$X^1\Sigma_g^+ \to {}^1\Pi_u \to {}^1\Delta_g$	J_b+1	$\frac{-3}{32}$	$\frac{3}{16J}$	$\frac{3}{32}$	$\frac{3}{160}$	$\frac{-3}{40}$	$\frac{3}{160}$
	J_b	$\frac{3}{16J}$	$\frac{-3}{8J^2}$	$\frac{-3}{16J}$	$\frac{-3}{40}$	$\frac{3}{10}$	$\frac{-3}{40}$
	J_b-1	$\frac{3}{32}$	$\frac{-3}{16J}$	$\frac{-3}{32}$	$\frac{3}{160}$	$\frac{-3}{40}$	$\frac{3}{160}$

7.6 Molecular Energy Level Information Provided by Selective Simplified Molecular Spectra

The analytical method discussed in this section for the spectral data from a selectively simplified molecular spectrum is suitable for the determination of the energy-level constants of the electronic ground state, the intermediate state or high-lying state by a laser-induced fluorescence spectrum, population modulation or polarization labelling, and two-step polarization labelling, respectively. The analytical steps can be summarized as follows:

(i) *Assignment of vibrational quantum numbers*:
There is a clear short-wavelength cut off in the v''-progression of a laser-induced fluorescence spectrum, a clear long-wavelength cut off of the probed v'-progression in a population modulation or polarization labelling spectrum. The cut-off positions correspond to the lowest vibrational levels in the respective final electronic states. Therefore the assignment can easily be made by counting the vibrational quantum numbers successively from zero value at the cut-off wavelength of a band progression.

(ii) *Identification of laser excitation transitions*:
It is direct and easy to identify the laser excitation transition from a laser-induced fluorescence spectrum. For example, Fig.7.2a shows clearly that the 488.8 nm Ar^+ laser excitation in the sodium dimers is the Q-branch starting from the lower level with $v'' = 3$; while the 476.4 nm line excitation is the P-branch with $v'' = 0$ (Fig.7.2b); Fig.7.3b exhibits that the equal-frequency two-photon excitation at 16601 cm^{-1} is the Q_2 branch starting from the lower level with $v'' = 2$, etc. Moreover, as the energy-level constants of the ground state are usually known, it will be not difficult to determine the rotatinal quantum number from the separation of R, P doublets in a fluorescence spectrum. By adding together the energy of the excitation transition and the energy of its lower level the energy of the upper level is calculated.

The identification of the pump transition in two-step polarization labelling requires more procedures. It can be achieved either by comparision of the calculated absorption spectrum according to the known molecular constants (including the Franck-Condon factors for the absorption intensity calculation) with the experimental traces, or by the help of complementary experiments, corresponding to the folded energy-level schemes of type 1 or 2 in Fig.7.7 [7.20]. The energy of the individual newly excited level relative to the bottom of the ground state is then calculated by adding together the energy of the probe transition, the energy of the pumped transition, and the energy of the lower level of the pumping transition.

(iii) *Determination of energy level constants*:
A single band progression from a chosen level clearly exhibits vibrational spacings at the chosen J value in the final electronic state. As an example, let us first see the analysis of a laser-induced fluorescence spectrum. Each signal line can be expressed by the term difference between the vibration-rotational levels in the upper and lower electronic states for the fluorescence emission:

$$\nu(v'', J''; v', J') = T(v', J') - T(v'', J'')$$
$$= (Te' - Te'') + [G'(v') - G''(v'')] + [F'_v(J') - F''_v(J'')] \,, \quad (7.47)$$

where Te is the bottom of the electronic state, in the fluorescence case Te" is usually zero; $G(v)$ and $F(J)$ are the vibrational and rotational terms, respectively, such as

$$G(v) = \omega_e(v + \tfrac{1}{2}) - \omega_e x_e(v + \tfrac{1}{2})^2 + \omega_e y_e(v + \tfrac{1}{2})^3 + \omega_e z_e(v + \tfrac{1}{2})^4 + \dots \,, \quad (7.48)$$

$$F(J) = B_v[J(J+1) - \Lambda^2] - D_v[J(J + 1) - \Lambda^2]^2 + \dots \,, \quad (7.49)$$

where the value of Λ can be determined according to the number of components in a simplified band with the relative intensity values, as listed in Tables 7.4 and 6. The rotational constants and the centrifugal constant containing higher-order corrections can be written by

$$B_v = B_e - \alpha_e(v + \tfrac{1}{2}) + \gamma_e(v + \tfrac{1}{2})^2 + \dots \,, \quad (7.50)$$

$$D_v = D_e + \beta_e(v + \tfrac{1}{2}) + \delta_e(v + \tfrac{1}{2})^2 + \dots \,. \quad (7.51)$$

Therefore the spacing between successive lines of a certain branch is

$$\Delta\nu_v(v'') = \nu(v'', J''; v', J') - \nu(v''+1, J''; v', J')$$
$$= [G''(v'' + 1) - G''(v'')] + [F''_{v+1}(J'') - F''_v(J'')] \,. \quad (7.52)$$

Substituting (7.48-51) into (7.52) one obtains [7.6]

$$\Delta\nu_v(v) = a_0 - a_1 v + a_2 v^2 + a_3 v^3 + \dots \,, \quad (7.53)$$

where

$$a_0 = \omega_e - 2\omega_e x_e + 3.25\,\omega_e y_e + 5\,\omega_e z_e + (-\alpha_e + 2\gamma_e)J(J + 1)$$
$$\quad - (\beta_e + 2\delta_e)J^2(J + 1)^2 \,,$$
$$a_1 = 2\omega_e x_e - 6\omega_e \gamma_e - 13\omega_e z_e - \gamma_e J(J + 1) + 2\delta_e J^2(J + 1)^2 \,, \quad (7.54)$$
$$a_2 = 3\omega_e y_e + 12\omega_e z_e \,,$$
$$a_3 = 4\omega_e z_e \,.$$

The intervals of the R,P doublets in a certain simplified progression have the relations of

$$\Delta\nu_r(v) = b_0 - b_1 v + b_2 v^2 \,, \quad (7.55)$$

where

$$b_0 = (4B_e - 2\alpha_e + \gamma_e - 6D_e - 3\beta_e - 1.5\delta_e)(J' + 1)$$
$$\quad - (8D_e + 4\beta_e + 2\delta_e)(J' + \tfrac{1}{2})^3 \,,$$
$$b_1 = (4\alpha_e - 4\gamma_e + 6\beta_e + 6\delta_e)(J' + \tfrac{1}{2}) + 8\beta_e(J' + \tfrac{1}{2})^3 \,, \quad (7.56)$$
$$b_2 = (4\gamma_e - 6\delta_e)(J' + \tfrac{1}{2}) - 8\delta_e(J' + \tfrac{1}{2})^3 \,.$$

Neglecting the higher-order terms in (7.54, 56), one finds the quadratic and linear dependence of a_0 and b_0 on J, respectively, which allows one to determine J values by two (or more) distinct excitation transitions. It allows one to calculate molecular constants including various correction coefficients with good accuracy through least square quadratic fitting computations, as was demonstrated by *Demtröder* et. al. [7.6].

The calculation method can be applied to other simplified spectra obtained either by population modulation labelling, polarization labelling or two-step labelling, provided the respective terms in (7.47) are substituted by those of the upper and lower levels of the probe transition. For the convenience of computer application it is much easier to summarize the results in terms of a Dunham expression

$$T(v, J) = \sum_{i, j} Y_{ij} (v + \tfrac{1}{2})^i [J(J + 1) - \Lambda^2]^j \tag{7.57}$$

instead of Te, G(v) and F(J) terms in (7.47).

(iv) *Calculation of molecular potential curve*:
If one substitutes the Morse potential into the Schrödinger equation, the exact solution for the eigenvalues is

$$G_v = \omega_e (v + \tfrac{1}{2}) - \omega_e x_e (v + \tfrac{1}{2})^2$$

without higher-order terms. However, as we have seen above, more accurate molecular constants can be obtained from the simplified molecular spectrum, which is the basis for a better description of the potential curve. In fact, an accurate potential curve can be calculated from the experimental constant B_v by using the Rydberg-Klein-Ress (RKR) method, that is to calculate the maximum and minimum values of the internuclear distances during classical vibration of the mass centre for each vibrational level according to the following relations

$$\begin{aligned} r_{max} &= \sqrt{f^2 + f/g} + f \, , \\ r_{min} &= \sqrt{f^2 + f/g} - f \, , \end{aligned} \qquad \text{where} \tag{7.58}$$

$$f(v) = \frac{\hbar}{(2kc\mu)^{1/2}} \int_{v_{min}}^{v} [G(v) - G(u)]^{-1/2} \, du \, ,$$

$$g(v) = \frac{(2kc\mu)^{1/2}}{\hbar} \int_{v_{min}}^{v} B(u) \, [G(v) - G(u)]^{-1/2} \, du \, . \tag{7.59}$$

The lower limit v_{min} is determined by

$$Y_{00} + G(v_{min}) = 0 \, ,$$

and is approximately equal to $-\frac{1}{2}$. Equation (7.59) means that it is no longer necessary to search for an analytical function for the fairly complicated molecular potential curve, and that it is sufficient to list the table of the r_{max} and r_{min} values via $G(v)$. The accuracy can then be 10^{-4} Å or better. Substituting the experimental term values of ground and excited electronic states into (7.58) separately, one can obtain the respective potential curves. The more bands in a observed progression the more accurate and complete is the potential curve obtained.

(v) *Determination of the Franck-Condon factors:*

There are different approaches for the determination of the Franck-Condon factors. One of them is the numerical calculation from the molecular constants obtained experimentally. In concrete terms, the following procedures are needed: *1*) simplified spectrum → molecular constants → potential curve → wavefunctions of vibrational levels (the last step can be found in [7.24]); *2*) repeat the procedure for both of the electronic states of the transitions of interest; *3*) calculate the overlap integral of the vibrational wave functions for each of the overtone transitions between the two electronic states; *4*) square the overlap integral to obtain the Franck-Condon factors. The approach has good reliability, which has been demonstrated by *Kaminsky* by comparing experimental data with the intensity distribution of a calculated absorption spectrum using the Franck-Condon factors so obtained. The agreement was within 1% [7.2].

Another approach for the determination of the Franck-Condon factors is to calculate it directly from the signal ratio of the band intensities in the progressions. For example, in the fluorescence spectrum case, the strength of spontaneous emission is

$$I_{ij} = N_i A_{ij} \hbar \nu_{ij} . \tag{7.60}$$

The Einstein coefficient A_{ij} can be expressed as

$$A_{ij} = 8\pi h \nu^3 R_{ij}/c^3 , \tag{7.61}$$

where the matrix elements R_{ij} for the electronic transition contain three strength factors, corresponding to the relative electronic, vibrational and rotational wave functions between the transition levels:

$$R_{ij} = M_e F_v H_r . \tag{7.62}$$

The middle factor in the above formula is the so-called Franck-Condon factor, which is equal to the vibration overlap integral between the transition levels.

The spectral lines belonging to the same branch in a laser-induced progression should have the same values of the electron-transition dipole moment and of the rotational wave-function overlap integral, so that their intensity ratio is

$$I_{v''_1 J''}/I_{v''_2 J''} = \nu_1^4 F_v(v', v''_1)/\nu_2^4 F_v(v', v''_2) , \tag{7.63}$$

244

which means that the intensity fluctuations in a certain branch of the simplified progression correspond to the variation in the Franck-Condon factor. Therefore, provided one is measuring the spectral frequencies and the intensities simultaneously, after eliminating the influence of the transmission characteristics of the spectrometer or spectrograph and spectral sensitivity of photodetector, one can easily calculate the Franck-Condon factor with (7.63). The method has been applied to laser-induced fluorescence spectra by *Demtröder* et. al. [7.6]. It certainly can be extended to any selective spectrum obtained by other laser-labelling techniques.

7.7 Comprehensive Identification of Equal-Frequency Molecular Two-Photon Transitions

All of the above-mentioned simplification techniques except one for the molecular spectrum take advantage both of the drastically reduced number of lines in a vibration-rotational band and the accompanying easier recognition. The molecular two-photon absorption spectrum does, however, give a greatly simplified spectrum but results in an ambiguous trace due to its usually incomplete fine structure and the alternations of the relative spectral distributions of the enhanced branches, as explained in Sect.4.5, especially for the perturbation shifted lines.

The identification of a molecular two-photon absorption line implies the finding of every quantum number for all of the three or four relevant levels, including the total electronic orbital angular momentum Λ, total electronic spin S, nuclear vibrations v_i, the fram rotation J, the resultant total angular momentum F and, in addition, the symmetry symbols for homonuclear diatomic molecules. It means already that identification of a molecular two-photon absorption spectrum is not an easy task. The irregular spectral structure makes the identification of molecular two-photon transition even more complicated. That is the reason for the suggestion of several complementary methods [7.25], as illustrated in the upper left corner in Fig.1.1 for examining some of the lines. Nevertheless, the explanations in Chap.4 are helpful for understanding the specific simplifications. The complementary identifications are suggested as the following steps, as demonstrated for the equal-frequency two-photon absorption spectra in molecular sodium:

7.7.1 Approximate Numerical Calculations for Predicting Observable Molecular Two-Photon Absorption Frequencies

1) *For the two-photon transitions between singlet states*

$$X^1\Sigma_g^+ + 2h\nu \quad \xrightarrow{A^1\Sigma_u^+} \quad (n)^1\Lambda_g \ , \tag{7.64}$$

where the initial level is in the groudn state $X^1\Sigma_g^+$, the final level is in the high-lying state $(n)^1\Lambda_g$, and the enhancing level is in the intermediate state $A^1\Sigma_u^+$. The way to predict a two-photon excitation spectrum is to calculate both of the electric-dipole-allowed transition bands from $X^1\Sigma_g^+$ to $A^1\Sigma_g^+$, and from the same levels in $A^1\Sigma_u^+$ to $(n)^1\Lambda_g$, to find the near equal-frequency transitions, as is illustrated in Fig.4.10b. According to the detection conditions the offset value of the intermediate level is limited to 2 cm^{-1} for the candidates of the observable two-photon signals with low-power cw laser sources. The molecular constants used for the calculations have been accurately determined for the $X^1\Sigma_g^+$, $A^1\Sigma_u^+$ and $(n)^1\Lambda_g$ states and have been published by *Kusch* et. al. [7.1] and, recently, by *Carlson* et. al. [7.21] and *Taylor* [7.20], respectively. A section of the calculated coarse structure of the three progressions and the fine structure of a molecular two-photon absorption band are presented by Fig.4.7 and the left side of Fig.4.10, respectively.

2) *For the two-photon transitions from singlet to triplet states such as*

$$X^1\Sigma_g^+ + 2h\nu \xrightarrow{\quad A^1\Sigma_u^+ \sim b^3\Pi_u \quad} (n)^3\Lambda_g ,\tag{7.65}$$

instead of Fig.4.10, we have to find the mixing center between $A^1\Sigma_u^+$ and $b^3\Pi_u$ before calculating the relevant enhancing centers, as shown in Fig.4.20a and b. Then the two-photon absorption coarse structure is expected as illustrated in Fig.7.22, which roughly agrees with the observed two-photon excitation spectrum depicted in Fig.4.19 with fluctuation intensities. The fine structure is exhibited in Fig.4.21. The advantage of the numerical calculation is that it gives the general appearance of the two-photon transitions of interest. Unfortunately, as thousands of vibration-rotational levels are involved in each of the molecular electronic states, it is possible that a few transitions will fall in the Doppler profile of a single two-photon absorption line. The known data, especially the upper states, are not accurate enough for assignment of the lines. In fact, after correct

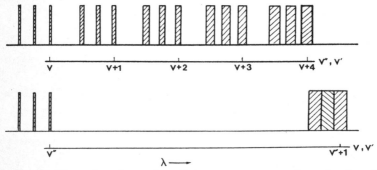

Fig.7.22. Illustration of a section of the coarse structure of a molecular spin-forbidden two-photon absorption spectrum enhanced by the mixing levels with a fixed vibrational quantum number v_A' in the $A^1\Sigma_u^+$ state. For the other section with variation of v_A' values the coarse structure is shifted.

identification of some of the lines one can use the above-mentioned calculation inversely to fit the observed data by computer to obtain the constants for the molecular Rydberg states.

7.7.2 Experimental Methods

The followign complementary methods may all be used alone or in combination to perform various identifications:

1) **Determination of the electronic orbital and spin quantum numbers of the upper state of a molecular two-photon transition:**

   a) By means of monitoring the two-photon transitions at particular fluorescence wavelengths:

Figures 7.23 and 5 show the low-resolution fluorescence spectra over a wide region from 320 to 800 nm for the two-photon transitions at 587.85, 643.4 and 654.7 nm in Na_2, recorded at a side window of a cross oven with the sodium vapour heated to 450° C. They all show band structures but with great disparity of the intensity distributions. Almost 90% of the signal is concentrated on the band around the excitation wavelength in Fig. 7.23, which shows a quite typical characteristic fluorescence band from an upper level (belonging to a singlet state) to the $A^1\Sigma_u^+$ and $B^1\Pi_u$ states and the cascade fluorescence band back to the $X^1\Sigma_g^+$ state. It then tells us that the upper state of the two-photon transition has the total electronic spin quantum number S equal to zero. On the other hand, the fluorescence band at 435.0 nm in Fig. 7.5b is significant. Its intensity is even comparable with the band around the excitation wavelength. According to the potential curves shown in Fig. 2.1, the fluorescence signals at about 435.0 nm and about 365.0 nm are the characteristic bands from the high-lying triplet states $(2)^3\Pi_g$ and $(3)^3\Pi_g$ to the lowest triplet state $a^3\Sigma_u^+$ [7.23, 26]. Owing to the lack of a 365.0 nm fluorescence band signal but a strong 435.0 nm signal, we may deduce that the upper level of this two-photon transition is a vibration-rotational level in the $(2)^3\Pi_g$ state. The fluorescence spectrum in Fig. 7.5a shows a significant band around 500.0 nm, which corresponds to the transi-

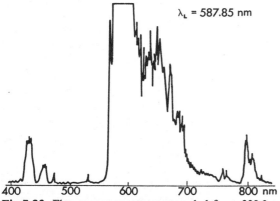

λ_L = 587.85 nm

Fig.7.23. Fluoresence spectrum recorded from 300.0 to 800.0 nm in Na_2 for the two-photon excitation at the wavelength of 587.85 nm

tions terminating in the $b^3\Pi_u$ state. Its integrated strength is even greater than the other two bands at 435.0 nm and 365.0 nm. This implies that the upper level is probably a mixture of a $^3\Delta_g$ state with $(2)^3\Pi_g$ and $(3)^3\Pi_g$ states.

The analyses of the wide region fluorescence spectra help us to organize the experimental set-up for the two-photon excitation spectrum by using narrow-band interference filters with desired center wavelengths. They can be used simultaneously or alternately. The double traces for the excitation spectra in Fig.4.22 were obtained by using the interference filters separately centered at 435.0 and 360.0 nm with a half width of 3.5 nm, inserted between the photomultipier tubes and the side windows of a four-ended oven. This arrangement efficiently provided information about the upper levels.

b) According to the fine structure of a fluorescence spectrum:

The fine structure of a fluorescence spectrum for a two-photon transition offers evidence for determining whether the upper level is in a triplet or a singlet state. While it is simply composed of two sets of doublet structure (for a $^1\Sigma$ - $^1\Sigma$ two-photon transition enhanced by a $^1\Sigma$ state, as shown in Fig.7.3) or a doublet combined with triplet structure (at least one of the electronic states with electronic orbital quantum number nonzero, but the total electronic spin quantum number equal to zero), etc. [7.27,28], the structure of the fluorescence spectrum at the excitation wavelength of a spin forbidden two-photon transition enhanced by mixing levels exhibits a much more complicated trace. Figure 7.24b displays the recorded fine structure of the fluorescence spectrum for a spin forbidden two-photon transition at 646.1 nm in molecular sodium [7.29]. The energy level scheme is illustrated in Fig.7.24a. Obeying the selection rules

$$\Delta\Omega = 0, \pm 1,$$

$$\Delta J = \begin{cases} 1, & \text{for } \Omega = 1 \longleftrightarrow \Omega' = 1, \\ 0, \pm 1, & \text{otherwise}, \end{cases} \tag{7.66}$$

the fluorescence lines should involve the cascade transitions from the upper level to all of the allowed lower levels, sharing the same vibrational levels either with the initial or the enhancing levels. It was interesting to see clearly that the fine structure of the fluroescence spectrum was composed of three sections. The center section with stronger intensities corresponds to the first step fluorescence transitions from the upper level of the two-photon absorption transition to the enhancing levels, whereas the two weaker side sections correspond to their cascade fluroescence transitions and to transitions from the upper level to another two intermediate fine electronic states. By means of more detailed analyses we were able to assign every line in Fig.7.24b. The subscripts of the R,Q,P branch transitions correspond to distinct lower states for A and with the Ω values of 0,1,2 for $b^3\Pi_{\Omega u}$ states. The three pairs of lines marked by P.Q.R with the subscripts of A and $b\Omega$ were considered to be the emission transitions to the $A^1\Sigma_u^+$ and $b^3\Pi_{\Omega u}$ levels, respectively. We saw that the separation between Q_{b1} and Q_A was

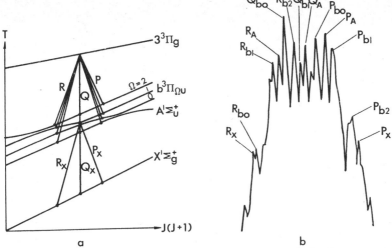

Fig.7.24. (a) Energy level scheme; (b) fine structure of the fluorescence spectrum from a vibration-rotatioanl level in $(3)^3\Pi_g$ as the upper level of a spin-forbidden two-photon transition in Na_2 [7.29]

smaller than that between R_{b1} and R_A, and between P_A and P_{b1}, which means that this two-photon transition was enhanced by one of the Q-branch mixing centers of the singlet-triplet couplings. Moreover, the separation between Q_{b1} and Q_A should be approximately equal to the perturbation shift. The presence of three similarly intense pairs of lines implies that the upper level must have Ω not equal to 1. With the aid of the wide--region fluorescence spectrum (Fig.7.5a) and the hyperfine structure recorded for the two-photon excitation transition, it could be shown that the upper level mainly belongs to the $(3)^3\Pi_{0g}$ state [7.29]. Thus the fluorescence transitions can take place from this level to twelve intermediate levels, including the $A^1\Sigma_u^+$ state and every fine electronic state of $b^3\Pi_{\Omega u}$ ($\Omega = 0,1,2$), and the corresponding cascade fluorescence transitions. Therefore the two-photon transition at 646.1 nm in Na_2 is derived to be

$$X^1\Sigma_g^+ \xrightarrow[\text{enhanced by } A^1\Sigma_u^+ \sim b^3\Pi_{1u}]{} (3)^3\Pi_{0g} \ . \qquad (7.67)$$

Moreover, the fine structure of the fluorescence spectrum provides information about the rotational quantum numbers and fruitful information about the rotational constants and their corrections, the mixing coefficients and perturbation shift, etc. (detailed later).

c) The line shape reveals the spin quantum number of the upper state of a two-photon transition. If the experimental conditions allow one to do Doppler-free spectroscopy, the linewidth, the symmetry of the line profile and the hyperfine structure of the Doppler-free parts will clearly show the nature of the upper state. As a comparison, the traces with singlet and triplet states as the final levels are exhibited in Fig.7.25a and b, respectively

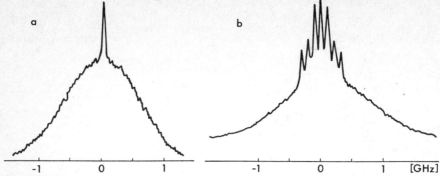

Fig.7.25. Doppler-free two-photon transition profiles with the upper levels belonging to (a) singlet state and (b) triplet state [7.9]

[7.10]. Since the hyperfine constant depends on rotational quantum numbers, the trace for the final level with a small rotational quantum number in a triplet state may not show such well-resolved hyperfine components. Nevertheless, it will at least show a much increased linewidth than that of the trace in Fig.7.25a. The possible serious assymmetries for the particular case of the spin-forbidden two-photon transitions have been detailed in Sect.4.9.

2) Determination of the vibrational quantum numbers

The vibrational quantum number of the initial level can be estimated by counting the resolved vibrational structures at the shorter-wavelength side of the excitation line in its surrounding fluorescence band system. For example, in Fig.7.5a there are three quite intense peaks at the blue side of the excitation wavelength of 643.4 nm with almost equal intervals, corresponding to the vibrational constant in the electronic ground state in Na_2. This means that this two-photon transition is near-resonantly enhanced with the offset of the intermediate level smaller than the Doppler width of the resonant transition from the initial level to the enhancing level, resulting in its being efficiently populated and in its forming such a strong fluorescence signal back to the ground state. On the other hand, the three strong peaks and a weak peak at 618.2 nm show that the vibrational quantum number v'' should be equal to or larger than 4. On acccount of the Franck-Condon factors some bands among the fluorescence signals would be very weak. Due to the serious overlapping of these signals with the fluorescence structures from the upper level to the intermediate states, the weak signals are difficult to gauge. This then is a disadvantage for the direct determination of v'' magnitude only by counting the vibrational peaks. In fact, it would be better to check the number of counted vibrational signals against the calculated Franck-Condon factors and numerical calculations of the transition frequencies from the ground state to the intermediate state.

Once the vibrational quantum number of the initial level is known, the vibrational quantum numbers of the enhancing levels and the final level of a two-photon transition can easily be determined by comparing the excitation frequency with the candidate transition frequencies.

3) Determination of the rotational quantum numbers

a) The following methods are used to identify all of the rotational quantum numbers involved. The rotational quantum number for the enhancing level of a two-photon transition can be determined by measuring the P-R branch intervals $\Delta\nu_{PR}$ in a branch-resolved two-photon fluorescence spectrum (Fig. 7.3) and calculating the J′ value according to

$$\Delta\nu_{PR} = (4J' + 2)B'' , \tag{7.68}$$

where B″ is the rotational constant of the initial state, which is usually known with fair accuracy.

This procedure is also suited to spin forbidden two-photon transitions by analyzing the fine structure of the fluorescence spectrum. With the perturbation shifts of the intermediate mixing levels in mind, one can recognize the cascade fluorescence lines among a dozen or so lines. As an example, the separation between line R_{x1} and P_{x1} in Fig. 7.24b gives a J value of 61.

b) Determine the J value of the two-photon transition by observing the signal ratio of the transition in the polarization-combined pulsed-laser field or the polarization-modulated excitation spectra with different phase setting for the cw source, and comparing them with the calculated signal ratios. By this means the transition branch can be identified by a few traces for dozens of equal-frequency molecular two-photon transitions [7.21], as described in Sect. 7.6.

c) The rotational quantum number of the final level could be determined by computer fitting of the well-resolved hyperfine structure of a two-photon vibration-rotational line if its upper level has considerable hyperfine splitting. Figure 7.26 shows the calculated partially resolve hyperfine structures for different rotational quantum numbers with the same linewidth and total separations. By comparison of the recorded trace in Fig. 7.25b with the curve in Fig. 7.26c and, furthermore, by computer fit-

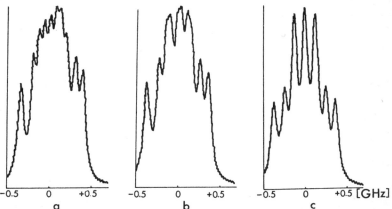

Fig. 7.26. Computed hyperfine structures with selected linewidth (40 MHz) and defined total separation of 460 MHz for (a) J = 10, (b) J = 30 and (c) J = 50

tings we determine that the two-photon transition at 646.1 nm in Na_2 has a J value of 61. Moreover, the number of hyperfine components indicates whether the rotational quantum number is even or odd [7.33].

4) Other determination

The F values decrease in steps of one from the higher to the lower frequency side of the Doppler-free peaks in Fig.7.25b, as described in Chap.5.

For homonuclear molecules such as Na_2, the symmetry of the electronic states (Sect.2.2) involved can be defined by g for the initial and final states, and by u for the enhancing state, which also suit to the perturbing states.

The symmetries for all of the related rotational levels can also be deduced according to the selection rules and the nuclear spin once J'' as determined, i.e., the symmetry of the initial level is known.

In addition to determining the rotational quantum numbers by means of analyzing the fine structure of a fluorescence spectrum, various molecular parameters can be estimated. Taking the example spectrum in Fig.7.24b, we may list the information obtained:

a) The perturbation shift near the mixing center due to the heterogeneous spin-orbital coupling, corresponding to the term plot intersection between $A^1\Sigma_u^+$ and $b^3\Pi_{1u}$, is about 2.3 cm^{-1} (e.g., the measurement of the separation between the fluorescence lines Q_{b1} and Q_A).

b) The separation between the fine electronic states $b^3\Pi_{0u}$ and $b^3\Pi_{2u}$, corresponding to the homogeneous spin-orbital coupling, is about 14.4 cm^{-1}. This means that the molecular constant A is about 7.2 cm^{-1} for conditions of large J (about 60) and large v (about 20), which is in contrast with 7.85 cm^{-1} for small J values, as determined by Li et al. [7.32].

c) The rotational constants of $A^1\Sigma_u^+$ are shown to be reduced here by more than the value expected by the constant corrections [7.14]. A 20% enlarged correction constant value Y_{02} will give much better fitting, which may be reasonable for such a large J levels even without perturbation shifts. The resulting rotational constant in $A^1\Sigma_u^+$ here is about 0.0928 cm^{-1}.

d) The resulting rotational constant in $b^3\Pi_{0u}$ with J of 60 and v of 20 is about 0.120 cm^{-1}. This implies that the correction term Y_{02} should have a similar value of Y_{02} in $A^1\Sigma_u^+$.

7.7.3 Examination of the Population Paths of a Fine Level for a Two-Photon Transition

As described in Sect.4.2, there are several ways to populate a high-lying level in a g-parity state (Fig.4.1). These can be distinguished by the following spectral characteristics:
a) excitation wavelengths,
b) excitation linewidths,
c) fluorescence spectra,
d) dependence on experimental conditions,
e) line shapes and hyperfine structures.

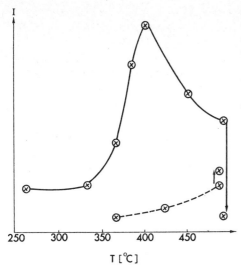

Fig.7.27. Signal dependences of the excitation pathways on experimental conditions. Upper curve corresponds to scheme (d) in Fig.4.1; lower curve corresponds to scheme (b); the arrows indicate the changing of the signals by decreasing the total pressure in the sample cell. It was noticed that different excitation pathways showing opposite features

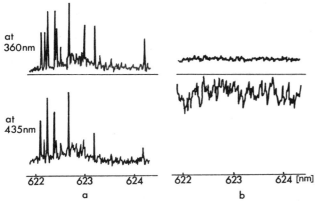

Fig.7.28. The signal dependences on the total pressure of (a) about 1 Torr; (b) about 100 Torr in a sodium oven heated to 450° C

The schemes a) and b) exhibits discrete lines in the two-photon excitation spectra with similarly narrow line widths of about ten to one or few hundred MHz; whereas schemes c) and d) should only give signals at atomic one-photon and two-photon resonance wavelengths (e.g., in sodium vapour they are 589.0, 589,6 and 588.7nm) with excitation linewidths of about 0.1 nm. In contrast, the schemes should show an almost continuous excitation spectrum. One can distinguish between a) and b) by using different interference filters for the monitor. The different fluorescence spectra are partially shown in Fig.7.5 and 23. The fluorescence spectrum omitted corresponded to scheme d) in Fig.4.1, which shows rather intense atomic emission transitions.

Figure 7.27 shows the different dependences of the signal intensity on the temperature (i.e., on the vapour density) and on buffer pressure for the scheme d) (upper curve) and scheme b) (lower curve). Figure 7.28 shows that the molecular energy pooling process can be greatly accelerated by filling the buffer gas to as high as 100 Torr (the continuous base). The dominant feature of the discrete lines on the continuous base in Fig. 7.28a corresponding to the direct two-photon excitations were replaced by quasicontineous ones in Fig. 7.28b, which showed a feature similar to a molecular one-photon absorption spectrum. The smooth upper trace in Fig. 7.28b in comparison with both the upper trace in Fig. 7.28a and the lower trace in Fig. 7.28b evidently demonstrate that the high-lying triplet state(s) $(n)^3\Delta_g$ in Na_2 could be hardly reached by molecular energy-pooling processes than the direct two-photon excitation via spin-orbital coupling mechanisms.

The discrimination of the excitation processes finally gives a population efficiency comparison for studying the hihg-lying g-parity triplet states in molecular sodium [7.30, 31].

References

Chapter 1

1.1   G. Herzberg: *Molecular Spectra and Molecular Structure I. Spectra of Diatomic Molecules* (Van Nostrand, Princeton, NJ 1950)

1.2   G. Herzberg: *Molecular Spectra and Molecular Structure II. Infrared and Raman Spectra of Polyatomic Molecules* (Van Nostrand, Priceton, NJ 1950)

1.3   G. Herzberg: *Molecular Spectra and Molecular Structure III. Electronic Spectra and Electronic Structure of Polyatomic Molecules* (Van Nostrand, Princeton, NJ 1966)

1.4   K.P. Huber, G. Herzberg: *Molecular Spectra and Molecular Structure; IV. Constants of Diatomic Molecules* (Van Nostrand, Princeton, New York 1979)

1.5   C.H. Townes, A.L. Schawlow: *Microwave Spectroscopy* (McGraw-Hill, New York 1955)

1.6   H.R. Xia, Z.G. Wang: Study of small but complete molecules - Na_2, in *Lasers, Spectroscopy and New Ideas*, ed. by W.M. Yen, M.D. Levenson, Springer Ser. Opt. Sci., Vol.54 (Springer, Berlin, Heidelberg 1987) p.174-182

1.7   V.S. Letokhov, V.P. Chebotayev: *Nonlinear Laser Spectroscopy*, Springer Ser. Opt. Sci., Vol.4 (Springer, Berlin, Heidelberg 1977)

1.8   W. Demtröder: *Laser Spectroscopy*, Springer Ser. Chem. Phys., Vol.5 (Springer, Berlin, Heidelberg 1981)
      W. Demtröder: *Laser Spektroskopie*, 2nd edn. (Springer, Berlin, Heidelberg 1991)

1.9   N. Bloembergen (ed.): *Nonlinear Spectroscopy*, Proc. Int. School Phys. ENRICO FERMI Course LXIV, Villa Monastero, Italian, June 30-July 12, 1975 (North-Holland, Amsterdam 1977)

1.10 K. Shimoda (ed.): *High Resolution Laser Spectroscopy*, Topics Appl. Phys., Vol.13 (Springer, Berlin, Heidelberg 1976)

1.11 H. Walther (ed.): *Laser Spectroscopy of Atoms and Molecules*, Topics Appl. Phys., Vol.2 (Springer, Berlin, Heidelberg 1976)

Chapter 2

2.1   G. Herzberg: *Molecular Spectra and Molecular Structure* I. Spectra of Diatomic Molecuels; II. Infrared and Raman Spectra of Polyatomic Molecules; and III. Electronic Spectra and Electronic Structure of Polyatomic Molecules (Van Nostrand, New York 1950; 1945; and 1966)

2.2   C.H. Townes, A.L. Schawlow: *Microwave Spectroscopy* (McGraw-Hill, New York 1955)

2.3   G. Jeung: J. Phys. B: At. Mol. Phys. **16**, 4289 (1983)

2.4   Li Li, S.F. Rice, R.W. Field: J. Mol. Spectrosc. **105**, 344 (1984)

2.5   Yao Zhao, Zhiye Shen, Qingshi Zhu, Cunhao Zhang: J. Mol. Spectrosc. **115**, 34 (1986)

2.6   I. Kovacs: Foundation of the theory of diatomic molecules. In *Atoms, Molecules and Lasers*, Lectures International Winter College (IAEA, Vienna 1974) p.377

H. Lefebvre-Brion: Perturbations in the spectra of diatomic molecules. In ibid. p.411

2.7   W.G. Sturrus, P.E. Sobol, S.R. Lumdeen: Phys. Rev. Lett. **54**, 792 (1985)

2.8   C.F. Barnett, J.A. Ray, A. Russek: Phys. Rev. **A5**, 2110 (1972)

2.9   S.M. Tarr, J.A. Schiavone, R.S. Freund: Phys. Rev. Lett. **44**, 1660 (1980)

2.10 G. Herzberg, Ch. Jungen: J. Mol. Spectrosc. **41**, 425 (1972)

2.11 R.A. Bernheim, L.P. Gold, T. Tipton: J. Chem. Phys. **78**, 3635 (1983)

2.12 M.W. McGeoch, R.E. Schlier: Chem. Phys. Lett. **99**, 347 (1983)

2.13 D. Eisel, W. Demtröder, W. Mueller, P. Botschwina: Chem. Phys. **80**, 329 (1983)

2.14 B. Hemmerling, S.B. Rai, W. Demtröder: Z. Phys. A **320**, 135 (1985)

2.15 N.W. Carlson, A.J. Taylor, K.M. Jones, A.L. Schawlow: Phys. Rev. A **24**, 822 (1981)

2.16 S. Martin, J. Chevaleyre, S. Valignat, J.P. Perrot, M. Broyer, B. Cabaud, A. Hoareau: Chem. Phys. Lett. **87**, 235 (1982)

2.17 S. Leutwyler, T. Heinis, M. Jungen, H.P. Haerri, E. Schumacher: J. Phys. **76**, 4290 (1982)

2.18 M. Broyer, J. Chevaleyre, G. Delacrelaz, S. Martin, L. Woeste: Chem. Phys. Lett. **99**, 206 (1983)

2.19 A.F.J. Van·Raan, J.E.M. Haverkort, B.L. Mehta, J. Korving: J. Phys. B **15**, L669 (1982)

2.20 M. Raab, G. Höning, W. Demtröder, C.R. Vidal: J. Chem. Phys. **76**, 4370 (1982)

2.21 Takanori Suzuki, Masao Kakimoto, Takahiro Kasuya: Reza Kagaku Kenjyu **4**, 13 (1982)

2.22 E.S. Chang, K. Yoshino: J. Phys. B **16**, L581 (1983)

2.23 S. Chung, C.C. Lin, E.T.P. Lee: J. Chem. Phys. **82**, 342 (1985)

2.24 A.R. Filippelli: Avail. Univ. Microfilms Int., Order No DA8421929. Diss. Abstr. Int. B**45**, 2587 (1985)

2.25 H. Sambe, D.E. Ramaker: Chem. Phys. **107**, 351 (1986)

2.26 S.E. Kupriyanov, A.A. Perov, A.Yu. Zayats, A.N. Stepanov: Pis'ma. Zh. Tekh. Fiz. **7**, 861, (1981)

2.27 A. Sur, C.V. Ramana, S.D. Colson: J. Chem. Phys. **83**, 904 (1985)

2.28 E.E. Eyler, F.M. Pipkin: Phys. Rev. **A27**, 2462 (1983)

2.29 P. Hvelplund, H.K. Haugen, H. Knudsen, L. Andersen, H. Damsgaard, F. Fukusawa: Phys. Scr. **24**, 40 (1981)

2.30 W.G. Sturrus, P.E. Sobol, S.R. Lundeen: Phys. Rev. Lett. **54**, 792 (1985)

2.31 G.V. Golubkov, G.K. Ivanov, I.E. Cherlina: Opt. Spektrosk. **54**, 427 (1983)

2.32 H. Takagi, H. Nakamura: J. Chem. Phys. **74**, 5808 (1981)

2.33 G. Herzberg, Ch. Jungen: J. Chem. Phys. **77**, 5876 (1982)

2.34 G.V. Golubkov, G.K. Ivanov: Zh. Eksp. Teor. Fiz. **80**, 1321 (1981)

2.35 J.A. Dagata: Avail. Univ. Microfilms Int., Order No. DA8409577. Diss. Abstr. Int. B **45**, 1479 (1984)

2.36 E.E. Eyler, R.C. Short, F.M. Pipkin: Phys. Rev. Lett. **56**, 2602 (1986)

2.37 H. Helm: Dynamic processes in molecular Rydberg states, in *Fundamentals of Laser Interactions*, ed. by F. Ehlotzky, Lect. Notes Phys., Vol.229 (Springer, Berlin, Heidelberg 1985) pp.208–220

2.38 N. Bjerre, R. Kachru, H. Helm: Phys. Rev. A **31**, 1206 (1985)

2.39 Li Jiaming, Lan Vo Ky: Commun. Theor. Phys. **2**, 1175 (1983)

2.40 F. Grein, A. Kapur: J. Chem. Phys. **77**, 415 (1982)

2.41 K. Kaufmann, C. Nager, M. Jungen: Chem. Phys. **95**, 385 (1985)

2.42 C. Henriet, G. Verhaegen: THEOCHEM **16**, 63 (1984)

2.43 Y. Houbrechts, I. Dubois, H. Bredohl: J. Phys. B **15**, 603 (1982)

2.44 Y. Houbrechts, I. Dubois, H. Bredohl: J. Phys. B **15**, 4551 (1982)

2.45 F. Grein, A. Kapur: J. Chem. Phys. **78**, 339 (1983)

2.46 C.Y.R. Wu: J. Chem. Phys. **76**, 4301 (1982)

2.47 Y. Ono, S.H. Linn, H.F. Prest, M.E. Gress, C.Y. Ng: J. Chem. Phys. 74, 1125 (1981)
2.48 A.Yu. Zayats, A.A. Perov, A.P. Simonov: Khim. Vys. Energ. 18, 182 (1984)
2.49 A.Yu Zayats, A.A. Perov, A.P. Simonov: Khim. Fiz. (3) 333–338 (1983)
2.50 G. Herzberg: J. Chem. Phys. 70, 4806 (1979)
2.51 G. Herzberg: Faraday Discuss. 71, 165 (1981)
2.52 S. Raynor, D.R. Herschbach: J. Phys. Chem. 86, 1214 (1982)
2.53 G. Herzberg, Ch. Jungen: J. Chem. Phys. 84, 1181 (1986)
2.54 S. Raynor, D.R. Herschbach: J. Phys. Chem. 86, 3592 (1982)
2.55 M.N. Ediger, N. Marwood: Avail. Univ. Microfilms Int., Order No. DA8600167. Diss. Abstr. Int. B 46, 3955 (1986)
2.56 P.M. Dehmer: J. Chem. Phys. 83, 24 (1985)
2.57 M.A. Baig, J.P. Connerade, J. Dagata, S.P. McGlynn: J. Phys. B 14, L25 (1981)
2.58 A.M. Woodward, S.D. Colson, W.A. Chupka, M.G. White: J. Phys. Chem. 90, 274 (1986)
2.59 J.M. Wiesenfeld, B.I. Benjamin: Phys. Rev. Lett. 51, 1745 (1983)
2.60 M.A.C. Nascimento, A. William: Chem. Phys. 53, 251 (1980)
2.61 M.A.C. Nascimento, A. William: Chem. Phys. 53, 265 (1980)
2.62 S.R. Prasad, A.N. Singh: Indian J. Phys. 59B, 1 (1985)
2.63 Ch. Jungen, E. Miescher: Can. J. Phys. 47, 1769 (1969)
      E. Miescher: ibid 54, 2074 (1976)
      P. Arrowsmith, W.J. Jones, R.P. Tuckett: J. Mol. Spectrosc. 86, 216 (1981)
2.64 K. Yoshino, D.E. Freeman: Can. J. Phys. 62, 1478 (1984)
      C.A. Mayhew, J.P. Connerade, M.A. Baigs, M.N.R. Ashfold, J.M. Bayler, R.N. Dixon, J.D. Prince: J. Chem. Soc. Faraday Trans. II, 83, Pt.2, 417 (1987)
      C.A. Mayhew, J.P. Connerrade, M.A. Baigs: J. Phys. B 19, 4149 (1986)
2.65 N.W. Carlson, F.V. Kowalski, R.E. Teets, A.L. Schawlow: Opt. Commun. 29, 302 (1979)
2.66 A.J. Taylor, K.M. Jones, A.L. Schawlow: Opt. Commun. 39, 47 (1981)
2.67 F.V. Kowalski, W.T. Hill, A.L. Schawlow: Opt. Lett. 2, 112 (1978)
2.68 P.M. Dehmer, W.A. Chupka: J. Chem. Phys. 65, 2243 (1976)
2.69 H.R. Xia, Z.G. Wang: In Laser, Spectroscopy and New Ideas, ed. by W.M. Yen, M.D. Levenson, Springer Ser. Opt. Sci., Vol.54 (Springer, Berlin, Heidelberg 1987) p.174
2.70 U. Fano: Phys. Rev. A2, 353 (1970)
2.71 M.J. Seaton: Rep. Prog. Phys. 46, 97 (1983)
2.72 U. Fano: Rep. Prog. Phys. 46, 167 (1983)
2.73 S. Martin, J. Chevalleyre, M.Chr. Bordas, S. Valignat, M. Broyer, B. Cabaud, A. Hoareau: In Laser Spectroscopy VI, ed. by H.P. Weber, W. Lüthy, Springer Ser. Opt. Sci. Vol.40 (Springer, Berlin, Heidelberg 1983) p.245
2.74 C. Bordas, M. Broyer, J. Chevaleyre, P. Labastie, B. Tribollet: Stud. Phys. Theor. Chem. 35, 451 (1985)
2.75 G.V. Golubkov, G.K. Ivanov: Zh. Eksp. Teor. Fiz. 80, 1321 (1981)
2.76 H. Takagi, H. Nakamura: J. Chem. Phys. 74, 5808 (1981)
2.77 G.V. Golubkov, G.K. Ivanov, I.E. Cherlina: Opt. Spektrosk. 54, 427 (1983)
2.78 J.M. Li, V.K. Lan: Commun. Theor. Phys. 2, 1175 (1983)
2.79 P.M. Dehmer: Comments At. Mol. Phys. 13, 205 (1983)
2.80 Ch.H. Greene, Ch. Jungen: In Adv. in Atomic and Molecular Physics, 21, 51 (Academic, New York 1985)
2.81 W.L. Glab, J.P. Hessler: Phys. Rev. A35, 2102 (1987)
2.82 W. Boehmer, R. Haensel, N. Schwentner, E. Boursey, M. Chergui: Chem. Phys. Lett. 91, 66 (1982)
2.83 W.G. Sturrus, P.E. Sobol, S.R. Lundeen: Phys. Rev. Lett. 54, 792 (1985)
2.84 W.G. Sturrus, E.A. Hessels, S.R. Lundeen: Phys. Rev. Lett. 57, 1863 (1986)
2.85 W. Demtröder, U. Diemer, H.J. Foth, R. Kullmer, G. Persch, H.J. Vedder, H.

Weickenmeier: High resolution molecular spectroscopy with Lasers. In *Colloq. Spectrosc. Int. XXIV-CSI XXIV* (Springer, Berlin, Heidelberg 1986) pp.109-115

2.86 K.M. Jones: Rydberg states in diatomic sodium. PhD Thesis, Stanford University (1983)

2.87 P. Labastie, M.C. Bordas, B. Tribollet, M. Broyer: Phys. Rev. Lett. 52, 1681 (1984)

2.88 Li Li, R.W. Field: J. Mol. Spectrosc. 117, 245 (1986)

2.89 G.P. Morgan, H.R. Xia, A.L. Schawlow: J. Opt. Soc. Am. 72, 315 (1982)

2.90 E.E. Eyler, R.C. Short, F.M. Pipkin: Phys. Rev. Lett. 56, 2602 (1986)

2.91 R.D. Knight, L.-G. Wang: Phys. Rev. Lett. 55, 1571 (1985)

2.92 A.J. Taylor, K.M. Jones, A.L. Schawlow: J. Opt. Soc. Am. 73, 994 (1983)

2.93 H.R. Xia, L.S. Ma, J.W. Xu, M. Yuan, I.S. Cheng: Study of g-parity triplet states by near-resonant equal-frequency two-photon transitions in Na_2, '87 Int'l Conf. on Laser, Xiamen, China (1987)

2.94 A.L. Schawlow: Spectroscopy in a new light. Nobel Prize Lecture. In *Les Prix Nobel* (Nobel Foundation, Stockholm 1982); Rev. Modern Phys. 54, 687 (1982)

2.95 N.W. Carlson: Identification of excited states of diatomic sodium by two-step polarization labeling. PhD Thesis, Stanford University (1980)

2.96 A.J. Taylor: Two-step polarization labeling spectroscopy of excited states of diatomic sodium. PhD Thesis, Stanford University (1982)

2.97 M.N.R. Ashfold: Mol. Phys. 58, 1-20 (1986)

2.98 H. Shinohara, K. Sato, Y. Achiba, N. Nishi, K. Kimura: Chem. Phys. Lett. 130, 231 (1986)

2.99 J.L. Hall, J.A. Magyar: High resolution saturated absorption studies of methane and some methyl-halides. In *High-Resolution Laser Spectroscopy*, ed. by K. Shimoda, Topics Appl. Phys., Vol.13 (Springer, Berlin, Heidelberg 1976) pp.174-199

2.100 M.D. Levenson, A.L. Schawlow: Phys. Rev. A 6, 10 (1972)

2.101 J.B. Atkinson, J. Becker, W. Demtröder: Chem. Phys. Lett. 87, 92 (1982)

2.102 J.B. Atkinson, J. Becker, W. Demtröder: Chem. Phys. Lett. 87, 128 (1982)

2.103 H.R. Xia, L.S. Ma, J.W. Xu, M. Yuan, I.S. Cheng: Opt. Commun., submitted

Chapter 3

3.1 G. Herzberg: *Molecular Spectra and Molecular Structure; I. Spectra of Diatomic Molecules* (Van Nostrand, Princeton, NJ 1950)

3.2 G. Herzberg: *Molecular Spectra and Molecular Structure; II. Infrared and Raman Sepctra of Polyatomic Molecules* (Van Nostrand, Princeton, NJ 1945)

3.3 G. Herzberg: *Molecular Spectra and Molecular Structure; III. Electronic Spectra and Electronic Structure of Polyatomic Molecules* (Van Nostrand, Princeton, NJ 1966)

3.4 C.H. Townes, A.L. Schawlow: *Microwave Spectroscopy* (McGraw-Hill, New York 1955)

3.5 Z.G. Wang, Y.C. Wang, G.P. Morgan, A.L. Schawlow: Opt. Commun. 48, 398 (1984)

3.6 J.B. Atkinson, J. Becker, W. Demtröder: Chem. Phys. Lett. 87, 128 (1982)

3.7 H.R. Xia, Z.G. Wang: Study of small but complete molecules - Na_2. In *Lasers, Sepctroscopy and New Ideas*, ed. by W.M. Yen, M.D. Levenson, Springer Ser. Opt. Sci., Vol.54 (Springer, Berlin, Heidelberg 1987) pp.174-182

3.8 R. Altkorn, F.E. Bartoszek, J. Dehaven, G. Hancock, D.S. Perry, R.N. Zare: Chem. Phys. Lett. 98, 212 (1983)

3.9 V.S. Letokhov: *Nonlinear Laser Chemistry*, Springer Ser. Chem. Phys., Vol.22 (Springer, Berlin, Heidelberg 1983)

3.10   H.B. Bechtel, C.J. Dasch, R.E. Teets: Combustion research with lasers. In *Laser Applications* 5, 129-212 (Academic, New York 1984)

3.11   R.L. Farrow, R.P. Lucht: High-resolution CARS for combustion diagnostics. In *Proc. 10th Int. Conf. on Raman Spectroscopy*, ed. by W.L. Peticolas, B. Hudson (Univ. Oregon, Eugene 1986) pp.27-28

3.12   L.A. Rahn, R.E. Palmer: Application of high-resolution coherent Raman spectroscopy to combustion measurements. In *Proc. 10th Int. Conf. on Raman Spectroscopy*, ed. by W.L. Peticolas, B. Hudson (Univ. Oregon, Eugene 1986) pp.25-26

3.13   H.R. Xia, S.J. Tan, Z.C. Xu, L.S. Ma, Y.D. Jiang, Z.D. Pan: J. Shanghai Normal University, Ser. of Science No.1, 30 (1978)

3.14   H.R. Xia, Z.X. Guo, L.S. Ma, P.P. Cai, L.E. Ding: ibid, No.2, 33 (1980)

3.15   C.J. Borde, J.L. Hall: Phys. Rev. Lett. 30, 1101 (1973)
       C.J. Borde, M. Ouhayoun, A. Van Lerberghc, C. Salomon, S. Avriller, C.D. Cantrell, J. Borde: High resolution saturation spectroscopy with CO_2 lasers. Application to the ν_3 bans of SF_6 and OsO_4. In *Laser Spectroscopy IV*, ed. by H. Walther, K.W. Rothe, Springer Ser. Opt. Sci., Vol.21 (Springer, Berlin, Heidelberg 1979) pp.142-153

3.16   S.N. Bagayev, L.S. Vasilenko, V.G. Goldort, A.K. Dmitriyev, A.S. Dychkov, V.P. Chebotayev: Appl. Phys. 13, 291 (1977)

3.17   K.K. Verma, J.T. Bahns, A.R. Rajaei-Rizi, C. Stwalley: J. Chem. Phys. 78, 3599 (1983)

3.18   F.P. Schäfer (ed.): *Dye Lasers*, 3rd edn., Topics Appl. Phys., Vol.1 (Springer, Berlin, Heidelberg 1990)

3.19   E.D. Hinkley, K.W. Nill, F.A. Blum: Infrared spectroscopy with tunable lasers. In *Laser Spectroscopy of Atoms and Molecules*, ed. by H. Walther, Topic Appl. Phys., Vol.2 (Springer, Berlin, Heidelberg 1976) pp.125-196

3.20   T.W. Hänsch, Appl. Opt. 11, 895 (1972)
       O.G. Peterson, S.A. Tuccio, B.B. Snavely: Appl. Phys. Lett. 17, 245 (1970)

3.21   R.M. Lees: Far infrared (FIR) and infrared (IR) spectroscopy of methanol applied to FIR laser assignments. Proc. SPIE 666, 158 (1986)

3.22   G. Taubmann, H. Jones, P.B. Davies: Appl. Phys. B, B41, 179 (1986)

3.23   J. Farhoomand, G.A. Blake, M.A. Frerking, H.M. Pickett: Proc. SPIE 598, 84 (1986)

3.24   C.M. Lovejoy, M.D. Schuder, D.J. Nesbitt: J. Chem. Phys. 85, 4890 (1986)

3.25   T. Nakanaga, T. Amano: Can. J. Phys. 64, 1356 (1986)

3.26   T. Nakanaga, T. Amano: Chem. Phys. Lett. 134, 195 (1987)

3.27   C.V. Shank: Investigation of nonthermal population distributions with femtosecond optical pulses. In *Methods of Laser Spectroscopy* (Plenum, New York 1986) pp.51-54

3.28   R.G. Bray: High vibration overtons spectroscopy of polyatomic molecules by intracavity dye laser photoacoustic spectroscopy. Proc. SPIE 286, 9 (1981)

3.29   C. Douketis, D. Anex, J.P. Reilly: High resolution overtone spectroscopy on a molecular beam. Proc. SPIE 669, 137 (1986)

3.30   J.B. Koffend, J.S. Holloway, M.A. Kwok, R.F. Heidner: J. Quant. Spectrosc. & Radiat. Transfer. 37, 449 (1987)

3.31   G. Gerber, R. Moller: AIP Conf. Proc. 146, 517 (1986)

3.32   C.F. Yu, F. Youngs, K. Tsukiyama, R. Bersohn, J. Preses: J. Chem. Phys. 85, 1382 (1986)

3.33   J.P. Pique, Y. Chen, R.W. Field, J.L. Kinsey: Phys. Rev. Lett. 58, 475 (1987)

3.34   Hajime Kato, Tooru Matsui, Chifuru Noda: J. Chem. Phys. 76, 5678 (1982)

3.35   Li Li, S.F. Rice, R.W. Field: J. Chem. Phys. 82, 1178 (1985)

3.36   Xingbin Xie, R.W Field: J. Chem. Phys. 83, 6193 (1985)

3.37   P. Kusch, M.M. Hessel: J. Chem. Phys. 63, 4087 (1975)

3.38   P. Kusch, M.M. Hessel: J. Chem. Phys. 68, 2591 (1978)

259

3.39 H.L. Dai, C.L. Korpa, J.L. Kinsey, R.W. Field: J. Chem. Phys. **82**, 1688 (1985)
3.40 Zhao Yao, Shen Zhiye, Zhu Qing-shi, Zhang Cun-hao: Chinese Phys. Lett. **1**, 69 (1984)
3.41 Yao Zhao, Zhiye Shen, Qingshi Zhu, Cunhao Zhang: J. Mol. Spectrosc. **115**, 34 (1986)
3.42 R. Anttila, R. Paso, G. Guelachvili: J. Mol. Spectrosc. **119**, 190 (1986)
3.43 Baoshu Zhang, Qingshi Zhu, Zhiye Shen, B.A. Thrush: J. Mol. Structure **159**, 77 (1987)
3.44 Bala S. Sankar, P.K. Panja, P.N. Ghosh: J. Mol. Struct. **157**, 339 (1987)
3.45 Peyroula E. Pebay, R. Jost: J. Mol. Spectrosc. **121**, 167 (1987)
3.46 J.C. Petersen, S. Saito, T. Amano, D.A. Ramsay: Can. J. Phys. **62**, 1731 (1984)
3.47 W.H. Weber, P.D. Maker, J.W.C. Johns, E. Weinberger: J. Mol. Spectrosc. **121**, 243 (1987)
3.48 R.C. Hilborn, Zhu Qingshi, D.O. Harris: J. Mol. Spectrosc. **97**, 73 (1983)
3.49 C.A. Mayhew, J.-P. Connerade, M.A. Baig, M.N.R. Ashfold, J.M. Bayley, R.N. Dixon, J.D. Prince: J. Chem. Soc. Faraday Trans. II **83**, 417 (1987)
3.50 E.D. Hinkley: Appl. Phys. Lett. **16**, 351 (1970)
3.51 K.W. Nill: Laser Focus **13**, 32 (1977)
3.52 R.V. Ambartzmian, Yu. A. Gorokhov, V.S. Letokhov, G.N. Makarov: Zh. Eksp. Teor. Fig. Pis'ma Red. **21**, 375 (1975)
3.53 R.J. Jensen, J.G. Marinzzi, C.P. Robinson, S.D. Rockwood: Laser Focus **12**, 51 (March 1976)
3.54 D.H. Cox, J. Elliott: Spectrosc. Lett. **12**, 275 (1979)
3.55 R.V. Ambartzumian, V.M. Apatin, A.V. Evseev, N.P. Furzikov: Kvantovaya Elektron. **7**, 1998 (1980)
3.56 C.A. Mayhew, J.P. Connerde, M.A. Baigs: J. Phys. B **19**, 4149 (1986)
3.57 G. Meijer, B. Jansen, J.J. Ter Meulen, A. Dymanus: Chem. Phys. Lett. **136**, 519 (1987)
3.58 S. Datta, R.W. Anderson, R.N. Zare: J. Chem. Phys. **63**, 5503 (1976)
3.59 H.C. Miller, J.W. Farley: J. Chem. Phys. **86**, 1167 (1987)
3.60 R.L. Robinson, Dz-Hung Gwo, D. Ray, R.J. Saykally: J. Chem. Phys. **86**, 5211 (1987)
3.61 D. Ray, R. Robinson, D.-H. Gwo, R.J. Saykally: Intracavity far infrared laser spectroscopy of supersonic jets: direct measurement of the vibrational motions in Van der Waals bonds. In *Laser Spectroscopy VII*, ed. by T.W. Hänsch, Y.R. Shen, Springer Ser. Opt. Sci., Vol.49 (Springer, Berlin, Heidelberg 1985) pp. 126–129
3.62 S.T. Pratt, P.M. Dehmer, J.L. Dehmer: AIP Conf. Proc. **146**, 493 (1986)
3.63 C.M. Lovejoy, M.D. Schuder, D.J. Nesbitt: J. Chem. Phys. **86**, 5337 (1987)
3.64 J.C. Drobits, I.M. Lester: J. Chem. Phys. **86**, 1662 (1987)
3.65 M.M. Doxtader, M.R. Topp: Chem. Phys. Lett. **124**, 39 (1986)
3.66 H. Kanamori, J.E. Butler, T. Minowa, K. Kawaguchi, C. Yamada, E. Hirota: Infrared laser kinetic spectroscopy. In *Laser Spectroscopy VII*, ed. by T.W. Hänsch, Y.R. Shen, Springer Ser. Opt. Sci., Vol.49 (Springer, Berlin, Heidelberg 1985) pp.115–117
3.67 C.B. Dane, W.B. Yan, D. Zeitz, J.L. Hall, R.F. Curl, Jr., J.V.V. Kasper, F.K. Tittel: AIP Conf. Proc. **146**, 458 (1986)
3.68 N. Bjerre, H. Helm: Chem. Phys. Lett. **134**, 361 (1987)
3.69 P.J. Sarre, J.M. Walmsley, C.J. Whitham: J. Chem. Soc. Faraday Trans. II **82**, 1243 (1986)
3.70 P.F. Bernath: AIP Conf. Proc. **146**, 443 (1986)
3.71 A. Bolovinos, S. Spyrou, P. Tsekerris, M. Compitsas: J. Mol. Spectrosc. **122**, 269 (1987)
3.72 M. Kawasaki, H. Sato, T. Kikuchi, A. Fukuroda, S. Kobayashi, T. Arikawa: J. Chem. Phys. **86**, 4425 (1987)

3.73 D.S. King, R.F. Wormsbecher: Proc. SPIE **286**, 111 (1981)
3.74 H.P. Yu: Quantitative laser spectroscopy ($\nu_3-\nu_1$) difference frequency band at 10.65 μm wavelength region in N_2O. Thesis, East China Normal University, Shanghai (1984)
3.75 P.W. Anderson: Phys. Rev. **76**, 647 (1949)
3.76 C.J. Tsao, B. Curnutte: J. Quant. Sepctrosc. Radiat. Transf. **2**, 41 (1962)
3.77 H.P. Yu, S.X. Shen, I.S. Cheng: Chinese J. Infrared Res. **3**, 334 (1984)
3.78 S.X. Shen, P.P. Cai, H.S. Zhang, I.S. Cheng: Chinese J. Infrared Res. **3**, 333 (1984)
3.79 W.S. Benedict, L.D. Kaplan: J. Chem. Phys. **30**, 388 (1959)
3.80 S.X. Shen, P.P. Cai, H.S. Zhang, I.S. Cheng: Int. J. Infrared and Millimeter Waves **6**, 423 (1985)
3.81 J.D. Xu: Study of temperature factor for self-broadening of CO_2 lines. Thesis, East China Normal University, Shanghai (1988)
3.82 S.X. Shen, P.P. Cai, H.S. Zhang, I.S. Cheng: Temperature factor of pressure broadening for vibration-rotational lines at 10 μm band of CO_2. Digest Annual Meeting of Chinese Optical Society, Shanghai (1985) p.294
3.83 W.G. Planet, G.L. Tettemer, J.S. Knoll: J. Quant. Spectrosc. Radiat. Transfer **20**, 547 (1978)
3.84 G.L. Tettemer, W.G. Planet: ibid, **24**, 343 (1980)
3.85 V. Malathy Devi, B. Fridovich, G.D. Jones, D.G. Snyder: J. Mol. Spectrosc. **105**, 61 (1984)
3.86 Eric Arie, Nelly Lacome, Armand Levy: Appl. Opt. **26**, 1636 (1987)
3.87 L.A. Gross and P.R. Griffiths: Appl. Opt. **26**, 2250 (1987)
3.88 C. Cousin, R. Le Doucen, C. Boulet, A. Henry: Appl. Opt. **24**, 3899 (1985)
3.89 J.H. Pierluissi, K. Tomiyama, R.B. Gomez: Appl. Opt. **18**, 1607 (1979)
3.90 S.L. Yu: Band model for quantitative studies of molecular absorption spectra. Thesis, East China Normal University, Shanghai (1987)
3.91 V.I. Moroz, L.V. Zasova: Kosm. Issled. **23**, 259 (1985) [Engl. transl.: Cosmic Res. **23**, 222 (1985)]
3.92 H.R. Xia, Z.X. Guo, L.S. Ma, P.P. Cai, L.E. Ding: J. Shanghai Normal Univ., Ser. Sci. No2, 33 (1980)
3.93 G.L. Loper, G.R. Sasaki, M.A. Stamps: Absorption spectra of toxic compounds at CO_2 laser wavelengths. In *Laser Spectroscopy for Sensitive Detection*, ed. by J.A. Gelbwachs, Proc. SPIE **286**, 2-8 (1981)
3.94 V.S. Letokhov, V.P. Zharov: *Laser Optoacoustic Spectroscopy*, Springer Ser. Opt. Sci., Vol.37 (Springer, Berlin, Heidelberg 1986)
3.95 P. Hess, J. Pelzl (eds.): *Photoacoustic and Photothermal Phenomena*, Springer Ser. Opt. Sci., Vol.58 (Springer, Berlin, Heidelberg 1988)
      J.C. Murphy, J.W. Maclachlan Spicer, L.C. Aamodt, B.S.H. Royce (eds.): *Photoacoustic and Photothermal Phenomena II*, Springer Ser. Opt. Sci., Vol.62 (Springer, Berlin, Heidelberg 1990)
3.96 P. Hess (ed.): *Photoacoustic, Photothermal, and Photochemical Processes in Gases*, Topics Curr. Phys., Vol.46 (Springer, Berlin, Heidelberg 1989)
3.97 Wendell T. Hill III, R.A. Abreu, T.W. Hänsch, A.L. Schawlow: Opt. Commun. **32**, 96 (1980)
3.98 Wendell T. Hill III, T.W. Hämsch, A.L. Schawlow: Appl. Opt. **24**, 3718 (1985)
3.99 D. Courtois, C. Thiebeaux, A. Delahaigue: Int. J. Infrared and Millim. Waves **8**, 103 (1987)
3.100 P. Menendez-Valdes Alvarez: Mundo Electron. **158**, 81 (1986)

Chapter 4

4.1   W. Kaiser, D.G.B. Garrett: Phys. Rev. Lett. **7**, 229 (1961)
4.2   L.S. Vasilenko, V.P. Chebotayev, A.V. Shishaev: JETP Lett. **12**, 113 (1970)
4.3   B. Cagnac, G. Grynberg, F. Biraben: J. Phys. **34**, 845 (1973)
4.4   M.D. Levenson, N. Bloembergen: Phys. Rev. Lett. **32**, 645 (1974)
4.5   T.W. Hänsch, K.C. Harvey, G. Meisel, A.L. Schawlow: Opt. Commun. **11**, 50 (1984)
4.6   A.H. Kung, R.H. Page, R.J. Larkin, Y.R. Shen, Y.T. Lee: Molecular spectroscopy by stepwise two-photon ion-pair production at 71 nm. In *Laser Spectroscopy VII*, ed. by T.W. Hänsch, Y.R. Shen, Springer Ser. Opt. Sci., Vol.49 (Springer, Berlin, Heidelberg 1985) pp.179-180
4.7   K.C. Harvey: Doppler-free two-photon spectroscopy. Ph.D. Thesis, Stanford University (1975)
4.8   J.P. Woerdman: Chem. Phys. Lett. **43**, 279 (1976)
4.9   H.R. Xia, G.Y. Yan, A.L. Schawlow: Opt. Commun. **39**, 153 (1981)
4.10 G.P. Morgan, H.R. Xia, A.L. Schawlow: J. Opt. Soc. Am. **72**, 315 (1982)
4.11 P.M. Dekmer, J.L. Dhemer, S.T. Pratt: AIP Con. Proc. **146**, 572 (1986)
4.12 B.K. Clark and I.M. Littlewood: AIP Conf. Proc. **146**, 472 (1986)
4.13 M.S. Dzhidzhoev, S.V. Ivanov, V.Ya. Panchenko, A.V. Chugunov: Kvantovaya Elektron. **13**, 740 (1986) [Engl. transl.: Sov. J. Quantum Electron. **16**, 481 (1986)]
4.14 M.B. Knickelbein, K.S. Haber, L. Bigio, E.R. Grant: Chem. Phys. Lett. **131**, 51 (1986)
4.15 W.K. Bischel, P.J. Kelly, C.K. Rhodes: Phys. Rev. Lett. **34**, 300 (1975)
4.16 J.A. Dagata, M.A. Scott, S.P. Mcglunn: J. Chem. Phys. **85**, 5401 (1986)
4.17 H.-H. Ritze: Ann. Phys. **43**, 545 (1986)
4.18 J.R. Lombardi, R. Wallenstein, T.W. Hänsch, D.M. Friedrich: J. Chem. Phys. **65**, 2357 (1976)
4.19 E. Riedle, H. Stepp, H.J. Neusser: Chem. Phys. Lett. **110**, 452 (1984)
4.20 M. Boiveneau, J. Le Calve, M.C. Castex, C. Jouvet: Chem. Phys. Lett. **128**, 528 (1986)
4.21 G.A. Bickel, K.K. Innes: J. Chem. Phys. **86**, 1752 (1987)
4.22 H. Shinohara, K. Sato, Y. Achiba, N. Nishi, K.Kimura: Chem. Phys. Lett. **130**, 231 (1986)
4.23 D.S. King, D.F. Heller, J. Krasinski, R.S. Bodaness: AIP Conf. Proc. **146**, 694 (1986)
4.24 N. Bloembergen, M.D. Levenson: Doppler-free two-photon absorption spectroscopy. In *High-Resolution Laser Spectroscopy*, ed. by K. Shimoda, Topics Appl. Phys., Vol.13 (Springer, Berlin, Heidelberg 1976) pp.315-369
4.25 T.W. Hänsch: Nonlinear high-resolution spectroscopy of atoms and molecules. In *Nonlinear Spectroscopy*, ed. by N. Bloembergen. Proc. Int. School Phys. "Enrico Fermi", Course LXIV (North-Holland, Amsterdam 1977) pp.17-86
4.26 V.S. Letokhov, V.P. Chebotayev: *Nonlinear Laser Spectroscopy*, Springer Ser. Opt. Sci., Vol.3 (Springer, Berlin, Heidelberg 1977)
4.27 E. Giacobinao, B. Cagnac: Doppler-free multiphoton spectroscopy. *Progress in Optics* **17**, 85-161 (North-Holland, Amsterdam 1980)
4.28 H.R. Xia, J.W. Xu, J.G. Cai, I.S. Cheng: Quantum Electronics (Chinese) **3**, 295 (1986)
4.29 L.S. Vasilenko, V.P. Chebotayev, A.V. Shishaev: JETP Lett. **12**, 113 (1970)
4.30 J.E. Bjorkholm, P.E. Liao: Phys. Rev. A**14**, 751 (1976)
4.31 H.R. Xia, J.W. Xu, I.S. Cheng: Quantum Electronics (Chinese) **3**, 293 (1986)
4.32 H.R. Xia, L.S. Ma, J.W. Xu, M. Yuan, I.S. Cheng: Paper presented at Int'l Conf. Lasers, Xiamen, China (1987)
4.33 P. Kusch, M.M. Hessel: J. Chem. Phys. **68**, 2591 (1978)

4.34   M.E. Kaminsky: Modulated population spectroscopy: identification of absorption lines by modulated lower level population: Spectrum of Na_2. Ph.D. Thesis, Stanford University (1976)

4.35   A.N. Taylor: Two-step polarization labeling spectroscopy of excited states of diatomic sodium. Ph.D. Thesis, Stanford University (1982)

4.36   N.W. Carlson, A.J. Taylor, K.M. Jones, A.L. Schawlow: Phys. Rev. A 24, 822 (1981)

4.37   N.W. Carlson, F.V. Kowalski, R.E. Teets, A.L. Schawlow: Opt. Commun. 18, 1983 (1979)

4.38   A.J. Taylor, K.M. Jones, A.L. Schawlow: Opt. Commun. 39, 47 (1981)

4.39   G. Jeung: J. Phys. B: At. Mol. Phys. 16, 4289 (1983)

4.40   G.Y. Yan, H.R. Xia: Scientia Sinica A (Chinese) 38, 504 (1985)

4.41   J.B. Atkinson, J. Becker, W. Demtröder: Chem. Phys. Lett. 87, 92 (1982)

4.42   Li Li, S.F. Rice, R.W. Field: J. Mol. Spectrosc. 105, 344 (1984)

4.43   Li Li, R.W. Field: J. Mol. Spectrosc. 117, 245 (1986)

4.44   H.R. Xia, Z.G. Wang: Study of small but complete molecules - Na_2, in *Lasers, Spectroscopy and New Ideas*, ed. by W.M. Yen, M.D. Levenson, Springer Ser. Opt. Sci., Vol.54 (Springer, Berlin, Heidelberg 1987) pp.174-182

4.45   J.W. Xu: Study on four-level coherent excitation of two-photon transitions. Thesis, East China Normal University, Shanghai (1987)

4.46   G. Herzberg: *Molecular Spectra and Molecular Structure; I. Spectra of Diatomic Molecules* (Van Nostrand, Princeton, NJ 1950) p.284

4.47   H.R. Xia, J.-W. Xu, I-S. Cheng: Lineshape variations of spin-forbidden two-photon transitions. *Proc. 4th Int'l Laser Science Conf.* 1988 (World Scientific, Singapore 1989) p.70

Chapter 5

5.1   T.W. Hänsch: Nonlinear high-resolution spectroscopy of atoms and molecules. In *Nonlinear Spectroscopy*, ed. by N. Bloembergen (North-Holland, Amsterdam 1977) pp.17-86

5.2   V.S. Letokhov, V.P. Chebotayev: *Nonlinear Laser Spectroscopy* (Springer, Berlin, Heidelberg 1977)

5.3   K. Shimoda (ed.): *High Resolution Laser Spectroscopy*, Topics Appl. Phys., Vol.13 (Springer, Berlin, Heidelberg 1976)

5.4   W. Demtröder: *Laser Spectroscopy* (Springer, Berlin, Heidelberg 1981)

5.5   V.P. Chebotayev: Superhigh resolution spectroscopy. In *Laser Handbook V*, ed. by M. Bass, M.L. Stitch (North-Holland, Amsterdam 1985) pp.289-404

5.6   T.W. Hänsch, M.D. Levenson, A.L. Schawlow: Phys. Rev. Lett. 27, 707 (1971)

5.7   P.W. Smith, T.W. Hänsch: Phys. Rev. Lett. 26, 740 (1971)

5.8   C. Borde: C.R. Acad. Sci. 271, 371 (1970)

5.9   M.D. Levenson, A.L. Schawlow: Phys. Rev. A6, 10 (1972)

5.10   M.E. Kaminsky: Modulated population spectroscopy: Identification of absorption lines by modulated lower level population: Spectrum of Na_2. PhD. Thesis, Stanford University (1976)

5.11   M.E. Kaminsky, R.T. Hawkins, F.V. Kowalski, A.L. Schawlow: Phys. Rev. Lett. 36, 671 (1976)

5.12   O.N. Kampanets, V.S. Letokhov: Zh. Eksp. Theor. Fiz. 62, 1302 (1972)

5.13   J.L. Hall, J.A. Magyar: High resolution saturated absorption studies of methane and some methyl-halides. In *High-Resolution Laser Spectroscopy*, ed. by K. Shimoda, Topics Appl. Phys., Vol.13 (Springer, Berlin, Heidelberg 1976) pp.173-199

5.14   E.V. Baklanov, V.P. Chebotayev: Usp. Fiz. Nauk 122, 513 (1977)

5.15   P. Minguzzi, M. Tonelli, A. Carrozzi, S. Profeti, A. Di Lieto: Doppler-free optoacoustic spectroscopy ammonia. In *Laser Spectroscopy VI*, ed. by H.P. Weber, W. Lüthy, Springer Ser. Opt. Sci., Vol.40 (Springer, Berlin Heidelberg 1983) pp.152-153

5.16   T. Suzuki, M. Kakimoto: J. Mol. Spectroscop. **93**, 423 (1982)

5.17   M.S. Sorem, A.L. Schawlow: Opt. Commun. **5**, 148 (1972)

5.18   F.V. Kowalski, W.T. Hill, A.L. Schawlow: Opt. Lett. **2**, 112 (1978)

5.19   R. Schieder: Opt. Commun. **26**, 113 (1978)

5.20   J.C. Tsai: Pressure effects measurements of $Na_2(X^1\Sigma_g^+ - B^1\Pi_u)$ with high resolution laser spectroscopy. PhD. Thesis, Stanford University (1980)

5.21   G.C. Bjorklund: Opt. Lett. **5**, 15 (1980)

5.22   G.C. Bjorklund, W. Lenth, M.D. Levenson, C. Ortiz: Frequency modulated (FM) spectroscopy. In *Laser Spectroscopy for Sensitive Detection*, ed. by J.A. Gelbwachs, Proc. SPIE **286**, 153-159 (1981)

5.23   J.L. Hall, T. Baer, L. Hollberg, H.R. Robinson: Precision spectroscopy and laser frequency control using FM sideband optical heterodyne techniques. In *Laser Spectroscopy V*, ed. by A.R.W. McKellar, T. Oka, B.P. Stoicheff, Springer Ser. Opt. Sci., Vol.30 (Springer, Berlin, Heidelberg 1981) pp.15-24

5.24   Ch. Breant, T. Baer, D. Nesbitt, J.L. Hall: State-dependent hyperfine Coupling of HF studied with a frequency-controlled color-center laser spectrometer. In *Laser Spectroscopy VI*, ed. by H.P. Weber, W. Lüthy, Springer Ser. Opt. Sci., Vol.40 (Springer, Berlin, Heidelberg 1983) pp.138-143

5.25   N.C. Wong, J.L. Hall: Servo control of amplitude modulation in FM spectroscopy: shot-noise limited measurement of water vapor pressure-broadening. In *Laser Spectroscopy VII*, ed. by T.W. Hänsch, Y.R. Shen, Springer Ser. Opt. Sci., Vol.49 (Springer, Berlin, Heidelberg 1985) pp.393-394

5.26   W.J. Childs, G.L. Goodman, L.S. Goodman, L. Young: J. Mol. Spectrosc. **119**, 166 (1986)

5.27   J.L. Hall, L. Hollberg, Ma Long-Sheng, T. Baer, H.R. Robinson: J. Physique-Colloque **42**, C8-59 (1981)

5.28   Z.Y. Bi: Phase modulation optical heterodyne for four-wave mixing. Thesis, East China Normal University (1986)

5.29   T.P. Duffey, D. Kammen, A.L. Schawlow, S. Svanberg, H.R. Xia, G.Q. Xiao, G.Y. Yan: Opt. Lett. **10**, 597 (1985)

5.30   C.E. Wieman, T.W. Hänsch: Phys. Rev. Lett. **36**, 1170 (1976)

5.31   R.E. Teets: Polarization Labeling Spectroscopy of Molecules. Thesis, Stanford University (1978)

5.32   R.E. Teets, F.V. Kowalski, W.T. Hill, N. Carlson, T.W. Hänsch: Laser polarization spectroscopy. In *Advances in Laser Spectroscopy*, ed. by Ahmed Zewail, SPIE **113**, 80 (1977)

5.33   Ming-Guang Li, Chongye Wang, Yingde Wang, Li Li: J. Mol. Spectrosc. **123**, 161 (1987)

5.34   T.W. Hänsch, D.R. Lyons, A.L. Schawlow, A. Siegel, Z.Y. Wang, G.Y. Yan: Opt. Commun. **37**, 87 (1981)

5.35   G.Y. Yan, H.R. Xia: Scientia sinica A**11**, 1043 (1984)

5.36   J.G. Cai, H.R. Xia, I.S. Cheng: Acta Optica Sinica **6**, 212 (1986)

5.37   M. Raab, W. Demtröder: Double-resonance polarization spectroscopy of the caesium-dimer (CS_2). In *Laser Spectroscopy V*, ed. by A.R. W. McKellar, T. Oka, B.P. Stoicheff, Springer Ser. Opt. Sci., Vol. 30 (Springer, Berlin, Heidelberg 1981) pp.126-129

5.38   B. Hemmerling, S.B. Rai, W. Demtröder: Z. Physik A**320**, 135 (1985)

5.39   M.D. Levenson, G.L. Eesley: Appl. Phys. **19**, 1 (1979)

5.40   M. Raab, A. Weber: J. Opt. Soc. Am. B**2**, 1476 (1985)

5.41   W.H. Weber, P.D. Maker, J.W.C. Johns, E. Weinberger: J. Mol. Spectrosc. **121**, 243 (1987)

5.42 A. Kiermeier, K. Dietrich, E. Riedle, H.J. Neusser: J. Chem. Phys. **85**, 6983 (1986)
5.43 L.S. Vasilenko, V.P. Chebotayev, A.V. Shishaev: JETP Lett. **12**, 113 (1970)
5.44 B. Cagnac, G. Grynberg, F. Biraben: J. Phys. (Paris) **34**, 845 (1973)
5.45 M.D. Levenson, N. Bloembergen: Phys. Rev. Lett. **32**, 645 (1974)
5.46 T.W. Hänsch, K.C. Harvey, G. Meisel, A.L. Schawlow: Opt. Commun. **11**, 50 (1974)
5.47 K.C. Harvey: Doppler-Free Two-Photon Spectroscopy. Thesis, Stanford University (1975)
5.48 J.P. Woerdman: Chem. Phys. Lett. **43**, 279 (1976)
5.49 H.R. Xia, G.Y. Yan, A.L. Schawlow: Opt. Commun. **39**, 153 (1981)
5.50 G.P. Morgan, H.R. Xia, A.L. Schawlow: J. Opt. Soc. Am. **72**, 315 (1982)
5.51 H.J. Neusser, E. Riedler: Comments At. Mol. Phys. **19**, 331 (1987)
5.52 E. Riedler, H. Stepp, H.J. Neusser: Doppler-free two-photon electronic spectra of large molecules with resolution near the natural linewidth. In *Laser Spectroscopy VI*, ed. by H.P. Weber, W. Lüthy, Springer Ser. Opt. Sci., Vol.40 (Springer, Berlin, Heidelberg 1983) pp.144-146
5.53 W.K. Bischel, P.J. Kelly, C.K. Rhodes: Phys. Rev. Lett. **34**, 300 (1975)
5.54 W. Demtröder, U. Diemer, H.J. Foth, R. Kullmer, G. Persch, H.J. Vedder, H. Weickenmeier: High resolution molecular spectroscopy with lasers. In *Colloquium Spectroscopicum International XXIV-CSI XXIV* (Springer, Berlin, Heidelberg 1986) pp.109-115
5.55 H.R. Xia, S.V. Benson, T.W. Hänsch: Laser Focus **17**, 54 (March 1981)
5.56 E.V. Baklanov, B.Ya. Dubetsky, V.P. Chebotayev: Appl. Phys. **9**, 171 (1976)
5.57 J.C. Bergquist, S.A. Lee, J.L. Hall: Phys. Rev. Lett. **38**, 159 (1977)
5.58 G. Kramer: J. Opt. Soc. Am. **68**, 1634 (1978)
5.59 M. Baba, K. Shimoda: Appl. Phys. **24**, 11 (1981)
5.60 C.J. Borde, S. Avrillier, A. van Lerberghe, C. Salomon, D. Bassi, G. Scoles: J. Physique Colloqut **42**, C8-15 (1981)
5.61 Ch. Salomon, S. Avrillier, A. van Lerberghe, Ch.J. Borde: Direct optical detection of Ramsey fringes in a supersonic beam of SF_6. In *Laser Spectroscopy VI*, ed. by H.P. Weber, W. Lüthy, Springer Ser. Opt. Sci., Vol.40 (Springer, Berlin, Heidelberg 1983) pp.159-160
5.62 A.G. Adam, T.E. Gough, N.R. Isenor, G. Scoles, J. Shelley: Phys. Rev. A**34**, 4803 (1986)
5.63 S.N. Bagayev, A.S. Dychkov, V.P. Chebotayev: Appl. Phys. **15**, 209 (1978)
5.64 M.M. Salour, C. Cohen-Tannoudji: Phys. Rev. Lett. **38**, 757 (1977)
5.65 V.P. Chebotayev, N.M. Dyuba, M.I. Skvortsov, L.S. Vasilenko: Appl. Phys. **15**, 319 (1978)
5.66 L.S. Vasilenko, N.N. Rubtsova, Kvantovaya Elektronika **9**, 2243 (1982)
5.67 L.S. Vasilenko, I.D. Matveyenko, N.N. Rubtsova: Opt. Commun. **53**, 371 (1985)
5.68 K.M. Evenson, J.S. Wells, F.R. Petersen, B.L. Danielson, G.E. Day: Appl. Phys. Lett. **22**, 192 (1973)
5.69 Ch.J. Borde, J. Borde, Ch. Breant, Ch. Chardonnet, A. van Lerberghe, Ch. Salomon: Internal dynamics of simple molecules revealed by the superfine and hyperfine structures of their infrared spectra. In *Laser Spectroscopy VII*, ed. by T.W. Hänsch, Y.R. Shen, Spriner Ser. Opt. Sci., Vol.49 (Springer, Berlin, Heidelberg 1985) pp.108-114
5.70 R. Teets, R. Feinberg, T.W. Hänsch, A.L. Schawlow: Phys. Rev. Lett. **37**, 683 (1976)
5.71 M. Raab, G. Höning, W. Demtröder, C.R. Vidal: J. Chem. Phys. **76**, 4370 (1982)
5.72 J.B. Atkinson, J. Becker, W. Demtröder: Chem. Phys. Lett. **87**, 92 (1982)
5.73 O.N. Kompanets, A.R. Kukudzhanov, V.S. Letokhov, V.G. Minogin, Ye.L. Mikhailov: Zh. Eksp. Teor. Fiz. **69**, 32 (1975)
5.74 C. Borde, J.L. Hall: Phys. Rev. Lett. **30**, 1101 (1973)

5.75 T.W. Meyer, J.F. Brilando, C.K. Rhodes: Chem. Phys. Lett. **18**, 382 (1971)
5.76 J.B. Atkinson, J. Becker, W. Demtröder: Chem. Phys. Lett. **87**, 128 (1982)
5.77 Y.D. Wang, C.Y. Wang, M.G. Li, Li Li: Chinese J. Laser **15**, 302 (1988)
5.78 H.R. Xia, Z.G. Wang: Study of small but complete molecules - Na_2. In *Laser, Spectroscopy and New Ideas*, ed. by W.M. Yen, M.D. Levenson, Springer Ser. Opt. Sci. Vol.54 (Springer, Berlin, Heidelberg 1987) pp.174-182
5.79 H.R. Xia, L.S. Ma, J.W. Xu, M. Yuan, I.S. Cheng: to be published
5.80 L. Li, R.W. Field: J. Mol. Spectrosc. **123**, 237 (1987)
5.81 E.V. Baklanov, M.V. Belyayev: Appl. Phys. **14**, 389 (1977)
5.82 K. Uehara, K. Shimoda: Japan J. Appl. Phys. **16**, 633 (1977)
5.83 K. Uehara: Optics Letters **6**, 191 (1981)
5.84 A.C. Luntz, R.G. Brewer, K.L. Foster, J.D. Swalen: Phys. Rev. Lett. **23**, 951 (1969)
5.85 A.C. Luntz, R.G. Brewer: J. Chem. Phys. **54**, 3641 (1971)
5.86 K. Uehara: J. Phys. Soc. Japan **34**, 777 (1973)
5.87 H. Calotti, G.D. Lonardo, A. Trombetti: J. Chem. Phys. **78**, 1670 (1983)
5.88 K. Takagi, K. Itoh, E. Miura, S. Tanimura: J. Opt. Soc. Am **B4**, 1145 (1987)
5.89 I. Kleiner, M. Godefroid, M. Herman: J. Opt. Soc. Am. **B4**, 1159 (1987)
5.90 G. Duxbury, J. McCombie: J. Opt. Soc. Am **B4**, 1197 (1987)
5.91 R. Wallenstein, J.A. Paisner, A.L. Schawlow: Phys. Rev. Lett. **32**, 1333 (1974)
5.92 E.E. Uzgiris, J.L. Hall, R.L. Barger: Phys. Rev. Lett. **26**, 289 (1971)
5.93 S.N. Bagayev, M.V. Balyayev, A.K. Dmitriyev, V.P. Chebotayev: Appl. Phys. **24**, 261 (1981)
5.94 T. Amano, K. Kawaguchi, M. Kakimoto, S. Saito, E. Hirota: J. Chem. Phys. **77**, 159 (1982)
5.95 K. Kawaguchi, T. Suzuki, S. Saito, E. Hirota: J. Opt. Soc. Am. **B4**, 1203 (1987)
5.96 E. Riedle, H.J. Neusser: J. Chem. Phys. **80**, 4686 (1984)
5.97 W.G. Harter: Phys. Rev. **A24**, 192 (1981)
5.98 J. Borde, Ch.J. Borde: Chem. Phys. **71**, 417 (1982)
5.99 J. Borde, Ch.J. Borde: J. Mol. Spectrosc. **78**, 353 (1979)
5.100 V.P. Chebotayev: Optical frequency standards. In *Metrology and Fundamental Constants* LXVIII Corso. Soc. Italiana di Fisica, Bologna, Italy, 1980, p.623
5.101 J.L. Hall, C.J. Borde: Appl. Phys. Lett. **29**, 788 (1976)
5.102 R. Feinberg, R.E. Teets, J. Rubbmark, A.L. Schawlow: J. Chem. Phys. **66**, 4330 (1977)
5.103 S.N. Bagayev, E.V. Baklanov, V.P. Chebotayev: Zh. Eksp. Teor. Fiz. Pisma. **16**, 15 (1972)
5.104 C.J. Borde: New sub-doppler interaction techniques. In *Laser Spectroscopy III*, ed. by J.L. Hall, J.L. Carlsten, Springer Ser. Opt. Sci., Vol.7 (Springer, Berlin, Heidelberg 1977) pp.121-134
5.105 A.T. Mattick, N.A. Kurnit, A. Javan: Chem. Phys. Lett. **38**, 176 (1976)
5.106 T.W. Meyer, C.K. Rhodes, H.A. Haus: Phys. Rev. **12**, 1993 (1975)
5.107 L.S. Vasilenko, V.P. Kochanov, V.P. Chebotayev: Opt. Commun. **20**, 409 (1977)
5.108 M.V. Belyayev, L.S. Vasilenko, M.N. Skvortsov, V.P. Chebotayev: Zh. Eksp. Teor. Fiz. **81**, 526 (1981)
5.109 O.N. Kompanets, A.P. Kukudzhanov, V.S. Letokhov, E.L. Mikhailov: Kvantovaja Electronika **16**, 28 (1973)
5.110 S.N. Bagayev, A.S. Duchkov, A.K. Dmitriyev, V.P. Chebotayev: Zh. Eksp. Teor. Fiz. **79**, 1160 (1980)
5.111 W.K. Bischel, P.J. Kelley, C.K. Rhodes: Phys. Rev. Lett. **34**, 300 (1975)
5.112 S.N. Bagayev, V.P. Chebotayev, A.S. Dychkov, S.V. Maltsev: J. Physique Colloque **42**, C8 (1981)
5.113 A.F.J. Van Raan, J.E.M. Haverkort, B.L. Mehta, J. Korving: J. Phys. B **15**, L669 (1982)
5.114 W.R. Fredrickson, C.R. Stannard: Phys. Rev. **44**, 632 (1933)

5.115 T. Carroll: Phys. Rev. **52**, 822 (1937)
5.116 P. Kusch, M.M. Hessel: J. Chem. Phys. **63**, 4087 (1975)
5.117 R. Teets, R. Feinberg, T.W. Hänsch, A.L. Schawlow: Phys. Rev. Lett. **37**, 683 (1976)
5.118 F. Engelke, H. Hage, C.D. Caldwell: Chem. Phys. **64**, 221 (1982)
5.119 K. Shimizu, F. Shimizu: J. Chem. Phys. **78**, 1126 (1983)
5.120 Li Li, S.F. Rice, R.W. Field: J. Mol. Spectrosc. **105**, 344 (1984)
5.121 R. Castell, W. Demtröder, A. Fischer, R. Kullmer, K. Wickert: Appl. Phys. B **38**, 1 (1985)
5.122 F.V. Kowalski, R.E. Teets, W. Demtröder, A.L. Schawlow: J. Opt. Soc. Am. **68**, 1611 (1978)
5.123 W. Preuss, G. Baumgartner: Z. Phys. A **320**, 125 (1985)
5.124 Z.C. Bao, L.S. Ma, L.E. Ding, I.S. Cheng: to be published
5.125 G. Herzberg: *Molecular Spectra and Molecular Structure. I. Sepctra of Diatomic Molecules* (Van Nostrand, Toronto 1950)
5.126 C.H. Townes, A.L. Schawlow: *Microwave Spectroscopy* (McGraw-Hill, New York 1955)
5.127 S. Chen, M. Takeo: Rev. Mod. Phys. **29**, 20 (1957)
5.128 P.R. Berman: Phys. Rev. A**13**, 2191 (1976)
5.129 R. Feinberg, R.E. Teets, J. Rubbmark, A.L. Schawlow: J. Chem. Phys. **66**, 4330 (1977)
5.130 G. Herzberg, Ch. Jungen: J. Mol. Spectrosc. **41**, 425 (1972)
5.131 H.G.M. Edwards, D.A. Long, G. Sherwood: Studies of the self-broadening and foreign gas-broadening coefficient dependence on temperature for diatomic molecules and monoatomic gases. In Proc.10th Int'l Conf. on Raman Spectroscopy, (Eugene, OR 1986) pp.6-7
5.132 G.J. Rosasco, W.S. Hurst: Nonlinear Raman spectroscopy with high-resolution CW laser. In Proc. 10th Int'l Conf. on Raman Spectroscopy, Eugene, OR 1986) pp.43-44
5.133 L.A. Rahn, R.E. Palmer: J. Opt. Soc. Am. B**3**, 1165 (1986)
5.134 J. Bonamy, D. Robert, C. Boulet: J. Quant. Spectrosc. Radiat. Transfer **31**, 23 (1984)
5.135 D.J. Clouthier, D.A. Ramsay, F.W. Birss: J. Chem. Phys. **79**, 5851 (1983)
5.136 S.M. Freund, A.G. Maki: J. Mol. Spectrosc. **93**, 433 (1982)
5.137 L. Nemes, A.R.W. McKellar, J.W.C. Johns: J. Opt. Soc. Am. **4**, 1165 (1987)
5.138 W.H. Weber (ed.): Stark and Zeeman techniques in laser spectroscopy. Special issue. J. Opt. Soc. Am. **4**, 1141-1226 (1987)
5.139 J.E. Lawler, A.I. Freguson, J.E.M. Goldsmith, D.J. Jackson, A.L. Schawlow: Phys. Rev. Lett. **42**, 1046 (1979)
5.140 D.R. Lyons, A.L. Schawlow, G.Y. Yan: Opt. Commun. **38**, 35 (1981)
5.141 J.A. Gelbwachs, C.F. Klein, J.E. Wessel: IEEE J. QE-**14**, 121 (1978)
5.142 J.E. Wessel, D.E. Cooper, C.M. Klimcak: Ultrasensitive molecular detection by multiphoton ionization spectroscopy. In *Laser Spectroscopy for Sensitive Detection*, ed. by J. A. Gelbwachs, Proc. SPIE **286**, 48-55 (1981)
5.143 D.H. Parker: Laser ionization spectroscopy and mass spectroscopy. In *Ultrasensitive Laser Spectroscopy*, ed. by D.S. Kliger (Academic, New York 1983)
5.144 M.J. Adams, D.R. Wake, N.M. Amer: Appl. Phys. Lett. **34**, 379 (1979)
5.145 A.C. Tam: Photoacoustics: Spectroscopy and their applications. In *Ultrasensitive Laser Spectroscopy*, ed. by D.S. Kliger (Academic, New York 1983)
5.146 G.L. Loper, G.R. Sasaki, M.A. Stamps: Absorption spectra of toxic compounds at CO_2 laser wavelengths. In *Laser Spectroscopy for Sensitive Detection*, ed. by J.A. Gelbwachs, SPIE **286**, 1 (1981)
5.147 T.W. Hänsch, A.L. Schawlow, P. Toschek: IEEE J. QE-**8**, 802 (1972)
5.148 W.T. Hill III, R.A. Abreu, T.W. Hänsch, A.L. Schawlow: Opt. Commun. **32**, 96 (1980)

5.149 W. Demtröder, F. Paech, R. Schmiedl: Chem. Phys. Lett. **26**, 381 (1974)
5.150 P. Jacquinot: Atomic beam spectroscopy. In *High Resolution Laser Spectroscopy*, ed. by K. Shimoda, Topics Appl. Phys., Vol.13 (Springer, Berlin, Heidelberg 1976)
5.151 W. Demtröder, H.J. Foth: Phys. Bl **43**, 7 (1987)
5.152 C. Douketis, D. Anex, J.P. Reilly: High resolution overtone spectroscopy on a molecular beam. In *Laser Applications in Chemistry*, Proc. SPIE Int. Soc. Opt. Eng., Vol.669, 1986, pp.137–142
5.153 T.C. Chang, M.V. Johnston: J. Phys. Chem. **91**, 884 (1987)
5.154 J.W. Xu: Study on four-level coherent excitation of two-photon transitions. Thesis, East China Normal University, Shanghai (1987)

Chapter 6

6.1   N. Bloembergen: *Nonlinear Optics* (Benjamin, New York 1965)A
6.2   M. Allegrini, G. Alzetta, A. Kopystynska, L. Moi, G. Orriols: Opt. Commun. **22**, 329 (1977)
6.3   J.P. Woerdman: Opt. Commun. **26**, 216 (1978); also Chem. Phys. Lett. **43**, 279 (1976)
6.4   A. Kopystynska, P. Kowalczyk: Opt. Commun. **28**, 78 (1979)
6.5   M. Allegrini, L. Moi: Opt. Commun. **32**, 91 (1980)
6.6   M.E. Koch, K.K. Verma, W.C. Stwalley: J. Opt. Soc. Am. **70**, 627 (1980)
6.7   C.Y.R. Wu, J.K. Chen: Opt. Commun. **44**, 100 (1982)
6.8   C.Y.R. Wu, J.K. Chen, D.L. Judge, C.C. Kim: Opt. Commun. **48**, 28 (1983)
6.9   J.T. Bahns, W.C. Stwalley: Appl. Phys. Lett. **44**, 826 (1984)
6.10 Z.G. Wang, L.S. Ma, H.R. Xia, K.C. Zhang, I.S. Cheng: Opt. Commun. **58**, 315 (1986)
6.11 Z.G. Wang, X.L. Tan, K.C. Zhang, L.J. Qin, I.S. Cheng: Scientia Sinica (Science in China) A **30**, 1045 (1987)
6.12 Z.G. Wang, B. Wellegehausen: 1987 Int'l Conf. on Laser, Xiamen, PR China (1987)
6.13 Z.G. Wang, K.C. Zhang, X.L. Tan, I.S. Cheng: Acta Opt. Sinica **6**, 1081 (1986)
6.14 D. Krökel, M. Hube, W. Luhs, B. Wellegehausen: Appl. Phys. B **37**, 137 (1985)
6.15 L.J. Qin, Z.G. Wang, K.C. Zhang, L.S. Ma, Y.Q. Lin, I.S. Cheng: In *Laser Spectroscopy VII*, ed. by T.W. Hänsch, Y.R. Shen, Springer Ser. Opt. Sci., Vol.42 (Springer, Berlin, Heidelberg 1985)
6.16 Z.G. Wang, L.J. Qin, K.C. Zhang, I.S. Cheng: Appl. Phys. B **41**, 125 (1986)
6.17 L.J. Qin, Z.G. Wang, K.C. Zhang, I.S. Cheng: Chinese Physics **6**, 986 (1986)
6.18 X.L. Tan, Z.G. Wang, K.C. Zhang, L.J. Qin, I.S. Cheng: Phys. Sinica (China) **37**, 1227 (1988)
6.19 J.L. Carlsten, A. Szöke, M.G. Raymer: Phys. Rev. A**15**, 1029 (1977)
6.20 W. Hartig: Appl. Phys. **15**, 427 (1978)
6.21 J.K. Chen, C.Y.R. Wu, C.C. Kim, D.L. Judge: Appl. Phys. B **33**, 427 (1984)
6.22 P.L. Zhang, Y.C. Wang, A.L. Schawlow: J. Opt. Soc. Am. B **1**, 9 (1984)
6.23 R.E. Johnson: *Introduction to Atomic and Molecular Collision* (Plenum, New York 1982)
6.24 R.B. Bernstein (ed.): *Atom-Molecule Collision Theory* (Plenum, New York 1979)
6.25 E.K. Kopeikina, M.L. Yanson: Opt. Spektrosk. **41**, 378 (1976) [Engl. transl.: Opt. Spectrosc. USSR **41**, 217 (1976)]
6.26 E.K. Kopeikina, M.L. Yanson: Opt. Spektrosk. **39**, 783 (1975) [Engl. transl.: Opt. Spectrosc. USSR **39**, 442 (1975)]
6.27 E.K. Kraulinya, E.K. Kopeikina, M.L. Janson: Chem. Phys. Lett. **39**, 565 (1976)
6.28 V.B. Grushevskii: Opt. Spectrosc. **42**, 572 (1977)

6.29 L.K. Lam, T. Fujimoto, A.C. Gallagher, M.M. Hessel: J. Chem. Phys. **68**, 3553 (1978)

6.30 Ya.P. Klyavinsh, M.L. Yanson: Opt. Spektrosk. **52**, 630 (1982) [Engl. transl.: Opt. Spectrosc. USSR **52**, 376 (1982)]

6.31 Z.G. Wang, L.J. Qin, L.S. Ma, Y.Q. Lin, I.S. Cheng: Opt. Commun. **51**, 155 (1984)

6.32 K.C. Zhang: Study of Stimulated Radiation Generated by Atom-Molecule Collision Energy Transfer, Thesis, East China Normal University (1986)

6.33 S.G. Dinev, I.G. Koprinkov, I.L. Stefanov: Appl. Phys. B **39**, 65 (1986)

6.34 R.L. Byer, R.L. Herbst, H. Kildal, M.D. Levenson: Appl. Phys. Lett. **20**, 463 (1972)

6.35 B. Wellegehausen, K.H. Stephan, D. Friede, H. Welling: Opt. Commun. **23**, 157 (1977)

6.36 J.B. Koffend, R.W. Field: J. Appl. Phys. **48**, 4468 (1977)

6.37 M.A. Henesian, R.L. Herbst, R.L. Byer: J. Appl. Phys. **47**, 1515 (1976)

6.38 H. Itoh, H. Uchiki, M. Matsuoka: Opt. Commun. **18**, 271 (1976)

6.39 B. Wellegehausen, S. Shahdin, D. Friede, H. Welling: Appl. Phys. **13**, 97 (1977)

6.40 H. Welling, B. Wellegehausen: In *Laser Spectroscopy III*, ed. by J.L. Hall, J.L. Carlsten, Springer Ser. Opt. Sci., Vol.7 (Springer, Berlin, Heidelberg 1977) pp.365-369

6.41 D.R. Guyer, S.R. Leone: 5th Conf. on Chemical and Molecular Lasers, St. Louis, MO (1977)

6.42 B. Wellegehausen, D. Friede, G. Steger: Opt. Commun. **26**, 391 (1978)

6.43 F.J. Wodarczyk, H.R. Schlossberg: J. Chem. Phys. **67**, 4476 (1977)

6.44 S.R. Leone, K.G. Kosnik: Appl. Phys. Lett. **30**, 346 (1977)

6.45 B. Wellegehausen, H. Welling, K.H. Stephan, H.H. Heitmann: 10th Int'l Quantum Electron. Conf., Atlanta, GA (1978)

6.46 W.P. West, H.P. Broida: Chem. Phys. Lett. **56**, 283 (1978)

6.47 P.L. Jones, U. Hefter, U. Gaubtz, K. Bergmann, B. Wellegehausen: Appl. Phys. B28, 196 (1982)

6.48 B. Wellegehausen, W. Luhs, H. Welling, A. Topouzkhanian: Appl. Phys. B **28**, 195 (1982)

6.49 A.R. Rajaei-Rizi, J.T. Bahns, K.K. Verma, W.C. Stwalley: Appl. Phys. Lett. **40**, 869 (1982)

6.50 C.N. Man-Pichot, A. Brillet: ACS Symp. Ser. **179**, 487 (1982)

6.51 C.N. Man-Pichot, A. Brillet: IEEE J. QE-16, 1103 (1980)

6.52 K.K. Verma, W.C. Stwalley, W.T. Zemke: J. Appl. Phys. **52**, 3821 (1981)

6.53 K.K. Verma, W.C. Stwalley, W.T. Zemke: J. Appl. Phys. **52**, 5419 (1981)

6.54 B. Wellegehausen, H.H. Heitmann: Appl. Phys. Lett. **34**, 44 (1979)

6.55 J.T. Bahns, K.K. Werma, A.R. Rajaei-Rizi, W.C. Stwalley: Appl. Phys. Lett. **42**, 336 (1983)

6.56 K.K. Verma, T.H. Vu, W.C. Stwalley: J. Mol. Spectrosc. **85**, 131 (1981)

6.57 S.I. Kanorsii, V.M. Saslin, O.F. Yakushev: Kvantovaya Electron. (Moscow) 7, 2201 (1980) [Engl. transl.: Sov. J. Quantum Electron. **10**, 1275 (1980)]

6.58 J.B. Koffend, R.W. Field, D.R. Guyer, S.R. Leone: In *Laser Spectroscopy III*, ed. by J.L. Hall, J.L. Carlsten, Springer Ser. Opt. Sci., Vol.7 (Springer, Berlin, Heidelberg 1977) p.382

6.59 J.B. Koffend, S. Goldstein, R. Bacis, R.W. Field, S. Ezekiel: Phys. Rev. Lett. **41**, 1040 (1978)

6.60 B. Wellegehausen: ACS Symp. Series **179**, 461 (1982)

6.61 B. Wellegehausen: IEEE J. QE-15, 1108 (1979)

6.62 W. Müller, I.V. Hertel: Appl. Phys. **24**, 33 (1981)

6.63 B. Wellegehausen, W. Luhs, A. Topouskhanian, J. d'Incan: Appl. Phys. Lett. **43**, 912 (1983)

6.64 Z.G. Wang, Y.C. Wang, G.P. Morgan, A.L. Schawlow: Opt. Commun. **48**, 398 (1984)
6.65 P. Bernage, P. Niay, H. Bocquet: J. Mol. Spec. **98**, 304 (1983)
6.66 S. Shahdin, B. Wellegehausen, Z.G. Ma: Appl. Phys. B29, 195 (1982)
6.67 P. Niay, P. Bernage: C.R. Acad. Sci. II **294**, 627 (1982)
6.68 R. Schmiele, W. Lüthy, H.P. Weber: J. Appl. Phys. **53**, 1356 (1982)
6.69 Z.G. Wang, H.R. Xia, L.S. Ma, Y.Q. Lin, I.S. Cheng, A.L. Schawlow: In *Proc. Int'l Conf. on Laser 1984*, ed. by K.M. Corcoran, D.M. Sullivan, W.C. Stwalley (STS, McLean, VA 1985) p.291
6.70 Z.G. Wang, H.R. Xia, L.S. Ma, Y.Q. Lin, I.S. Cheng: Appl. Phys. B **37**, 233 (1985)
6.71 S.G. Dinev, I.G. Koprinkov, I.L. Stefanov: J. Phys. B **19**, 1 (1986)
6.72 K.-P. Huber, G. Herzberg: *Constants of Diatomic Molecules* (Van Nostrand-Reinhold, New York 1979)
6.73 J. Verges, C. Effantin, J. d'Incan, A. Topouzkhanian, B.F. Barrow: Chem. Phys. Lett. **94**, 1 (1983)
6.74 P. Kusch, M.M. Hessel: J. Chem. Phys. **68**, 2591 (1978)
6.75 K.K. Verma, T.H. Vu, W.C. Stwalley: J. Mol. Spectrosc. **91**, 325 (1982)
6.76 N.W. Carlson: Ginzburg Lab. Report No. 3114, Stanford Univ. (1980)
6.77 J. Verges, C. Effantin, J. d'Incan, D.L. Cooper, R.F. Barrow: Phys. Rev. Lett. **53**, 46 (1984)
6.78 J.V.V. Kasper, G.C. Pimentel: Appl. Phys. Lett. **5**, 231 (1964)
6.79 R. Burnham: Appl. Phys. Lett. **30**, 132 (1977)
6.80 D.J. Ehrlich, J. Maya, R.M. Osgood: Appl. Phys. Lett. **33**, 931 (1978)
6.81 J.C. White: Appl. Phys. Lett. **33**, 325 (1978)
6.82 H. Hemmati, G.J. Collins: Appl. Phys. Lett. **34**, 844 (1979)
6.83 T.F. Deutsch, D.J. Ehelich, R.M. Osgood: Opt. Lett. **4**, 378 (1979)
6.84 D.J. Ehrlich, R.M. Osgood: Appl. Phys. Lett. **34**, 378 (1979)
6.85 H. Hemmati, G.J. Collins: IEEE J. QE-16, 594 (1980)
6.86 P.P. Sorokin, J.R. Lankard: J. Chem. Phys. **54**, 2184 (1971)
6.87 H. Hemmati, G.J. Collins: IEEE J. QE-16, 1014 (1980)
6.88 D.G. Cunningham, D. Denvir, I. Duncan, T. Morrow: Optica Acta 31, 1321 (1984)
6.89 W. Lüthy: Appl. Phys. B40, 121 (1986)
6.90 P. Andresen, G.S. Ondrey, B. Titze: Phys. Rev. Lett. **50**, 486 (1983)
6.91 D.E. Johnson, J.G. Eden: IEEE J. QE-18, 1836 (1982)
6.92 S.G. Dinev, H.U. Daniel, H. Walter: Opt. Commun. **41**, 117 (1982)
6.93 M.N. Edinger, A.W. McCown, J.G. Eden: Appl. Phys. Lett. **40**, 99 (1982)
6.94 A.W. McCown, M.N. Ediger, J.G. Eden: Opt. Commun. **40**, 190 (1982)
6.95 A.W. McCown, J.G. Eden: Appl. Phys. Lett. **39**, 371 (1981)
6.96 E. Gerck, E. Fill: IEEE J. QE-17, 2140 (1981)
6.97 W.M. Trott, J.K. Rice, J.R. Woodworth: J. Chem. Phys. **74**, 518 (1981)
6.98 E.E. Fill, W. Skrlac, K.-J. Witte: Opt. Commun. **37**, 123 (1981)
6.99 V.N. Kurzenkov: Sov. J. Quantum Electron. **11**, 588 (1981)
6.100 W. Lüthy, P. Burkhard, T.E. Gerber, H.O. Weber: Opt. Commun. **38**, 413 (1981)
6.101 D.J. Ehrlich, R.M. Osgood, jr.: IEEE J. QE-16, 257 (1980)
6.102 P. Burkhard, W. Lüthy, T. Gerber: Opt. Commun. **28**, 451 (1980)
6.103 E.J. Schimitschek, J.E. Celto, J.A. Trias: Appl. Phys. Lett. **31**, 608 (1977)
6.104 J. Maya: IEEE J. QE-15, 579 (1979)
6.105 J.C. White, D. Henderson: Phys. Rev. A **25**, 1226 (1982)
6.106 J.C. White, D. Henderson: Opt. Lett. **7**, 204 (1982)
6.107 K. Ludewigt, K. Birkman, B. Wellegehausen: Appl. Phys. B **33**, 133 (1984)
6.108 J.C. White, D. Henderson: Opt. Lett. **8**, 520 (1983)
6.109 J.C. White, D. Henderson: IEEE J. QE-20, 462 (1984)

6.110 B. Wellegehausen, K. Ludewigt, H. Welling: Proc. Soc. Photo-Opt. Instrum. Eng. **492**, 10 (1985)
6.111 K. Ludewigt, R. Dierking, B. Wellegehausen: Opt. Lett. **10**, 606 (1985)
6.112 K. Ludewigt, R. Dierking, W. Pfingsten, B. Wellegehausen: IEEE J. QE-22, 1967 (1986)
6.113 K. Ludewigt, W. Pfingsten, C. Möhlmann, B. Wellegehausen: Opt. Lett. **12**, 39 (1987)
6.114 N.J.A. VanVeen, M.S. DeVries, T. Batler, A.E. DeVries: Chem. Phys. **55**, 371 (1981)
6.115 T.Y. Chang, T.J. Bridges: Opt. Commun. **1**, 423 (1970)
6.116 N.G. Douglas:*Millimetre and Submillimetre Wavelength Lasers*, Springer Ser. Opt. Sci., Vol.61 (Springer, Berlin, Heidelberg 1989)
6.117 M. Fourrier, A. Kreisler: Appl. Phys. B **41**, 57 (1986)
6.118 J.C. Petersen, G. Duxbury: Appl. Phys. B **37**, 209 (1985)
6.119 D. Pereira, E.C.C. Vasconcellos, A. Scalarin, K.M. Evenson, F.R. Petersen, D.A. Jennings: Int. Infrared Millimeter Waves **6**, 877 (1985)
6.120 S. Kon, T. Kachi, Y. Tsunawaki, M. Yamanaka: *Review of Infrared and Millimeter Waves* (Plenum, New York 1984) p.159
6.121 S.F. Dyubko, V.A. Svich, L.D. Fesenko: Zh. Tekhn. Fiz. **43**, 1772 (1973) [Engl. transl.: Sov. Phys.-Tech. Phys. **18**, 1121 (1974)]
6.122 S. Kon, E. Hagiwara, T. Yano, H. Hirose: Japan J. Appl. Phys. **14**, 731 (1975)
S. Kon, T. Yano, E. Hagiwara, H. Hirose: Japan J. Appl. Phys. **14**, 1861 (1975)
6.123 D.M. Watson: In *Galactic and Extragalactic Infrared Spectroscopy*, ed. by M.F. Kessler, J.P. Phillips (Reidel, Dordrecht 1984) pp.195-219
6.124 F.R. Peterson, K.M. Evenson, D.A. Jennings: In *Laser Spectroscopy IV*, ed. by H. Walther, K.W. Rothe, Springer Ser. Opt. Sci., Vol.21 (Springer, Heidelberg, Berlin 1979) pp.39-48
6.125 D.J.E. Knight, G.J. Edwards, P.R. Pearce, N.R. Cross, T.G. Blaney, B.W. Jolliffe: In *Proc. 6th Vavilov Conf. on Nonlin-Physics*, Pt.I, Novosibirsk (1979) pp.112-119
6.126 T.Y. Chang: In *Nonlinear Infrared Generation*, ed. by Y.-R. Shen, Topics Appl. Phys., Vol.16 (Springer, Heidelberg, Berlin 1977) pp.215-272
6.127 Xu. Fuyong, Chen Menyao, Zhao Keyu, Du Li: Chin. Journ. Lasers **13**, 341 (1986)
6.128 Fu Engshen, Huang Gong: Infrared Physics and Technique (Chinese) **48**, 43 (1980)
6.129 T.A. De Temple: *Infrared and Millimeter Waves* **1**, 129 (Academic, New York 1979)
6.130 J.O. Henningsen: *Infrared and Millimeter Waves* **5**, 29 (Academic, New York 1982)
6.131 F. Strumia, M Inguscio: *Infrared and Millimeter Waves* **5**, 130 (Academic, New York 1982)
6.132 T.A. De Temple, E.J. Danielewicz: *Infrared and Millimeter Waves* **7**, 1 (Academic, New York 1983)
6.133 K. Walzel: *Infrared and Millimeter Waves* **7**, 119 (Academic, New York 1983)
6.134 F. Strumia, N. Iolia, A. Moretti: In *Physics of New Laser Sources*, ed. by N.B. Abraham, F.T Arecchi, A. Moorradian, A. Sona (Plenum, New York 1985) p.217
6.135 M. Inguscio, G. Moruzzi, K.M. Evenson, D.A. Jennings: J. Appl. Phys. **60**, 161 (1986)
6.136 N. Ioli, A. Moretti, G. Moruzzi, P. Roselli, F. Strumia: J. Mol. Spectr. **105**, 284 (1984)
6.137 C. Rolland, J. Reid, B.K. Garside: Appl. Phys. Lett. **44**, 725 (1984)
6.138 A.Z. Grasiuk, V.S. Letokhov, V.V. Lobko: Prog. Quant. Electr. **6**, 245 (1980)
6.139 C.R. Jones: Laser Focus **14**, 68 (1978)

6.140 R.G. Harrison, H.N. Rutt: In *Physics of New Laser Sources*, ed. by N.B. Abraham, F.T. Arecchi, A. Moorradian, A. Sona (Plenum, New York 1985) p.201
6.141 H.N. Rutt: Opt. Commun. **34**, 434 (1980)
6.142 J.J. Tiee, C. Witting: Appl. Phys. Lett. **30**, 420 (1977)
6.143 R.R. Jacobs, D. Prosnitz, W.K. Bischel, C.K. Rhodes: Appl. Phys. Lett. **29**, 710 (1976)
6.144 P. Wazen, J.M. Lourtisz: Appl. Phys. B **32**, 105 (1983)
6.145 J. Telle: IEEE J. QE-19, 1469 (1983)
6.146 N.N. Rutt: Infrared Physics **24**, 535 (1984)
6.147 H.R. Schlossberg, H.R. Fetterman: Appl. Phys. Lett. **26**, 316 (1975)
6.148 R.M. Osgood, jr.: Appl. Phys. Lett. **32**, 564 (1978)
6.149 A.H. Bushnell, C.R. Jones, M.I. Buchwald, M. Gundersen: IEEE J. QE-15, 208 (1979)
6.150 J. Telle: IEEE J. QE-19, 1469 (1983)
6.151 H. Kidal, T.F. Deutch: Appl. Phys. Lett. **27**, 500 (1975)
6.152 T.Y. Chang, O.R. Wood II: Appl. Phys. Lett. **24**, 182 (1974)
6.153 T.Y. Chang, O.R. Wood II: IEEE J. QE-13, 907 (1977)
6.154 T.Y. Chang, O.R. Wood II: Appl. Phys. Lett. **23**, 370 (1973)
6.155 H. Tashiro, K. Suzuki, K. Toyoda, S. Namba: Appl. Phys. **21**, 237 (1980)
6.156 T. Yoshida, N. Yamabayashi, K. Miyazaki, K. Fujisawa: Opt. Commun. **26**, 410 (1978)
6.157 S.M. Fry: Opt. Commun. **19**, 320 (1976)
6.158 P.K. Gupta, A.K. Kar, M.R. Taghizadeh, R.G. Harrison: Appl. Phys. Lett. **39**, 32 (1981)
6.159 T.Y. Chang, J.D. McGee: Appl. Phys. Lett. **28**, 256 (1976)
6.160 C. Rolland, B.K. Garside, J. Reid: Appl. Phys. Lett. **40**, 655 (1982)
6.161 C. Rolland, J. Reid, B.K. Garside: Appl. Phys. Lett. **44**, 725 (1984)
6.162 C. Rolland, J. Reid, B.K. Garside: Appl. Phys. Lett. **44**, 380 (1984)
6.163 G.B. Hocker, C.L. Tang: Phys. Rev. **184**, 356 (1969)
6.164 R.L. Shoemaker, E.W. Van Stryland: J. Chem. Phys. **64**, 1733 (1976)
6.165 A.G. Adam, T.E. Gough, N.R. Isenor, G. Scoles: Phys. Rev. A32, 1451 (1985)
6.166 G.B. Hocker, C.L. Tang: Phys. Rev. Lett. **21**, 591 (1968)
6.167 T.W. Mossberg, S.R. Hartmann: Phys. Rev. A 23, 1271 (1981)
6.168 P.B. Berman, J.M. Levy, R.G. Brewer: Phys. Rev. A 11, 1668 (1975)
6.169 K.L. Foster, S. Stenholm, R.G. Brewer: Phys. Rev. A 10, 2318 (1974)
6.170 R. Beach, S.R. Hartmann: Phys. Rev. Lett. **53**, 663 (1984)
6.171 J.A. Kash, E.L. Hahn: Phys. Rev. Lett. **47**, 167 (1981)
6.172 J.A. Kash, Sun Tao-Heng, E.L. Hahn: Phys. Rev. A 26, 2682 (1982)
6.173 J.E. Golub, T.W. Mossberg: J. Opt. Soc. Am. B 3, 554 (1984)
6.174 H. Nakatsuka, M. Tomita, M. Fujiwara, S. Asaka: Opt. Commun. **52**, 150 (1984)
6.175 N. Morita, T. Yajima: Phys. Rev. A 30, 2525 (1984)

Chapter 7

7.1 P. Kusch, M.M. Hessel: J. Chem. Phys. **68**, 2591 (1978)
7.2 M.E. Kaminsky: Modulated Population Spectroscopy: Identification of Absorption Lines by Modulated Lower Level Population: Spectrum of Na_2. Ph.D. Thesis, Stanford University (1976)
7.3 R.W. Wood, E.L. Kinsay: Phys. Rev. **30**, 1 (1927); **31**, 793 (1928)
7.4 F.P. Schäfer (ed.): *Dye Lasers*, 3rd. edn., Topics Appl. Phys., Vol.1 (Springer, Berlin, Heidelberg 1990)
7.5 N.W. Carlson, K.M. Jones, G.P. Morgan, A.L. Schawlow, A.J. Taylor, H.R. Xia, G.Y. Yan: Selective Spectrum Simplification by Laser Level Labeling. In *Laser Spectroscopy V*, ed. by A.R. W. McKellar, T. Oka, B.P. Stoicheff, Springer Ser. Opt. Sci., Vol.30 (Springer, Berlin, Heidelberg 1981)

7.6   W. Demtröder, M. McClintock, R.N. Zare: J. Chem. Phys. 51, 5495 (1969)
7.7   J.P. Woerdman: Chem. Phys. Lett. 43, 279 (1976)
7.8   J.P. Woerdman, M.F.H. Schuurmans: Opt. Commun. 21, 243 (1977)
7.9   Li Li, S.F. Rice, R.W. Field: J. Mol. Spectrosc. 105, 344 (1984)
7.10 H.R. Xia, Z.G. Wang: Study of Small but Complete Molecules. In *Laser, Spectroscopy and New Ideas*, ed. by W.M. Yen, M.D. Levenson, Springer Ser. Opt. Sci., Vol.54 (Springer, Berlin, Heidelberg 1987) pp.174-182
7.11 H.R. Xia, F.H. Yao, J.G. Cai, I.S. Cheng: Appl. Lasers (China) 6, 177 (1986)
7.12 M.E. Kaminsky, R.T. Hawkins, F.V. Kowelski, A.L. Schawlow: Phys. Rev. Lett. 34, 1073 (1975)
7.13 P. Kusch, M.M. Hessel: J. Chem. Phys. 63, 4087 (1975)
7.14 A.L. Schawlow: Physica Scripta 25, 333 (1982)
7.15 R.E. Teets: Polarization Labeling Spectroscopy of Molecules. Ph.D. Thesis, Stanford Univ. (1978)
7.16 A.R. Edmonds: *Angular Momentum in Quantum Mechanics* (Princeton Univ. 1974)
7.17 A.J. Taylor: Two-Step Polarization Labeling Spectroscopy of Excited States of Diatomic Sodium. Ph.D. Thesis, Stanford Univ. (1982)
7.18 R.E. Teets, R. Feinberg, T.W. Hänsch, A.L. Schawlow: Phys. Rev. Lett. 37, 683 (1976)
7.19 W. Demtröder: *Laser Spectroscopy*, Springer Ser. Chem. Phys., Vol.5 (Springer, Berlin, Heidelberg 1981)
      W. Demtröder: *Laser Spektroskopie*, 2nd edn. (Springer, Berlin, Heidelberg 1990)
7.20 N.W. Carlson, A.J. Taylor, K.M. Jones, A.L. Schawlow: Phys. Rev. A24, 822 (1981)
7.21 J.G. Cai, H.R. Xia, I.S. Cheng: Acta Optica Sinica 6, 212 (1986); Chinese Phys. 7, 185 (1987)
7.22 R.N. Zare: J. Chem. Phys. 40, 1934 (1964)
7.23 D. Eisel, D. Zevgolis, W. Demtröder: J. Chem. Phys. 71, 2005 (1979)
7.24 J.W. Cooley: Math. Compotation 15, 363 (1961)
7.25 H.R. Xia, J.G. Cai, I.S. Cheng: Bull. Am. Phys. Soc. 31 (11) (1985)
7.26 Li Li, S.F. Rice, R.W. Field: J. Chem. Phys. 82, 1178 (1985)
7.27 G. Morgan, H.R. Xia, A.L. Schawlow: J. Opt. Soc. Am. 72, 315 (1982)
7.28 G.Y. Yan, B.W. Sterling, A.L. Schawlow: J. Opt. Soc. Am. B 5, 2305 (1988)
7.29 H.R. Xia, J.W. Xu, I.S. Cheng: In *Proc. of the Topical Meeting on Laser Materials and Laser Spectroscopy*, ed. by W. Zhizian, Z. Zhiming (World Scientific, Singapore 1989) p.181
7.30 H.R. Xia, J.W. Xu, I.S. Cheng: Quantum Electronics (China) 3, 293 (1986)
7.31 H.R. Xia, J.W. Xu, I.S. Cheng: Quantum Electronics (China) 3, 295 (1986)
7.32 Li Li, R.W. Field: J. Mol. Spectrosc. 117, 245 (1986)
7.33 H.R. Xia, L.S. Ma, J.W. Xu, M. Yuan, I.S. Cheng: Optics Commun., submitted

Subject Index

Raman beat 208,209
Ramsey fringes 137-139
Recoil effect elimination 142
Reduced mass 6,15,31
Regular perturbations 6
Relaxation processes 58,143,207
Rigid-rotator 14,16,17,42
Rotation-wave approximation 70
Rotational angular momentum 20
Rotational constants 14,16-18,23,32,40,
  50,112,242,251,252
Rotational fine structure 42,50,249,251
Rotational levels 14-18,20,21,29,52,242
Rotational quantum number 15,25,84,
  95,223,242,243,251
Rotational spectra 42,43,50
Rydberg-Klein-Ress (RKR) method 243
Rydberg states 33-41,62,112,143,157,
  212

S branch 83-88,99,132,133,215,249
S-tetrazine 157,160
S_2 188
SbI_3 197
Schrödinger equation 5,6,10,14,67
Second-order Doppler effects 143
Selection rules 9,12,15,16,44,45,47,48,
  50,51,64,93,212,215,248
Selective excitation 211
Selective simplification 210
Self-broadening 59,60,155
Separated field spectroscopy 134,
  137-140
SF_6 57,62,112,139,140,142,143,151,153,
  154
Short-lived molecules 57
SiCl 57
Side band 118
SiF 34
Signal intensity 229,233,237,238,240,253
Signal phase 120,125,131,245
Simplification of molecular spectra 76,
  81,82,96,99,211,214,216,221,229,231
Simplified band progressions 214,215,
  231,242
Singlet state 98,245,246
Small-signal gain 163
SnI_2 197
SO_2 135,142
Spatially separated fields 137,140
Spectral density 99
Spectral effects of external fields 142,
  150,157,158
Spectral resolution 137,141

Spectral strength factor 223,228,230,237,
  238
Spectral terms 11,14-17,75-78,94,95,
  210,241-243
Spherical-top molecules 18
Spin angular momentum 20
Spin-forbidden transitions 93-100,250,
  251
Spin-orbit interaction 6,24-26,252,254
Spin-rotation interaction 6,112
Spontaneous emission 54,244
Stark effect 142,157,158
Stark-switch technique 207
Stimulated diffuse band radiation
  169-174
Stimulated electronic Raman scattering
  (SERS) 162,163,165,177-180,183
Stimulated hyper-Raman scattering
  (SHRS) 162,165,168
Stimulated radiation 161,174
Successive perturbation solutions 70-74,
  92,102,103,108,110,164-168
Sum-frequency coherent excitation 103
Superhyperfine structure 152
Supersensitive spectroscopy 117,120,
  123,134
Supersonic molecular beam 138,144,159
Susceptibility 114,115,179,180-182
Symmetric-top molecules 16,17,28,151,
  202,203
Symmetry breaking 143
Symmetry species 52,58,203

Te_2 188
Term values 5,24
Thermal population 210
Third-order polarization 163,164,168
Three-level system 16,66-74,89-93,
  162-165,222
Time-of-flight broadening 134,137,143
Time separated fields 138,140
TII 197,200
Total angular momentum 2,20,26,122
Total electronic angular momentum 9
Total electronic orbital angular momen-
  tum 7
Total electronic spin angular momen-
  tum 8
Total energy 5
Total wave function 5
Trans-butadiene 35
Trans-1,3,5-hexatriene 35
Transient molecules 57
Transit broadening 136

Transition relaxation 207
Transition strength 135
Triatomic molecules 13,43
Triplet states 94-99,144-149,246,250
Two-level system 110,138
Two-photon absorption 36,62,66,76,
  245-254
Two-photon band progression 74-83
Two-photon band sequence 82
Two-photon branches 83-88,99,132,133,
  215,249
Two-photon coarse structure 74-83
Two-photon enhancing factor 72,102,
  103,105,106
Two-photon excitation 64,66,166,169,
  170,172,196-198,205,247
- mechanisms 89,93,101-104
Two-photon fine structure 83-89
Two-photon lineshapes 89-93,101-107,
  130-132,250,251
Two-photon polarization labelling
  236-240
Two-photon spectroscopy 62-107,
  129-136,236-240
Two-photon transition 3,38,40,62-107,
  128-136
Two-step excitation 35,36,232-236
Two-step hybrid four-wave mixing 182
Two-step hybrid off-resonance pumping
  177
Two-step hybrid resonance 174,176,
  181-185
Two-step molecule-atom hybrid reso-
  nance laser 175
Two-step polarization labelling spectros-
  copy 35-40,232-236

UF_6 57
Ultrafast relaxation processes 209

Valence electrons 29,141
Van der Waals molecules 35,57
Vibration-rotation interaction 6,19,45
Vibration-rotation interaction constant
  18,32,242
Vibration-rotational bands 45
Vibration-rotational energy level 18,19
Vibration-rotational spectra 44
Vibrational angular momentum 44
Vibrational band systems 48-50
Vibrational bands 43,47,49
Vibrational constants 11-14,32,39,242
Vibrational energy 11
Vibrational levels 10,12-14,26,50,52,242
Vibrational quantum number 11,18,214,
  241,243,250
Vibrational wave function 244
Voigt profile 58

Wave function 7,10,14
Wave-function mixing 95,104,142
Weak intramolecular coupling 208

Xe_2 57
XeCl 62

Zeeman effect 143,150,157,158

Printing: Mercedesdruck, Berlin
Binding: Buchbinderei Lüderitz & Bauer, Berlin